# PETROLOGY OF THE OCEAN FLOOR

*Elsevier Oceanography Series, 33*

# PETROLOGY OF THE OCEAN FLOOR

R. HEKINIAN

*Centre National pour l'Exploitation des Océans (CNEXO),
Centre Océanologique de Bretagne, Brest, France*

ELSEVIER SCIENTIFIC PUBLISHING COMPANY
Amsterdam — Oxford — New York          1982

ELSEVIER SCIENTIFIC PUBLISHING COMPANY
1, Molenwerf
P.O. Box 211, 1000 AE Amsterdam, The Netherlands

*Distributors for the United States and Canada:*

ELSEVIER SCIENCE PUBLISHING COMPANY INC.
52, Vanderbilt Avenue
New York, N.Y. 10017

Library of Congress Cataloging in Publication Data

Hékinian, R. (Roger), 1935-
    Petrology of the ocean floor.

    (Elsevier oceanography series, ISSN 0-444-41623-4 ;
33)
    Bibliography: p.
    Includes index.
    1. Submarine geology.  2. Petrology.  I. Title.
II. Series.
QE39.H43          551.46'08        82-2468
ISBN 0-444-41967-5                 AACR2


ISBN 0-444-41967-5 (Vol. 33)
ISBN 0-444-41623-4 (Series)

Printed in The Netherlands

*"This book is dedicated to Ginny, Aram, Diran and Anna and to the memory of the Armenian People who perished during the 1894—1896 and 1914—1921 periods".*

# PREFACE

"Ocean floor petrology" is concerned primarily with the study of volcanism and related processes involving the creation and the transformation of crustal material which takes place at contact with seawater. The field of ocean floor petrology has developed and will continue to evolve rapidly with man's effort to penetrate and explore greater depths through the use of more sophisticated technology. The present work was initially intended to be a comprehensive review of the major mineralogical and geochemical investigations carried out in the study of oceanic crust. However, during the process of gathering the data together, new information has been added in order to further increment our knowledge of the oceanic crust.

It is not the pretense of this work to give a complete coverage of ocean-floor genesis, but to give insights into the compositional diversities of basement rock samples associated with various geological settings. Since the petrological studies of the ocean floor are interrelated with other disciplines, an attempt has been made throughout this work to cover some general aspects of the morphology and the structure of oceanic features for which data and samples are available. Both the sampling and the gathering of other geological data from the ocean floor have been found to be an expensive and difficult task. Submersibles and unmanned bottom-navigated instruments have demonstrated their efficiency and are prerequisites for sea-floor exploration in order to implement further detailed studies of magmatism, hydrothermalism and structural processes related to different types of geological features.

This book has been divided into twelve chapters which can be separated into two main parts. The first part, including Chapters 1 through 8, deals with the composition of the oceanic basement associated with the various types of structures thus far recognized and sampled. Since oceanic ridges are one of the main features and also the most accessible for sampling, they are the best known. Even so, we are far from having a clear picture of the deep-seated processes involved in the creation of new crust. Only a few segments of oceanic ridges have been studied in detail using advanced technology. The FAMOUS project (Franco-American Mid-Ocean Undersea Study) and project RITA (study of the *Ri*vera and *Ta*mayo Fracture Zones) provided new structural and petrological data for the understanding of plate boundaries accreting at different rates in the Atlantic and Pacific Oceans. Deep-sea drilling has proved to be invaluable in recovering samples in heavily sedimented areas hitherto inaccessible by other methods of sampling.

Other major oceanic structures such as aseismic ridges, which are thought to be chains of elevated edifices active during the early stage of the separation of the continental masses, are still inadequately sampled. The crustal composition of oceanic trenches is also poorly known. More sampling of the outer island arcs or ocean-side regions of the trenches is necessary in order to have more insight into crustal behavior prior to the plunging of the lithosphere.

Other poorly known regions, located at the plate boundaries and characterized by both extensional and compressional plate motions, have been tentatively called thrust-faulted regions and these are also discussed.

The aspect of subaerial ophiolites and their inferred occurrences in an oceanic environment has not been discussed in depth in this book. Obviously, this subject is of great interest in understanding the early emplacement of ocean-floor ultramafic and mafic complexes at accreting plate boundaries. The lack of a continuous section through layers 2 and 3 of the oceanic crust has, up to now, prevented marine geologists from making any meaningful correlation between subaerial ophiolites and similar oceanic types of rock associations.

The second part of the book (Chapters 9, 10, and 11) deals mainly with the alteration of oceanic crust after its creation when exposed to the progressive effects of weathering, hydrothermalism, and metamorphism.

So far, the economic aspects of deep-seated oceanic environments have been centered on the study of polymetallic manganese nodules (made up of Mn, Fe, Ni, Cu). However, the mineral wealth of the ocean floor is still not well assessed and it was only recently (1979) that massive sulfide deposits with higher metal concentrations than those of the manganese nodules were seen to be formed at depths of 2600 meters. It is necessary that future exploration be oriented towards a systematic interdisciplinary study of oceanic ridges and associated fracture zones in order to obtain more insight into the genesis of ore deposits in both oceanic and subaerial regions.

The last chapter (Chapter 12) deals with subcrustal and upper mantle processes related to accreting plate boundary regions. This chapter is based on inferences made from major experimental petrological work and geophysical investigations carried out on oceanic rocks and environments.

As we increase our knowledge in oceanic petrology, the questions become more complex, the problems more complicated. Oversimplification is replaced by a feeling of confusion, albeit closer to the "truth". It is essentially due to the increasing complexity of the field of oceanic petrology that an interdisciplinary approach is vital in future work.

ROGER HEKINIAN
St. Renan, 30 December 1980

# ACKNOWLEDGEMENTS

This book would not have been possible were it not for the previous studies by the innumerable scientists whose names appear in the bibliography.

I am also indebted to my colleagues, and in particular to D. Bideau, W.B. Bryan, M. Fevrier, J. Francheteau, H.D. Needham, V. Renard, and R. Smith, who contributed in both numerous discussions and the field-work operations which permitted the clarification of various aspects of ocean floor studies.

I would like to thank Drs. W.B. Bryan and W.J. Morgan who reviewed the manuscript.

I am also very grateful for the skillful technical capabilities of M. Bohn in performing the microprobe analyses, of P. Cambon and J. Etoubleau for their X-ray fluorescence analyses of bulk-rock samples, of G. Floch for the thin-section preparation, and of D. Carre, R. Thirion, A. Grotte, J. Guichardot-Wirmann, and S. Monti for their draftmanship. Some of the bibliographical search was done by S. Marques.

I wish to thank C. Robart, N. Guillo, M. Le Menn, M.L. Quentel, B. Berthe, and my wife, V. Hékinian, who have helped considerably in the typing and editing of the manuscript.

Encouragement and support were given by the administration of the Centre National pour l'Exploitation des Océans (CNEXO; President: G. Piketty), by the administration of the Centre Océanologique de Bretagne (C.O.B.; Director: J. Vicariot) and by the members of the Department of Geology and Geophysics of C.O.B. (Department Chairman: G. Pautot). The X-ray fluorescence performed at C.O.B. was supervised by H. Bougault.

Also, I gratefully acknowledge the various Institutions which provided some of the ocean floor data used in this study. In particular I would like to thank R. Capo and F.W. McCoy from Lamont-Doherty Geological Observatory, Palisades, N.Y.; T. Simkin of the Smithsonian Institution (Oceanographic Sorting Center), Washington, D.C.; R.L. Fisher, F.H. Spiess and J. Hawkins from Scripps Institution of Oceanography, La Jolla, Calif.; W.A. Riedel and the Staff of the Deep Sea Drilling Project at Scripps; J. Edmond from the Massachussets Institute of Technology; and H.D. Holland from Harvard University, Cambridge, Mass. I am grateful to the captains, crewmen, and scientists who participated in numerous oceanographic expeditions.

# ABBREVIATIONS

| | |
|---|---|
| DSDP | Deep Sea Drilling Project |
| FAMOUS | Franco American Mid-Ocean Undersea Survey |
| IPOD | International Phase of Ocean Drilling |
| JOIDES | Joint Oceanographic Institutions for Deep Earth Sampling |
| RITA | An acronym derived from the names of the Rivera and Tamayo Fracture Zones on the East Pacific Rise |
| TAG | Trans-Atlantic Geotraverse |
| | |
| ALV | "Alvin" (submersible) |
| AR, ARP | "Archimède" (bathyscaphe) |
| CY, CYP | "Cyana" (diving saucer) |

# CONTENTS

XIV

CHAPTER 1

# MINERALOGY AND CHEMISTRY OF OCEAN FLOOR ROCKS

ULTRAMAFIC ROCKS

Most of the ultramafic rocks found in the ocean floor are serpentinized to various degrees. It is also observed that in many cases the serpentinized peridotites collected from various structures of the ocean floor are accompanied by gabbros and sometimes by minor amounts of anorthosite. Basaltic rocks are also intermixed in various proportions within the dredge hauls, suggesting that some kind of interrelationship might exist between these different rock types. The distribution of ultramafics from the ocean floor is shown on a world map in Fig. 1-1.

*Serpentinized peridotite*

It is often difficult to establish the primary composition of altered peridotite. Assuming that the serpentinization of oceanic peridotite proceeds under isochemical conditions, we can therefore convert the silicate analyses to anhydrous residue in order to obtain the primary normative composition of the peridotitic material. Chemical analyses of some serpentinized peridotites are shown in Table 1-1.

Serpentinization is a process of alteration of an original mineral, usually pyroxene or olivine, which gives rise to different types of serpentine minerals and structures. Often the original rock structure and mineral outlines are preserved, and it is sometimes easy to reconstruct the ghost minerals. Two examples where these may be readily recognized are the formation of bastite from pyroxene and the preserved sharp outlines formed by Fe-oxide minerals replacing many olivine crystals.

The texture of serpentinized peridotite varies. There are massive varieties with allotriomorphs of polygonal olivine relics and crystals, and xenomorphic randomly oriented tabular pyroxene. Other textural features, such as linear orientation or a contorted appearance of mineral grains are often encountered in specimens of serpentinized material. Some pyroxene minerals show deformed twin lamellae and broken-up plagioclase crystals, and an effect of granulation on mafic minerals may also occur. These later textural features are attributed to cataclastic metamorphism (see Chapter 9).

Because of the different degrees of rock serpentinization, some authors (Dmitriev and Sharaskin, 1975) think that there are at least two stages of serpentinization. The first stage consists of uniform serpentinization of minerals (pseudomorphism) with no signs of alteration of the primary rock fabric, while the second stage consists of the formation of veins and veinlets plus many new mineral phases due to the recrystallization of serpentine.

TABLE 1-1

Chemical analyses of serpentinized peridotite from the Mid-Atlantic Ridge near 53°N and from 30°N (Hékinian and Aumento, 1973; Miyashiro et al., 1969a). The data from the Kane and from the Vema Fracture Zones are from Miyashiro et al. (1969a) and from Melson and Thompson (1971). Samples 97 and 93 are lherzolite from the Indian Ocean (C.G. Engel and Fisher, 1969)

| | Mid-Atlantic Ridge, 53°N CM4-DR2-1 | Mid-Atlantic Ridge, 30°N A150-6 AM-3 | Kane Fracture Zone V25-8-T20 | Vema Fracture Zone AII-20-9-1 | Romanche Fracture Zone CH80-DR10-15 | Rodriguez Fracture Zone 97-W | Mid-Indian Oceanic Ridge, 12°S 93-3 |
|---|---|---|---|---|---|---|---|
| $SiO_2$ (wt.%) | 37.80 | 41.76 | 40.68 | 39.71 | 43.00 | 4.5 | 38.29 |
| $Al_2O_3$ | 3.75 | 2.30 | 2.86 | 2.59 | 1.08 | 3.59 | 2.74 |
| $Fe_2O_3$ | 10.73* | 5.66 | 4.69 | 7.31 | 10.29 | 1.60 | 4.65 |
| FeO | – | 3.14 | 3.43 | 1.30 | 1.12 | 5.23 | 2.25 |
| MnO | 0.10 | 0.11 | 0.10 | 0.12 | 0.12 | 0.10 | 0.06 |
| MgO | 32.29 | 33.83 | 33.20 | 33.15 | 31.87 | 34.84 | 35.32 |
| CaO | 1.87 | 1.53 | 2.05 | 1.52 | 0.08 | 5.50 | 3.75 |
| $Na_2O$ | 0.21 | 0.23 | 0.12 | 0.28 | 0.29 | 0.19 | 0.20 |
| $K_2O$ | 0.52 | 0.03 | 0.03 | 0.06 | 0.03 | 0.20 | 0.20 |
| $TiO_2$ | 0.22 | 0.14 | 0.12 | 0.06 | 0.05 | 0.03 | 0.01 |
| $P_2O_5$ | – | 0.02 | 0.01 | 0.02 | 0.08 | – | – |
| $H_2O$ | 12.95 | 11.41 | 12.03 | 13.88 | 12.19 | 4.52 | 11.46 |
| Total | 100.44 | 100.16 | 99.32 | 100.00 | 100.20 | 98.78 | 98.75 |
| Cr (ppm) | n.d. | 3700 | 3600 | 3300 | 2520 | 6400 | 4800 |
| Ni | n.d. | 1500 | 2500 | 2300 | 2238 | 3500 | 3300 |
| Co | n.d. | | | | | | |

n.d. = not determined.

*Indicates total Fe calculated as $Fe_2O_3$.

Tremolite, at the expense of pyroxene, was found in veins showing the effect of strong directional stress (Aumento and Loubat, 1971). However, it is not excluded that many transitional stages of serpentinization might have occurred between the beginning and the end of peridotitic alteration. It is observed that the final stage of serpentinization, with the formation of iddingsite, hematite and carbonate minerals, occurs at a lower confining pressure (Aumento and Loubat, 1971).

Various types of serpentinized peridotite have been described; however, the most common ultramafics from the ocean floor are divided according to their mineral constituents into four major types: the dunites, the harzburgites, the lherzolites, and the ophicalcites (Fig. 1-2). Other minor types of ultramafics include pyroxenites, wehrlites, and websterites (see Fig. 1-3).

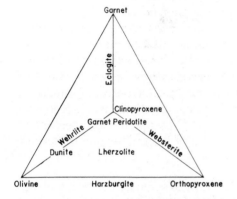

Fig. 1-2. Tetrahedon of the garnet-olivine-clinopyroxene-orthopyroxene association showing the compositional distribution of the various types of ultramafics (Yoder, 1975).

*Dunite.* Dunite consists almost entirely of olivine which has undergone various degrees of serpentinization. Among the ultramafics, it is the least commonly encountered and is usually associated with harzburgites and lherzolites. Dunites were found on the Carlsberg Ridge and on the western branch of the Mid-Indian Ocean Ridge (Dmitriev and Sharaskin, 1975). Others occur in the central and north Atlantic regions (Bonatti et al., 1970; Aumento and Loubat, 1971). Mineralogically, dunite differs from both harzburgite and lherzolite by its higher olivine content. In addition, it contains spinel, which is usually chromiferous. When altered by secondary processes, which is commonly the case in oceanic environments, the dunites also contain chlorite, serpentine (lizardite), talc, and hydrogarnet. Garnet was also seen in veinlets with a serpentine-chlorite association (Aumento and Loubat, 1971). Chemically, no major differences are detected among the dunites and the other serpentinized peridotites (harzburgite and lherzolite); the Indian Ocean dunite reported by Dmitriev and Sharaskin (1975), however, shows a distinctly lower total iron content (<7%) than do the other associated serpentinized peridotites.

*Harzburgite.* This variety of serpentinized peridotite is characterized by a relatively high amount of bastite (25—30% of the bulk rock). Both modal and normative counts show 60—80% olivine, 20—40% pyroxene, and 0.5—2% spinel. Theoretically, harzburgite should contain more orthopyroxene than clinopyroxene when compared to the lherzolite variety of serpentinized peridotite. The harzburgite variety also differs from the lherzolite by its generally higher total iron content recalculated as FeO (8—11%), its higher MgO content (35—55%), and its lower CaO content (<3%).

Many serpentinized peridotites from the central equatorial Atlantic consist of the following assemblage: olivine (85—95%), enstatite with clinopyroxene exsolution lamellae, and Cr-spinel (Bonatti et al., 1970).

*Lherzolite.* When recalculated to be water-free and normalized to 100%, lherzolite contains relatively higher amounts of normative pyroxene (35—45%) and olivine (58—70%) than does harzburgite. Relics of orthopyroxene and clinopyroxene are usually recognizable. Identification between the two pyroxenes is made on the cleavage planes of bastite. Exsolution lamellae of diopside occur in enstatic lamellae (Aumento and Loubat, 1971). Relics of diopsidic-augite or augite, occasionally surrounded by patches and veinlets of Fe-oxide material, are found in abundance in the lherzolite-type of serpentinized peridotite.

*Ophicalcite.* The term ophicalcite was coined by Brongniart in 1813 and was assigned to a calcitic type of serpentinite. Ophicalcites dredged in an equatorial Atlantic fracture zone ("Charcot" cruise 80, 1977) consist of brownish-red foliated or massive rocks. The major constituents are calcite, serpentine, Fe-oxides, altered orthopyroxene and clinopyroxene, chlorite, and talc. The reddish color of the specimens is attributed to the oxidized nature of the iron-bearing minerals. The chemistry of these ophicalcites differs from that of other serpentinized peridotites by its higher CaO (7—10%) and its lower MgO contents (<30%) (Table 1-2).

*Mineralogy of ultramafic rocks*

Because of the pronounced degree of alteration of most ocean floor serpentinized peridotites, it is difficult to assess the original composition of the various mineral phases. The formation of secondary minerals such as serpentine, chlorite, pargasite, hydromica, talc, brucite, çalcite, apatite, amphibole, mica, prehnite, and hematite is prominent. However, primary Cr-Al-spinel, clinopyroxene, orthopyroxene, and olivine relics were found to occur in these rocks and probably represent the major constituents of the unaltered peridotite.

TABLE 1-2

Chemical analyses of ophicalcite from the Romanche Fracture
Zone at 0°05′S and 18°29′W at about 5400 m depth

|  | CH80-DR10-13 | CH80-DR10-12 |
| --- | --- | --- |
| $SiO_2$ (wt.%) | 40.00 | 35.07 |
| $Al_2O_3$ | 2.34 | 1.14 |
| $Fe_2O_3$ | 6.06 | 10.35* |
| FeO | 1.17 | — |
| MnO | 0.09 | 0.10 |
| MgO | 25.59 | 27.05 |
| CaO | 10.24 | 8.22 |
| $Na_2O$ | 0.11 | 0.23 |
| $K_2O$ | 0.01 | 0.02 |
| $TiO_2$ | 0.10 | 0.06 |
| $P_2O_5$ | 0.08 | 0.13 |
| $H_2O$ | 13.71 | 16.46 |
| Total | 99.50 | 98.83 |
| Cr (ppm) | 2043 | 2932 |
| Ni | 1507 | 2100 |
| Co | 101 | 90 |

*Indicates total Fe calculated as $Fe_2O_3$.

*Serpentine.* This is a hydrous mineral which is formed by the reaction of
water with primary mafic mineral phases, especially pyroxene and olivine.
The hypothetical reaction between forsterite and water, giving rise to serpentine and brucite, may take place at a temperature of about 350—400°C
(Johannes, 1968):

$$2Mg_2SiO_4 + 3H_2O \rightarrow Mg_3Si_2O_5(OH)_4 + Mg(OH)_2$$
(olivine        $\rightarrow$ serpentine      + brucite)

Assuming a ratio of olivine to orthopyroxene of about 1:1, the following
reaction can be written:

$$Mg_2SiO_4 + MgSiO_2 + H_2O \rightarrow Mg_3Si_2O_5(OH)_4$$
(olivine    + orthopyroxene    $\rightarrow$ serpentine)

Various types of serpentine minerals are found within a single specimen of
serpentinized peridotite from the ocean floor: lizardite, chrysotile, and
antigorite. Lizardite and antigorite show plate-like textural features, while
chrysotile is fibrous. The recognition of various types of serpentine can be of
value in establishing the pressure/temperature conditions present during
serpentinization (Coleman, 1977). Lizardite and chrysotile can form at
temperatures from 350°C down to ambient temperatures (Barnes and O'Neil,
1969), whereas antigorite is stable up to temperatures slightly above 500°C.

TABLE 1-3

Chemical analyses of spinel and pyroxene found in ultramafic rocks collected from the ocean floor (total Fe calculated as FeO except for sample 5324)

| | Spinel | | | | | | Pyroxene in serpentinized peridotites | | | | |
| --- | --- | --- | --- | --- | --- | --- | --- | --- | --- | --- | --- |
| | | | | | | | V30-DR5 | | CH77-DR10-96 | | A150-DR20 |
| | 5324 | 165-8 | 159-10 | 156-10 | 165-1 | 165-2 | cpx | opx | cpx | opx | cpx |
| $SiO_2$ (wt.%) | 2.11 | n.d. | n.d. | n.d. | n.d. | n.d. | 52.36 | 55.67 | 51.85 | 55.65 | 51.58 |
| $Al_2O_3$ | 18.69 | 20.6 | 25.0 | 32.4 | 23.9 | 24.1 | 3.38 | 2.33 | 4.58 | 3.54 | 4.57 |
| $Fe_2O_3$ | 2.71 | — | — | — | — | — | — | — | — | — | — |
| FeO | 14.70 | 26.9 | 22.9 | 14.1 | 21.3 | 21.4 | 2.76 | 6.14 | 2.93 | 5.82 | 2.66 |
| MnO | 0.33 | n.d. | n.d. | n.d. | n.d. | n.d. | 0.08 | 0.13 | 0.25 | 0.15 | 0.08 |
| MgO | 13.26 | 11.6 | 11.7 | 15.2 | 14.2 | 14.4 | 17.07 | 32.85 | 17.28 | 33.30 | 17.98 |
| CaO | — | n.d. | n.d. | n.d. | n.d. | n.d. | 21.12 | 1.68 | 21.76 | 1.56 | 20.49 |
| $Na_2O$ | 0.05 | n.d. | n.d. | n.d. | n.d. | n.d. | 0.37 | 0.03 | 0.25 | 0.01 | 0.25 |
| $K_2O$ | 0.02 | n.d. | n.d. | n.d. | n.d. | n.d. | 0.00 | 0.00 | 0.00 | 0.00 | 0.00 |
| $TiO_2$ | 0.12 | n.d. | n.d. | n.d. | n.d. | n.d. | 0.57 | 0.21 | 0.13 | 0.04 | 0.08 |
| $Cr_2O_5$ | 46.78 | 40.7 | 40.8 | 38.7 | 39.7 | 39.9 | 1.34 | 0.75 | 1.15 | 0.61 | 1.40 |
| Total | | | | | | | 99.42 | 100.23 | 99.67 | 100.62 | 99.10 |
| Wo | — | — | — | — | — | — | 44.84 | 3.25 | 45.92 | 3.00 | 43.02 |
| En | — | — | — | — | — | — | 50.45 | 87.42 | 49.99 | 88.32 | 52.50 |
| Fs | — | — | — | — | — | — | 4.71 | 9.34 | 4.32 | 8.87 | 4.48 |

Sample 5324 (polygon X) is from a serpentinized peridotite dredged from the western branch of the Mid-Indian Oceanic Ridge near 28°22'S, 62°36.7'E (Dmitriev, 1975). Samples 165-8, 159-10, and 156-10 are serpentinized peridotites from 45°N (Aumento and Loubat, 1971) and 165-1 and 165-2 are dunites from 45°N (Aumento and Loubat, 1971). Sample V30-DR05 is from the Romanche Fracture Zone, 01°31'N, 19°06'W (4125 m); CH77-DR10-96 is from the Vema Fracture Zone, 10°41.50'N, 42°40.80'W (4125 m), and A150-DR20 is from the Atlantis Fracture Zone (Camebax microprobe analyses, courtesy of R. Hebert).

*Olivine.* This mineral is rarely preserved and generally occurs as relic constituents in the form of small crystals surrounded by dust-like Fe-oxide minerals. The olivine is colorless with high relief, has an index of refraction of $N_\gamma = 1.685-1.690$ and $n_\alpha = 1.645-1.662$, and has an optical angle $2V$ of about $90°$. The forsterite content of olivine is usually around $Fo_{90}$.

Two main types of serpentinized peridotite found on the ocean floor are lherzolite and harzburgite. The difference between the two is that the lherzolite variety has a higher content of clinopyroxene and usually less olivine than does the harzburgite variety. Evidence of the relative mineral abundance is calculated from both visual observation and modal analyses recalculated to the anhydrous phase.

*Spinel.* This mineral usually occurs as a mixture of chromite and common spinel with a euhedral or anhedral shape. The major chemical constituents are $Cr_2O_3$ (35—41%) and $Al_2O_3$ (20—33%) (Table 1-3). It was found (Aumento and Loubat, 1971) that the chrome spinel from harzburgite has a lower FeO and $Cr_2O_3$ content, with a cell parameter ranging in the lower range from 8.20 to 8.23 Å, than does a chrome spinel from dunite, wehrlite and olivine gabbros, which have a larger cell parameter (8.23—8.29 Å). Spinel may occur either as cumulated grains or dispersely associated with olivine relics.

*Pyroxene.* Ortho- and clinopyroxene occur usually as anhedral crystals with contorted cleavage planes and partially altered by phyllosilicates. The orthopyroxene is usually colorless with extinction angle $C \wedge \gamma = 0°$ to $7-10°$. The index of refraction is $N_\gamma = 1.665-1.682, N_\alpha = 1.650-1.665$, and $2V$ is $75-90°$. The optical properties and the chemical analyses of some orthopyroxenes indicate that they belong to the enstatite variety. The clinopyroxene is also colorless, has a similar refractive index to that of orthopyroxene, but has a higher extinction angle: $C \wedge \gamma = 20-50°$.

GABBROIC ROCKS

The compositional field of oceanic gabbros in the AMF ternary diagram $(Na_2O + K_2O)$-MgO-FeO (total Fe calculated as FeO) shows a scattering of the plots along the MgO-FeO side of the diagram, and follows a similar trend of variability to oceanic basalt in general (Fig. 1-3). The gabbros from both the Mid-Atlantic Ridge and Mid-Indian Oceanic Ridge, showing a wide range in the FeO/MgO ratio (0.30—2.90), suggest a marked trend of fractionation (Fig. 1-3). The trend is accompanied by a decrease in the anorthite content of the plagioclase from about $An_{79}$ to $An_{47}$ from the gabbro located nearer to the MgO corner down to the ferrogabbros (Miyashiro et al., 1970b). The gabbroic rocks are found mostly in association with the ultramafic group of the serpentinized peridotite suites. In general, the gabbros when unaltered

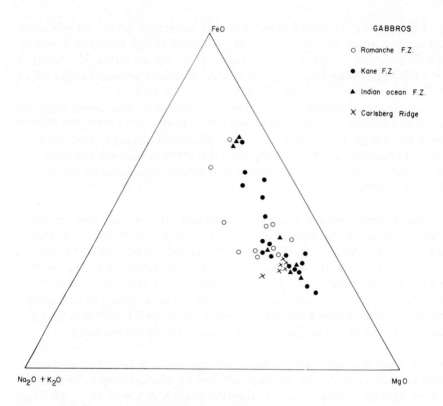

Fig. 1-3. (Na$_2$O + K$_2$O)-MgO-FeO ternary diagram of gabbros collected from various oceanic structural settings. For references see Table 1-5.

consist of plagioclase, olivine, clinopyroxene, orthopyroxene, and accessory minerals such as sphene, hornblende, apatite, and titanomagnetite (Table 1-4). The relative proportions of the various minerals make it possible to distinguish different types of gabbros.

*Normal gabbros* consist essentially of plagioclase and clinopyroxene. The plagioclase varies in composition between An$_{45}$ and An$_{70}$ and may account for up to 40—60 vol.% of the bulk rock. The clinopyroxene is usually xenomorphic and could account for 25—35 vol.% of the rock. The clinopyroxene is a diopside or an augite with $2V$ varying between 38 and 56°. The minor constituents are mainly colorless olivine (5 vol.%). These gabbros are also characterized by a relatively low K$_2$O content (0.06—0.15%) and a low TiO$_2$ content (0.2—0.5%) when compared to ocean ridge basalts (Table 1-4).

TABLE 1-4

Chemical analyses of selected gabbroic rocks from the ocean floor

| | Romanche Fracture Zone | Romanche Fracture Zone | Mid-Atlantic Ridge, 30°N | Mid-Atlantic Ridge, 24°N | Carlsberg Ridge |
|---|---|---|---|---|---|
| | alkaline gabbro P6707-25F | pyroxene-plagioclase gabbro CH80-KS-5-1 | gabbro GE260 | troctolite 1 | gabbros (ave. of 4) |
| $SiO_2$ (wt.%) | 46.32 | 50.58 | 48.20 | 46.42 | 49.81 |
| $Al_2O_3$ | 16.84 | 16.45 | 14.30 | 23.68 | 16.64 |
| $Fe_2O_3$ | 2.50 | 1.14 | — | 0.52 | 1.63 |
| FeO | 4.60 | 4.07 | 11.20 | 3.22 | 3.67 |
| MnO | 0.28 | 0.13 | n.d. | 0.06 | 0.13 |
| MgO | 8.46 | 7.81 | 7.30 | 11.49 | 9.90 |
| CaO | 7.60 | 11.68 | 10.50 | 11.47 | 14.28 |
| $Na_2O$ | 4.03 | 3.41 | 2.54 | 1.81 | 2.39 |
| $K_2O$ | 0.08 | 0.16 | 0.17 | 0.03 | 0.12 |
| $TiO_2$ | 0.43 | 0.47 | 1.75 | 0.05 | 0.40 |
| $P_2O_5$ | 0.05 | 0.02 | 0.15 | 0.01 | n.d. |
| $H_2O$ | 5.80 | 2.73 | n.d. | 1.15 | 0.88 |
| Total | 99.92 | 98.64 | 96.11 | 99.91 | 99.85 |

The gabbros from the Romanche Fracture Zone are from Bonatti et al. (1971), and from the present study (CH80-KS-5-1; 0°02'N, 12°34'W; depth 3760 m). The Mid-Atlantic Ridge samples (GE260) are from Kay et al. (1970), and sample 1 is from Miyashiro et al. (1970b). The Carlsberg Ridge samples are from Dmitriev and Sharaskin (1975).

*Gabbros with orthopyroxene* were often found to have undergone various degrees of alteration. Hypersthene is usually less abundant (8—17%) than the clinopyroxene. The hypersthene is usually altered into antigorite. Laminated hypersthene-bearing gabbros with parallel oriented cumulates of plagioclase were found in the Romanche Trench (Melson and Thompson, 1971; Bonatti et al., 1971), in the St. Paul Fracture Zone (Bonatti et al., 1971), and in a fracture zone of the Indian Ocean (C.G. Engel and Fisher, 1969).

*Troctolite or olivine-enriched gabbros* are often uralitized. This is due to the formation of amphibole pseudomorphs after olivine. The olivine is also altered into serpentine. Other products of mineral alteration are chlorite, talc, tremolite, and antophyllite. The olivine content of these rocks is usually higher than 16%. Chemical analyses show that the rocks usually consist of a low $SiO_2$ content (<47%), a relatively high MgO content (>12%), and high $Al_2O_3$ content (>20%) (Table 1-4). Both the high magnesia and alumina contents reflect the high plagioclase and olivine contents of these rocks.

14

*Norites* are a variety of gabbro in which the orthopyroxene, usually hypersthene, is a constituent dominant over the clinopyroxene. This type of rock consists essentially of calcic plagioclase and hypersthene. The secondary minerals are chlorite, antophyllite, hornblende, grunerite, epidote, and zoisite. Chemical analyses of norite shown in Table 1-5 indicate a higher FeO (10.02%) and $Fe_2O_3$ (9.09%) and a higher $TiO_2$ content (4.0%) with respect to the other types of gabbros.

TABLE 1-5

Chemical analyses of anorthosite (C.G. Engel and Fisher, 1969), pyroxenite (Hékinian, 1970), and norite (Bonatti et al., 1971) from the Indian Ocean (Indian Oceanic Ridge and Ob Trench) and from the equatorial Atlantic, respectively

|  | Anorthosite 97-X | Pyroxenite V16-97-8 | Norite P6707-5M |
|---|---|---|---|
| $SiO_2$ (wt.%) | 56.43 | 49.20 | 43.38 |
| $Al_2O_3$ | 26.10 | 3.92 | 11.89 |
| $Fe_2O_3$ | 0.51 | 8.17* | 9.09 |
| FeO | 0.63 | — | 10.02 |
| MnO | 0.1 | — | 0.28 |
| MgO | 0.92 | 17.13 | 5.41 |
| CaO | 8.34 | 18.05 | 9.97 |
| $Na_2O$ | 6.36 | 0.35 | 3.00 |
| $K_2O$ | 0.7 | 0.01 | 0.14 |
| $TiO_2$ | 0.18 | 0.16 | 4.00 |
| $P_2O_5$ | tr. | n.d. | 1.25 |
| $H_2O$ | 0.24 | n.d. | 1.42 |
| Total | 99.79 | 97.99 | 99.95 |
| Ba (ppm) | <2 |  |  |
| Co | 370 |  |  |
| Cr | 4 |  |  |
| Cu | 13 |  |  |
| Ni | 34 |  |  |

tr. = trace.

*Indicates total Fe calculated as $Fe_2O_3$.

*Alkali gabbros* are so called because they contain modal and/or normative nepheline (Honnorez and Bonatti, 1970). They were recovered in the Romanche Fracture Zone of the equatorial Mid-Atlantic Ridge. They are holocrystalline rocks with ophitic and miarolitic texture. The major mineral constituents are plagioclase ($An_{20}$—$An_{70}$), nepheline (when present), titanaugite or diopside. The minor components consist of natrolite, barkevikite, biotite, hornblende, Fe-oxide minerals, sphene, pyrite, apatite, and zircon.

Some of the gabbroic rocks collected from the ocean floor are metamor-phosed. Contorted lamellae of clinopyroxene are often encountered in the gabbros from the Ob Fracture Zone in the Indian Ocean (Hékinian, 1970). Metagabbros were drilled in the Gorringe Bank (Atlantic Ocean) and they consist of pyroxene which is partially replaced by actinolite. The cataclastic type of gabbro dredged from the Gorringe Bank by the R.V. "Conrad" is made up of labradorite, diallage Fe-oxide minerals, varying amounts of secondary amphibole, chlorite, and quartz.

*Rodingites.* In the study of ocean floor rocks, the name rodingite was given to gabbroic (Honnorez and Kirst, 1975) or to ultramafic (Aumento and Loubat, 1971) rocks which have undergone various stages of alteration. Amphibolitization, prehnitization, and serpentinization are the main types of primary mineral replacement encountered in the rocks. Hence, the most altered specimens usually consist of serpentine, chlorite, amphibole, prehnite, and hydrogarnet. It was found that in some specimens prehnite made up as much as 80% of the bulk rock (Honnorez and Kirst, 1975). Relatively fresh specimens of rodingite from central equatorial fracture zones are comprised of plagioclase, diopsidic augite, hypersthene (2%), prehnite, and hydrogarnet (Honnorez and Kirst, 1975). In continental ophiolitic provinces the term rodingite has been used to indicate extremely altered rocks of different types such as gabbros, basalts, graywackes, granites, dacites, and shales. Hence, in ophiolitic complexes rodingites are developed from diverse rock types associated with serpentinites (Coleman, 1977).

TABLE 1-6

Composition of amphiboles found in rocks from the Mid-Atlantic Ridge

|  | Pargasite in peridotite from St. Paul's Rocks (Melson et al., 1967) | Pargasite in serpentinite (Miyashiro et al., 1969a) | Amphibole in gabbro DSDP Leg 37, 334-22-1/17—23 (Symes et al., 1977) |
|---|---|---|---|
| $SiO_2$ (wt.%) | 43.00—47.00 | n.d. | 44.44 |
| $Al_2O_3$ | 10.00—12.00 | 8.00—13.20 | 7.64 |
| $TiO_2$ | 0.20— 0.50 | n.d. | 0.46 |
| FeO* | 3.60— 7.80 | n.d. | 9.65 |
| MgO | 17.00—20.00 | n.d. | 15.96 |
| CaO | 11.00—13.00 | 11.00—12.00 | 10.83 |
| $Na_2O$ | n.d. | 1.00— 2.60 | 2.00 |
| $K_2O$ | n.d. | n.d. | 0.22 |

*Total Fe as FeO.

16

*Hornblendite.* Early recognition of rocks with a composition close to a hornblendite was made by Lacroix (1917). Experimentally, Yoder and Tilley (1962) have suggested a possible similarity between hornblendites and tholeiites.

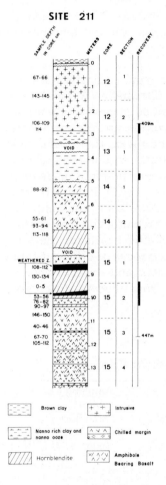

Fig. 1-4. Schematic diagram of a drilled core (DSDP Leg 22, Site 211) containing hornblendite from the northeastern Indian Ocean at 09°46.53'S, 102°41.95'E and from a water depth of 5535 m (subbottom depth of 409 m) (Hékinian, 1974a).

A hornblendite found during a deep-sea drilling operation in the eastern Indian Ocean at about 500 miles from Christmas Island is interlayered between flows of altered basaltic rocks (Figs. 1-4, 1-5A). This hornblendite consists of various flows with chilled margins. The inner part of the flow is primarily made up of amphiboles (79%), mica (4%), olivine (5.6%), and glass (11.4%). The amphiboles occur as tabular and elongated needle-like crystals which are

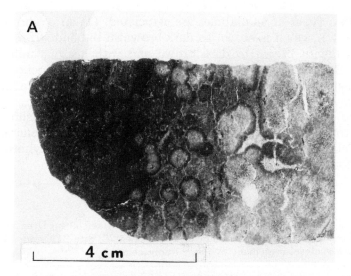

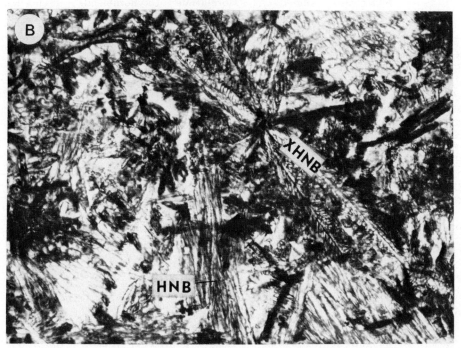

Fig. 1-5. A. Contact zone between the hornblendite (dark area) and the basalt (light area) in DSDP Leg 22, Site 211 (core 15, section 2, 81—89 cm). The circular patches consist of plagioclase, chalcedony and altered olivine. B. Microphotograph of hornblendite from Site 211 (15-1, 130—134 cm). Arborescent oxyhornblende (XHNB), common hornblende (HNB) and mica occur.

radially arranged. Two types of amphiboles are associated: (1) an oxyhorn-blende ($2V = 52\text{—}64°$), which is brown to dark-brownish red and strongly pleochroic from yellowish brown to dark brown and with nearly parallel extinction (Fig. 1-5B), and (2) a common hornblende which is pleochroic from pale yellowish-green to greenish light brown ($2V = 72\text{—}80°$) and tabular in shape (Fig. 1-5B).

Chemical analyses of these two hornblendes, reported in Table 1-7, show differences in their total Fe-oxide contents (calculated as FeO). The common hornblende has a FeO content of less than 10% while that of the oxyhorn-blende could have double this value (Table 1-7).

TABLE 1-7

Electron microprobe analyses of amphiboles from hornblendite recovered during Deep Sea Drilling Project operations in the northeastern Indian Ocean (Leg 22, Site 211, core 15, section 2). The bulk rock analyses were made by X-ray fluorescence

|  | Bulk rock 0—5 cm | Glass 40—49 cm (ave. of 3) | Hornblende 40—49 cm | Oxyhornblende 40—49 cm | Plagioclase 40—49 cm |
|---|---|---|---|---|---|
| $SiO_2$ (wt.%) | 40.95 | 45.41 | 46.52 | 30.25 | 67.22 |
| $Al_2O_3$ | 16.54 | 20.01 | 17.56 | 14.25 | 19.30 |
| $Fe_2O_3$ | 4.10 | — | — | — | — |
| FeO | 5.00 | 9.25* | 8.23* | 19.29* | — |
| CaO | 8.75 | 4.05 | 8.30 | 10.29 | 0.27 |
| MgO | 4.32 | 4.92 | 6.67 | 14.81 | — |
| $Na_2O$ | 3.24 | 5.58 | 4.82 | 1.53 | 10.86 |
| $K_2O$ | 2.01 | 1.43 | 1.33 | 0.14 | 0.26 |
| $TiO_2$ | 2.51 | 2.37 | 2.49 | 6.21 | — |
| $P_2O_5$ | 0.69 | 0.54 | 1.11 | 0.34 | n.d. |
| $CO_2$ | 2.38 | n.d. | n.d. | n.d. | n.d. |
| LOI | 7.72 | n.d. | n.d. | n.d. | n.d. |
| Total | 98.21 | 93.02 | 97.03 | 97.11 | 97.90 |

LOI = loss on ignition.

*Total Fe calculated as FeO.

## BASALTS

Basalt is the general term used by marine geologists to define the major type of solidified lava erupted on the ocean floor. It is important to find and define variations among the different types of basalts in order to obtain more insight into magmatic evolutions and evaluate different volcanic events associated with different structural settings.

Other prerequisites of rock classification are (1) the recognition of a major group of basaltic rocks as products of various magma types, and (2) modal compositions of the members of the major groups leading to their more specific classification.

## Classification of basaltic rocks

Since the term basalt, which is of uncertain origin but probably derived from the Greek word meaning touchstone (Johansen, 1939), was invented, there has been a large number of variants of rocks called basalts. Through a series of outstanding studies of basaltic provinces throughout the world, remarkable progress was made by Richey and Thomas (1930), Kennedy (1933) and others in the recognition of two fundamental basaltic magma types, such as the alkali basalt and the tholeiites. Chiefly through the work of Tilley (1950), Kuno et al. (1957) and Kuno (1959), the recognition of the main petrographic differences between tholeiites and alkali basaltic rock types was made. Yoder and Tilley (1962) suggested that all the elements for chemical definition of the two basaltic magma types are present in the system silica-nepheline-forsterite-diopside. It is clear that a major part of the explored structure of the ocean floor is made up of tholeiites, which are defined by

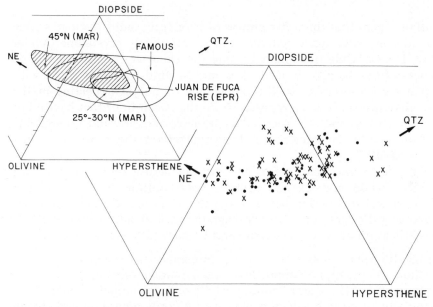

Fig. 1-6. Nepheline-diopside-olivine-hypersthene-quartz normative diagram (Yoder and Tilley, 1962). Crosses indicate the rock samples collected by the submersibles "Archimède" and "Cyana". The black dots represent 94 analyses of selected basalt samples from the Mid-Atlantic Ridge (Cann, 1971a). The inset shows the fields for (1) both dredged and submersible samples from near 36—37°N on the Mid-Atlantic Ridge (Bougault and Hékinian, 1974); (2) the Juan de Fuca Rise (Kay et al., 1970); (3) 25—30°N on the Mid-Atlantic Ridge (Miyashiro et al., 1969b); and (4) 45°N on the Mid-Atlantic Ridge (Muir et al., 1964; Aumento, 1968).

Yoder and Tilley (1962) as a rock type consisting of augite or sub-calcic augite, labradorite, and Fe-oxide. From petrographic observations it is noticed that the ocean-floor basalts can be defined within the triangular system of plagioclase-olivine-pyroxene.

The scheme which follows for the classification of continental, island, and ocean floor basalts is primarily that which was proposed by Kuno (1960), Yoder and Tilley (1962), and G.A. Macdonald and Katsura (1964), and is based on rocks from Japan and Hawaii. Yoder and Tilley (1962), using mainly the Hawaiian alkali and tholeiitic series and considering the various properties of the normative tetrahedron, have defined five types of basaltic rocks (Fig. 1-6):

(1) Oversaturated tholeiite with normative quartz and hypersthene.
(2) Saturated tholeiite with normative hypersthene.
(3) Undersaturated olivine tholeiite with normative hypersthene and olivine.
(4) Olivine tholeiite with normative olivine.
(5) Alkali basalt with normative nepheline.

These authors claim that these five groups of basalt are readily distinguishable by the modal minerals, provided the rock is completely crystalline (Yoder and Tilley, 1962). While this might be true for island volcanics, it is not so for the ocean-floor basalts. Indeed, it is rare to find tholeiitic rocks with modal quartz corresponding to the percentage of quartz appearing in the norm. No tholeiites have been reported from the ocean floor that have a fair amount of modal hypersthene. Hence, the presence of quartz and hypersthene derived from purely chemical data does not reflect the most crystalline ocean floor basalts.

Another type of classification for the Hawaiian basaltic rocks was proposed by G.A. Macdonald and Katsura (1964). These authors have also used a primarily chemical distinction between the tholeiites and the alkali suites as seen by plotting the rocks on a Harker diagram (wt.% alkalies versus $SiO_2$).

The classification of G.A. Macdonald and Katsura (1964) is as follows:

(1) Tholeiitic suite: rocks with tholeiitic mineral composition —
    (a) tholeiitic basalt, containing less than 5% modal olivine;
    (b) tholeiitic olivine basalt, containing 5% or more modal olivine;
    (c) oceanic (picritic basalt of oceanic type), containing abundant pheno-crysts of olivine and less than 30% feldspar.
(2) Alkalic suite: rocks with alkalic mineral composition —
    (a) alkalic basalt, containing less than 5% modal olivine;
    (b) alkalic olivine basalt, containing 5% or more modal olivine, and less than 5% normative nepheline;
    (c) basanite, containing more than 5% normative nepheline and with both modal nepheline and feldspar;

(d) basanitoid, containing more than 5% normative but no modal nepheline;

(e) ankaramite (picritic basalt of ankaramite type), containing very abundant phenocrysts of olivine and augite and less than 30% total feldspar.

The main contribution of Kuno (1960) in basalt classification was to introduce the high-alumina basalt as a primary magma, transitional in its chemistry and mineralogy between the tholeiite and alkali olivine basalt magmas. The non-porphyritic high-alumina basalt which Kuno considered as a primary magma is not considered in the same way by Yoder and Tilley (1962), who feel that the high-alumina content of these basalts is a consequence of the proximity of the determined derivative liquid lines to plagioclase conditions. As discussed later on, the high-alumina basalts of the ocean floor consist primarily of an abnormal accumulation of plagioclase, and non-porphyritic high-alumina basalts are not observed. The high-alumina type of basalt is not only found among rocks of the tholeiite suites but also occurs in alkali basalt types and in the calc-alkali basalt suites of the orogenetic basalts. G.D. Nicholls (1965) showed that glassy basalts from the ocean floor representing the compositions of magmatic liquids have variable $Al_2O_3$ contents which are higher than 17% in some specimens. The high $Al_2O_3$ content would not necessarily be due to the accumulation of plagioclase phenocrysts.

Kuno (1960), in his study of basalts from Honshu and Izu Islands in Japan, found three types of basalts: (1) the tholeiite with low $Al_2O_3$ and alkalies, (2) the alkali basalt with variable $Al_2O_3$ and higher alkalies, and (3) the high-alumina basalt with higher $Al_2O_3$ and intermediate alkalies.

The high-alumina basalt (with $Al_2O_3$ >16.5%) is also considered by Kuno (1960) to be a type transitional between the tholeiite and the alkali basalt. These high-alumina basalts from Japan are aphyric rocks and hence are not considered by Kuno (1960) to be the result of an accumulation of plagioclase.

Aphyric high-alumina basalts (with $Al_2O_3$ >17%) were also found near 29°N in the Mid-Atlantic Ridge (G.D. Nicholls, 1965). The high-alumina basalt from Japan occurs in a zone between the tholeiite and the alkali province. This structural setting and the transitional chemistry of the high-alumina basalt suggested to Kuno (1960) that it is produced by a partial melting of the mantle at an intermediate depth (about 200 km). The term high-alumina basalt is often used for ocean-floor basalts which have an abnormal accumulation of early-formed plagioclase. This results in a depletion of Mg-olivine and the extraction of other non-feldspathic components. The high-alumina groups of Kuno (1960) show many of the rocks to be depleted in their MgO content (<8%), while those described by G.D. Nicholls (1965) from the Mid-Atlantic Ridge with a MgO content of about 10% are more likely to be true aphyric high-alumina basalts. Experimental studies (D.H. Green and Ringwood, 1967) have shown that at pressures between 8 and 10 kbar (about 30 km depth) an olivine tholeiite melt will fractionate and the residual liquid will be enriched in $Al_2O_3$, CaO, and $Na_2O$, thus giving rise to a plagioclase-enriched rock. Alternatively, a fractional melting of 20—25% of hypothetical pyrolitic mantle will segregate under appropriate pressure/temperature conditions (8—10 kbar, 1150°C), resulting in magma having the composition of a high-alumina basalt (D.H. Green and Ringwood, 1967).

Few detailed rock descriptions are found in the literature concerning the basalts of the ocean floor. Shand (1949), Muir et al. (1964), Aumento (1968), Miyashiro et al. (1969b), and Hékinian and Aumento (1973) have described in some detail the basaltic rocks encountered in the ocean floor. From these petrographic data it is possible to gain more insight into the relative proportion and order of crystallization of the basaltic tholeiites. Thus, the order of crystallization of minerals in tholeiites was used to define two types of rocks on the basis of whether the first generation of minerals is plagioclase or olivine: the olivine tholeiites and the plagioclase tholeiites (Miyashiro et al., 1969b; Shido et al., 1971). Based purely on petrographic data, the order of mineral crystallization, and the relative mineral proportions, six types of basaltic rocks were recognized along the Mid-Atlantic Ridge (Hékinian and Aumento, 1973; Bougault and Hékinian, 1974; Hékinian et al., 1976):

(1) Picritic basalt.
(2) Highly-phyric plagioclase basalt (HPPB).
(3) Moderately-phyric plagioclase basalt (MPPB).
(4) Plagioclase-olivine-pyroxene basalt (POPB).
(5) Olivine basalt.
(6) Plagioclase-pyroxene basalt (PPB).

This type of classification was made on samples collected from the Mid-Atlantic Ridge near 53°N and 36°N. The mineralogical and chemical criteria used here to classify ocean-floor basalt is also applied to other volcanic provinces such as basalts from the East Pacific Rise near 21°N (Chapter 2). Mineral proportion data based on both the glassy margins and the most crystalline interior of rock fragments were used to identify the various rock types. The difference between this classification and those previously mentioned (Yoder and Tilley, 1962; G.A. Macdonald and Katsura, 1964) is that more weight is given to the observed mineral phases and this grouping best reflects the bulk rock mineral content. The olivine basalts, the picritic basalt and the MPPB and HPPB were already mentioned by Yoder and Tilley (1962), G.A. Macdonald and Katsura (1964) and Kuno (1960) as olivine tholeiites, oceanite and high-alumina basalts, respectively. Other types such as the plagioclase-pyroxene basalts correspond to the oversaturated tholeiites of Yoder and Tilley (1962).

As already mentioned, the basalt classification adopted here is mainly based on the observed mineral distribution in the rock, which should theoretically reflect the chemistry. However, some difficulties could arise when the rock is mostly made up of glass. In this case it was found (Hékinian et al., 1976) that the proportion of the early-formed minerals present in the glass is diagnostic and representative of the bulk sample. This is in agreement with the fact that the early-formed minerals control the path of the remaining liquid during crystallization. The order of crystallization with diminishing temperature is:

Liquid → megacryst → phenocryst → microphenocryst → matrix → mesostasis

Such an order of crystallization is also accompanied by chemical changes attributed to the liquid-crystal reaction during solidification.

*The picritic basalts* contain megacrysts (mgcr), phenocrysts (ph), micro-phenocrysts (mcrph), and matrix (mtx). Their principal mineral phases with their order of crystallization are:

ol (mgcr) + ol, pl (ph) + ol, pl, sp (mcrph) + ol, pl, cpx + sp (mtx)

ol = olivine, pl = plagioclase, sp = spinel, and cpx = clinopyroxene.

The early-formed mineral phases as defined from the study of pillow-lava glassy rims show a phenocrystic ratio ol/(ol + pl + cpx) of 0.64—0.80, while the same ratio for their bulk mineral ratio is around 0.30. In both glassy margins and in the crystalline interior of the pillow lavas, the amount of early-formed plagioclase and clinopyroxene is subordinate to that of olivine (Fig. 1-7). Spinel, usually chromite, is a minor constituent of this type of rock and it occurs as euhedral light-brown crystals included in large olivine crystals and in the groundmass (Fig. 1-8B).

The olivine occurs throughout the whole range of crystal sizes and is represented by megacrysts (>2 mm in diameter), phenocrysts (0.5—2 mm in diameter), and microphenocrysts (<0.5 mm in diameter). The term megacryst applies here to large crystals ranging between 2 and 10 mm in diameter. The olivine megacrysts are often resorbed with rounded edges and they show undulated extinction. Sometimes the large olivine crystals are surrounded by tiny plagioclase laths showing fluidal texture or by a more glassy rim than the rest of the bulk-rock groundmass. It is likely that the megacrysts of olivine were crystallized before extrusion and brought to the surface by flottage.

*The highly-phyric plagioclase basalts (HPPB).* These are the equivalent of the picritic basalt, but instead of having olivine as the major phase, it is the plagioclase which is the most abundant mineral (Fig. 1-7). Porphyritic with megacrysts, phenocrysts and microphenocrysts of plagioclase, the groundmass is ophitic, subophitic and intersertal (Fig. 1-9B). The principal mineral phases

A

B

and their order of crystallization are summarized as follows:

pl (mgcr, ph, mcrph, mtx) + cpx (ph, mcph, mtx) + ol (ph, mcrph, mtx)

The megacrystic plagioclases are often rounded and resorbed. They also show rims of more sodic plagioclase. These early-formed plagioclases were probably brought towards lower pressure zones by flottage and reacted with the melt during their ascent. Olivine may also occur as an early mineral phase but it is believed to be a xenocrystic mineral rather than being in equilibrium with the bulk rock (Fig. 1-9C). The early-formed mineral ratio, ol/(ol + pl + cpx) is 0.01—0.22 (Table 1-9). The $TiO_2$ and the FeO/MgO weight ratio of the bulk rock are 0.6 and 0.9, respectively. The $Al_2O_3$ content is higher than 17% (Table 1-9). These rocks may also contain spinel which is always included within the olivine xenocrysts.

*The moderately-phyric plagioclase basalts (MPPB)* are porphyritic rocks which differ from the HPPB by their smaller amount of plagioclase phenocrysts and by the lack of megacrystic resorbed plagioclase. The major mineral association is (Fig. 1-8A):

pl (ph, mcrph, mtx) + ol (ph + mcrph + mtx) + cpx (ph, mcrph, mtx)

The MPPB have lower $Al_2O_3$ contents (15—17%) than the HPPB. The olivine occurs both as phenocrysts (<1%) and in the matrix (<10%). Cr-spinel is scarce or absent. The $TiO_2$ content of these rocks is higher (1.1—1.3%) than in both the HPPB and the picritic basalts (Table 1-9). For the East Pacific Rise rocks (21°N) the range in variability of $TiO_2$ and FeO/MgO is higher (>1.3% and >1.2, respectively) than that encountered in rocks from the Mid-Atlantic Ridge (Tables 1-9, 2-13).

*The plagioclase-olivine-pyroxene basalts (POPB)* differ from the previous types of rocks in their depletion in plagioclase phenocrysts and their relatively higher pyroxene content (Fig. 1-9D). Also the POPB contain moderate amounts of olivine phenocrysts. The rocks are sometimes characterized by having pyroxene reaction rims around microphenocrysts and matrix olivine. Their mineral constituents are:

ol (ph, mcrph, mtx) + pl (ph, mcrph, mtx) + cpx (ph, mcrph, mtx)

The glassy margins of these rocks contain a fair amount of plagioclase (3—12%), some olivine (2—4%), and generally some clinopyroxene (0.3—2%).

---

Fig. 1-7. A. Fragment of a picritic basalt (pillow flow; CYP78-10-3) showing megacrysts of olivine set in a dark cryptocrystalline groundmass. B. Sample CH31-DR2-352 is a highly-phyric plagioclase basalt (HPPB) with plagioclase megacrysts. Both samples were collected from the Rift Valley near 36°N on the Mid-Atlantic Ridge (Figs. 2-5, 2-6).

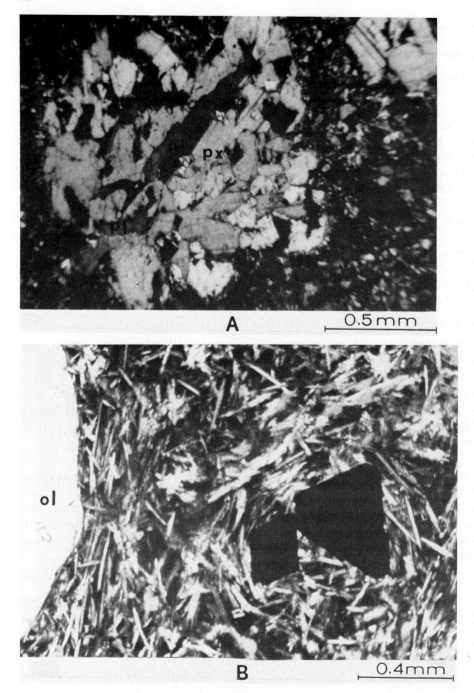

Fig. 1-8. Microphotographs of basaltic rocks from the Rift Valley and from Transform Fault "A" near 37°N on the Mid-Atlantic Ridge. The sample locations are shown in Figs. 2-6 and 7-2. A. A moderately-phyric plagioclase basalt (CYP74-31-39) showing "mirmekitic-like" texture of plagioclase included in a large clinopyroxene crystal (*px*). B. An olivine basalt (CYP73-10-03) showing chrome spinel (black). The groundmass is made up of tiny olivine (*ol*) and plagioclase laths with traces of Cr-spinel (Rift Valley). C. An altered

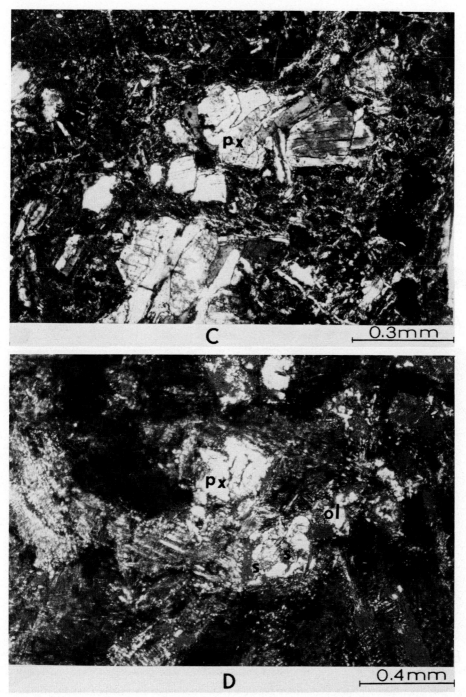

pyroxene basalt (CYP74-19-5) showing broken laths of plagioclase set in a groundmass of chlorite, serpentine, Fe-hydroxide material. Veinlets of phyllosilicates abound throughout. D. A pyroxene basalt (CYP74-26-15B) showing microlites of olivine altered into serpentine and clay (*s*). The groundmass is cryptocrystalline with traces of serpentine material. This sample was taken in association with the Mn-Fe hydrothermal deposit (Arcyana, 1975).

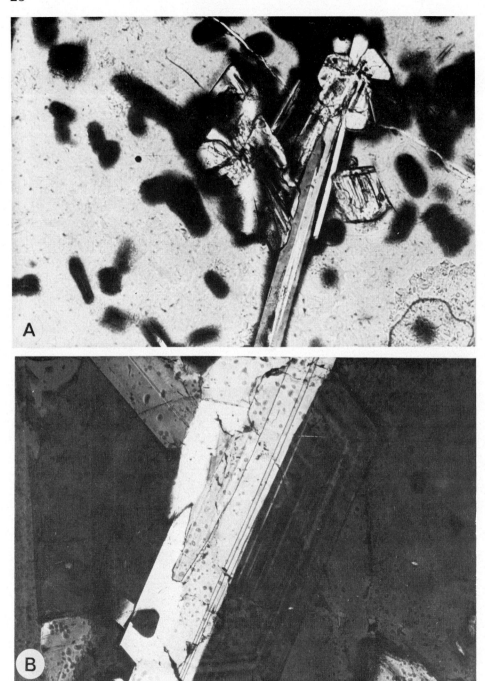

Fig. 1-9. Microphotograph of various basaltic suites collected from the East Pacific Rise near 21°N. A. Sample GL-DR-8-1 is an olivine basalt showing olivine microphenocrysts and plagioclase both set in a glassy matrix. B. A megacryst of a zoned plagioclase is set in a glassy matrix (sample CYP78-10-16) (nicols at 20°). C. A highly-phyric plagioclase basalt

29

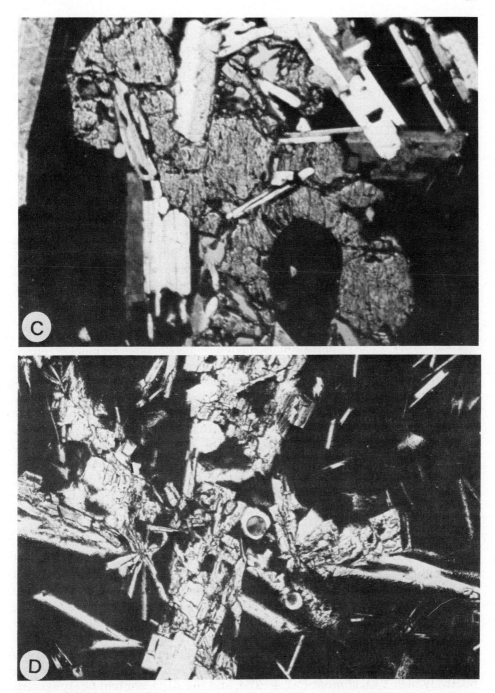

(sample CYP78-10-10) shows intergrowths of partially resorbed olivine (gray colored) with plagioclase laths (light color, under crossed nicols). D. A sample of plagioclase-olivine-pyroxene basalt (CYP78-02-03) shows pyroxene and plagioclase intergrowth set in a glassy groundmass (crossed nicols). Magnification ×63.

The occurrence of clinopyroxene phenocrysts is noticed in some specimens collected from near 36°N in the Atlantic Ocean. The ol/(ol + pl + cpx) value of the early-formed mineral phase is 0.2—0.6, while that of the bulk mineral constituents is 0.01—0.2. Both the $TiO_2$ content and the weight ratio of FeO/MgO are greater than 1.0 (Table 1-8). This type of rock is generally deprived of Cr-spinel, with the exception of a spinel with scarce chromite which was noticed near 36°N (Atlantic Ocean) and on the East Pacific Rise.

The POPB from the East Pacific Rise are generally more fractionated than their counterparts from the Mid-Atlantic Ridge. They have higher $TiO_2$ (1.3—2.0%) and higher FeO/MgO ratios (1.3—1.6) than Mid-Atlantic Ridge volcanics (Table 1-8, see also Table 2-11 and Fig. 2-34).

*The olivine basalts* differ from the POPB and MPPB in their relative abundance of olivine phenocrysts (1—4%) and their relative depletion of the plagioclase phenocryst-microphenocryst association (<1%) (Figs. 1-8B, 1-9A). They also differ from the picritic basalts by the lack of olivine megacrysts of a resorbed nature and in their smaller amount of olivine content (10—25%). Their major mineral association is comparable to that of the picritic basalts:

ol (ph, mcrph, mtx) + pl (ph, mcrph, mtx) + cpx (mtx) + sp (mtx)

Their order of crystallization is olivine → plagioclase → spinel → clinopyroxene. The phenocrystic ratio ol/(ol + pl + cpx) of the pillow lava glassy margin is 0.78—1.00. The same ratio for the bulk-rock mineralogy is 0.18—0.30. The olivine basalts are distinguishable from all other types except the picritic basalts by the predominance of olivine among the early formed crystals. Usually the olivine basalts have less than 3% early-formed plagioclase phases. These rocks have a narrow range of $Al_2O_3$ variability (15—16%), a low $TiO_2$ content (0.7—1.0%), and a relatively high MgO content (9—10%) for the Mid-Atlantic Ridge rocks. The East Pacific Rise olivine basalts have a higher $TiO_2$ and FeO/MgO ratio (>1% and 0.9—1, respectively) than their counterparts found along the Mid-Atlantic Ridge north of latitude 32°N (Figs. 2-24, 2-31, Tables 1-10, 2-12).

*Plagioclase-pyroxene basalt (PPB)* is essentially deprived in olivine and spinel (Fig. 1-8C, D). When olivine occurs, it forms a minor constituent of less than 2% of the bulk rock, and has crystallized at the same time as both pyroxene and plagioclase. The texture of the crystalline rocks is subophitic to ophitic, while the glassy margins of the pillow flow often show a glamoporphyritic texture. The glassy portion of most PPB is characterized by the early occurrence of both clinopyroxene and plagioclase. The content of clinopyroxene in these rocks is usually higher than 50%. The chemistry of the PPB shows a relatively high $TiO_2$ content (>1.4—2%), and the $Al_2O_3$ content rarely exceeds 15% by weight (Table 1-8).

TABLE 1-8

Classification of oceanic basalts using some chemical parameters. Plagioclase-olivine-pyroxene basalts (POPB) and plagioclase-pyroxene basalts (PPB) from the Mid-Atlantic Ridge Rift Valley near 36°50′—37°N in the FAMOUS area

| Sample No.* | C.I.** | $\dfrac{Fe_2O + FeO}{MgO}$ | $\dfrac{K_2O \times 100}{K_2O + Na_2O}$ | $TiO_2$ | $Al_2O_3$ |
|---|---|---|---|---|---|
| Plagioclase-olivine-pyroxene basalt | | | | | |
| 10—2 | 41.23 | 2.00 | 9.29 | 1.58 | 14.55 |
| 7—5 | 46.12 | 1.29 | 7.24 | 1.43 | 14.72 |
| 7—6 | 46.13 | 1.25 | 7.46 | 1.44 | 14.61 |
| 7—7 | 47.18 | 1.24 | 5.50 | 1.42 | 14.41 |
| 7—8 | 47.47 | 1.16 | 8.70 | 1.25 | 14.45 |
| 7—9 | 45.97 | 1.13 | 8.04 | 1.26 | 14.05 |
| 8—10 | 46.63 | 1.35 | 6.17 | 1.16 | 14.70 |
| 13—22 | 49.62 | 1.24 | 6.45 | 1.06 | 14.91 |
| 13—23 | 49.76 | 1.22 | 6.58 | 1.06 | 14.92 |
| 13—24 | 49.50 | 1.22 | 7.22 | 1.05 | 14.91 |
| 18—1 | 49.89 | 0.94 | 8.19 | 1.11 | 14.40 |
| Plagioclase-pyroxene basalt | | | | | |
| 31—38 | 51.01 | 1.19 | 7.01 | 0.98 | 14.48 |
| 14—26 | 51.90 | 1.24 | 7.59 | 1.04 | 14.98 |
| 14—30 | 51.42 | 1.28 | 8.05 | 1.03 | 14.64 |
| 17—40 | 49.54 | 1.24 | 11.12 | 1.11 | 15.37 |
| 17—41 | 47.87 | 1.27 | 7.35 | 1.21 | 14.70 |
| 17—42 | 48.80 | 1.34 | 7.53 | 1.15 | 14.81 |
| 18—44 | 48.00 | 1.22 | 5.37 | 1.05 | 14.55 |
| 5 | 46.68 | 0.92 | 5.91 | 1.07 | 15.03 |
| 6 | 50.89 | 1.08 | 6.28 | 1.05 | 15.13 |
| 30 | 43.97 | 1.45 | 6.05 | 1.48 | 14.16 |
| 31A | 48.25 | 0.99 | 4.52 | 1.13 | 14.17 |
| 31C | 49.09 | 1.05 | 5.43 | 1.10 | 14.40 |

*Sample numbers indicate the last 4 digits of the "Archimède" and "Cyana" submersible dives (ARP73, ARP74, and CYP74, respectively). The location of these samples is shown in Figs 2-6, 7-2 and 7-9.
**Crystallization index (Poldervaart and Parker, 1964).

In certain provinces of the Pacific Ocean, basaltic rocks high in $TiO_2$ (>2%) and FeO (>12%, where total Fe is calculated as FeO), are also as enriched in clinopyroxene as are the PPB (see Chapter 2). However, these Pacific Ocean rocks contain a clinopyroxene which is richer in iron than that of the PPB. The name "ferrobasalt" is most appropriate for this latter type of rock.

*Ternary system pyroxene-plagioclase-olivine.* In order to visualize and study the trend of mineral crystallization, a ternary diagram comprising the major

TABLE 1-9

Chemical characteristics of the plagioclase-rich basalts from the Mid-Atlantic Ridge near 36°50′—36°57′N

| Sample No. | C.I. | $\dfrac{Fe_2O_3 + FeO}{MgO}$ | $\dfrac{K_2O \times 100}{K_2O + Na_2O}$ | $TiO_2$ | $Al_2O_3$ |
|---|---|---|---|---|---|
| *Highly-phyric plagioclase basalt* | | | | | |
| 31—36 | 67.56 | 0.87 | 4.46 | 0.55 | 20.58 |
| 31—37 | 66.97 | 0.85 | 5.55 | 0.55 | 20.65 |
| 14—31 | 68.05 | 0.85 | 4.20 | 0.43 | 19.20 |
| 14—32 | 65.34 | 0.85 | 3.82 | 0.48 | 17.77 |
| 14—33 | 67.71 | 0.82 | 2.08 | 0.45 | 18.97 |
| | | | | | |
| *Moderately-phyric plagioclase basalt* | | | | | |
| 11—18 | 52.93 | 0.96 | 4.89 | 1.00 | 15.51 |
| 30—33 | 49.49 | 1.04 | 5.08 | 1.14 | 15.33 |
| 9—13 | 48.23 | 1.36 | 6.60 | 1.23 | 15.80 |
| 31—39 | 51.14 | 1.13 | 6.01 | 1.14 | 15.80 |
| 31—40 | 52.61 | 1.16 | 5.46 | 1.16 | 16.30 |
| 16—37 | 54.18 | 1.09 | 8.98 | 1.04 | 16.80 |

See footnotes to Table 1-8.

mineral phases (plagioclase, olivine, and pyroxene) of the ocean floor basalts is used (Fig. 1-10). This diagram is similar to the system diopside-forsterite-anorthite of Osborn and Tait (1952) and was also used by Shido et al. (1971), and Miyashiro et al. (1969b), to show the distribution of plagioclase and olivine tholeiites in the Mid-Atlantic Ridge (Fig. 1-10). This ternary diagram is part of the more complex diopside-forsterite-nepheline-quartz diagram of the iron-free basalt system of Yoder and Tilley (1962). The plagioclase-olivine-pyroxene system corresponds theoretically to the critical plane of silica undersaturation.

The ternary plagioclase-olivine-pyroxene diagram is used to show both the modal and normative composition and the order of mineral crystallization. The relationship between the early-formed mineral phases and the bulk rocks with more than 50% of identifiable crystalline material was used to construct the plots of Fig. 1-11. The mineral proportion is shown by a tie-line going from the bulk rock to its corresponding phenocryst-microphenocryst assemblage (Fig. 1-11). Three distinct groupings with three different assemblages having crystallized in various proportions are seen (Fig. 1-11). Thus, the plagioclase corner shows the field where a melt has first crystallized plagioclase with very little early clinopyroxene and olivine phases (Fig. 1-11). Both early and late crystalline phases of the HPPB and MPPB consist essentially of plagioclase and clinopyroxene (Fig. 1-11). Some of the MPPB plot in this latter field and show a simultaneous crystallization of plagioclase and clino-

TABLE 1-10

Chemical characteristics of olivine-rich basalts from the Mid-Atlantic Ridge near 36°50'—
36°57'N

| Sample No. | C.I. | $\dfrac{Fe_2O_3 + FeO}{MgO}$ | $\dfrac{K_2O \times 100}{Na_2O + K_2O}$ | $TiO_2$ | $Al_2O_3$ |
|---|---|---|---|---|---|
| *Picritic basalt* | | | | | |
| 30—32 | 66.48 | 0.61 | 3.76 | 0.64 | 14.58 |
| 30—34 | 65.06 | 0.60 | 3.72 | 0.63 | 14.51 |
| 10—3 | 60.81 | 0.42 | 3.46 | 0.45 | 10.80 |
| 10 | 61.82 | 0.73 | 1.70 | 0.56 | 15.85 |
| *Olivine basalt* | | | | | |
| 7—1 | 52.08 | 0.92 | 5.24 | 0.88 | 15.20 |
| 13—4 | 53.44 | 1.01 | 5.33 | 0.87 | 15.25 |
| 9—12 | 57.45 | 0.78 | 4.34 | 0.87 | 15.05 |
| 10—14 | 54.09 | 0.83 | 4.56 | 0.84 | 15.14 |
| 10—15 | 52.15 | 0.87 | 6.34 | 1.02 | 15.02 |
| 10—16 | 52.84 | 0.80 | 4.83 | 0.82 | 15.05 |
| 11—17 | 50.14 | 0.95 | 6.74 | 1.14 | 15.04 |
| 31—35 | 55.86 | 0.79 | 5.10 | 0.86 | 15.13 |
| 12—19 | 54.07 | 0.93 | 5.00 | 1.00 | 15.00 |
| 16—36 | 53.38 | 0.94 | 6.25 | 1.09 | 15.20 |
| 9 | 59.59 | 0.92 | 5.01 | 0.91 | 15.64 |
| 11 | 59.31 | 0.91 | 4.43 | 0.85 | 15.49 |
| 8 | 57.42 | 0.89 | 5.74 | 1.05 | 14.72 |
| 20 | 54.67 | 0.97 | 5.47 | 1.14 | 14.80 |
| 10A | 57.25 | 0.92 | 6.17 | 0.88 | 15.64 |
| 10B | 58.05 | 0.86 | 5.95 | 0.90 | 15.56 |
| 10C | 56.72 | 0.89 | 6.71 | 0.90 | 15.65 |
| 10D | 57.29 | 0.85 | 5.02 | 0.91 | 15.68 |
| 21 | 52.44 | 0.93 | 8.23 | 1.12 | 14.63 |
| 28 | 52.16 | 0.87 | 10.65 | 1.14 | 14.42 |
| 26 | 52.24 | 0.99 | 8.77 | 1.16 | 14.77 |

See footnotes to Table 1-8.

pyroxene (Fig. 1-11). This fact is supported by the occurrence of myrmekitic and interpenetrative intergrowth of plagioclase in the clinopyroxene host mineral. The pyroxene basalts falling near the plagioclase-clinopyroxene line contain early phases of pyroxene and plagioclase. This is represented by a tie-line going towards the field of the plagioclase-enriched rocks (HPPB and MPPB, Fig. 1-11). Other pyroxene basalts without any tie-lines are depleted in phenocrystic and microphenocrystic material and are intergranular holocrystalline rocks. Both the plagioclase- and the pyroxene-enriched rocks (HPPB, MPPB, and PPB) contain very little olivine (<5%), and in most cases the olivine is part of the late crystalline phase found in the groundmass. The plagioclase-olivine-pyroxene basalts containing almost equal amounts of early-

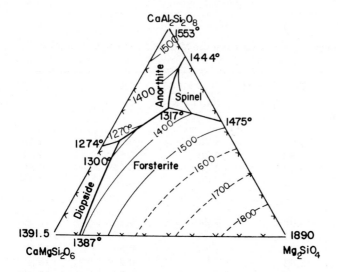

Fig. 1-10. Phase diagram for the system diopside-forsterite-anorthite (after Osborn and Tait, 1952), showing the temperature curves for different melts. The crystallization of all three phases (olivine, pyroxene and plagioclase) occurs in the piercing point at about 1270°C (eutectic), where all three phases crystallize simultaneously.

formed plagioclase and olivine fall along the plagioclase-olivine line, while their bulk composition related by a tie-line is close to the eutectic point $E$ on the ternary diagram of Fig. 1-11. The plagioclase-olivine-pyroxene basalts have none of the early-formed clinopyroxene. These types of basalts differ from the pyroxene basalts in their showing the effect of an olivine-clino-pyroxene reaction. The olivine-microphenocrysts and matrix are sometimes surrounded completely or partially by a rim of clinopyroxene. This could be interpreted as due to a continued crystallization of the olivine in the matrix, and its reaction with the residual melt has given rise to the clinopyroxene. A similar type of reaction relationship between olivine and clinopyroxene is seen within the olivine basalt. The olivine basalts and the picritic basalts have a bulk composition approaching that of the plagioclase-olivine-pyroxene basalts, but their abundance in early-formed olivine plots them towards the olivine corner.

This experimental plagioclase-pyroxene-olivine ternary diagram shows a trough corresponding to a low-temperature cotectic line along which two phases, olivine and plagioclase, crystallize simultaneously. Such a cotectic line in natural basaltic glass corresponds to a zone where the normative major mineral phases plot (Fig. 1-12). This is a broad zone rather than a line because other compounds and variables (pressure, temperature) intervene in a natural system. The broad zone of natural glass composition may also reflect the

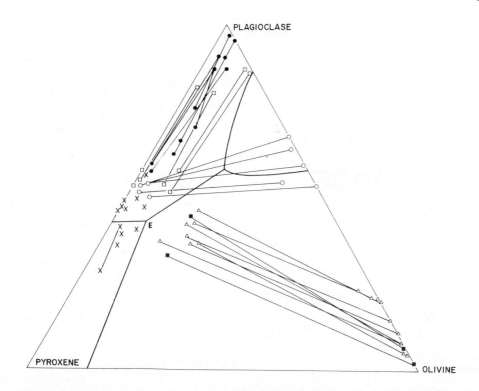

Fig. 1-11. Pyroxene-plagioclase-olivine (modal vol. %) ternary diagrams of basaltic rocks collected on the Mid-Atlantic Ridge near 36°50′N (FAMOUS area) and on the East Pacific Rise near 9°N. The tie-lines relate phenocrystal phases with their corresponding bulk rock composition. The full circles indicate the highly-phyric plagioclase basalts (HPPB); the empty circles are the plagioclase-olivine-pyroxene basalts (POPB); the triangles indicate the olivine basalts; the squares are the picritic basalts; the crosses are the plagioclase-pyroxene basalts (PPB); and the filled squares are the moderately-phyric plagioclase basalts (MPPB). The samples with no tie-lines are those deprived of phenocrysts and micropheno-crysts. E indicates a eutectic point located on the experimentally-derived reaction lines of Osborn and Tait (1952) as seen in Fig. 1-10; it is interesting to note that most of the final products of crystallization (bulk-rock compositions) plot near the line of reaction from pl = 60%, ol = 20%, cpx = 20% to pl = 40%, ol = 10%, cpx = 50%, at which point plagio-clase, olivine and clinopyroxene crystallize simultaneously.

amount of crystals which have separated from the melt during cooling. The bulk-rock, modal mineral data of basalt crystallized up to at least 60% of its volume also show a tendency to plot in the general area of the normative glass data. The two main paths of melt-crystal reaction are represented by the depletion of (1) early-formed plagioclase, and (2) early-formed olivine. The fact that a group of rocks have their early mineral phases plotting close to the plagioclase-olivine line is explained by the nature of their melt, which has an almost equal proportion of plagioclase to olivine (Fig. 1-11). The absence

36

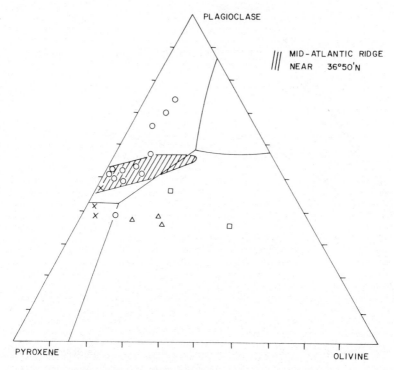

Fig. 1-12. Plagioclase-pyroxene-olivine normative ternary diagram of basaltic glass from both the Atlantic and Pacific Oceans. The hypothetical reaction lines dividing the various fields are the same as in Fig. 1-10. Symbols as in Fig. 1-11.

of spinel in these rocks is due to the reaction of spinel and olivine giving rise to plagioclase. Hence, it is likely that the temperature of the melt which gave rise to the early-formed olivine and plagioclase must have been lower than that of the spinel solidus.

*Mineralogy of basaltic rocks*

The primary mineral phases of unaltered basaltic rocks from the ocean floor can be summarized in five major constituents: pyroxene, olivine, plagioclase, opaques (Fe-Ti oxides), and spinel. The mineralogical description presented here deals primarily with samples from accreting plate boundary regions and seamounts. Details on the mineralogy of altered basement rocks will be given in separate chapters.

*Pyroxene.* Clinopyroxene, usually colorless or with a faint, light-green shade, is one of the main constituents of most mid-oceanic ridge basalts. The composition of clinopyroxene usually corresponds to a diopsidic augite and an

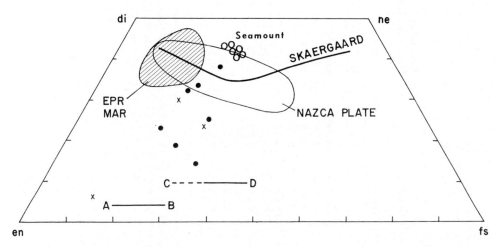

Fig. 1-13. Wollastonite-enstatite-ferrosilite ternary diagram of pyroxene composition from basaltic rocks collected from both the Mid-Atlantic Ridge and from the East Pacific Rise on a crust of less than 10 Ma old. The composition of pyroxene from a basin region (Nazca plate) is from G. Thompson et al. (1974). Some (×) of the pyroxene data from the East Pacific Rise (Leg 54) is from Schrader et al. (1980). The sub-calcic augite reported here (•) are from the Minia Seamount near 53°N on the Mid-Atlantic Ridge. The composition of the pyroxene at various latitudes along the East Pacific Rise and the Mid-Atlantic Ridge were determined by Camebax electron microprobe analyses (J. Morel and M. Bohn). The Skaergaard intrusion trend is from Muir (1951). Line *A-B* indicates the crystallization of primary orthopyroxene (Deer et al., 1963). The empty circles indicate the samples of trachyte (RC10-241, 260 cm) from the Mathematicians seamount chain area (Tables 1-12, 1-15).

augite. Sub-calcic augite and Mg-pigeonite are rarely detected in ocean floor basaltic rocks (Fig. 1-13, Tables 1-11, 1-12). The compositions of the clino-pyroxenes from most segments of the East Pacific Rise and the Mid-Atlantic Ridge are similar (Fig. 1-13). Most oceanic ridge basalts have a clinopyroxene composition falling in a limited area of the wollastonite-enstatite-ferrosilite ternary diagram (Fig. 1-13). The composition of these pyroxenes is within the range of: Wo = 35—44%, En = 43—55%, and Fs = 8—20%. The increase in ferrosilite content is most prominent in the groundmass clinopyroxene.

Compositional variations of clinopyroxene were used (Kushiro, 1960; LeBas, 1962; Schweitzer et al., 1979) to differentiate between alkali basalts and tholeiites. The elements which are the most sensitive to variation are Cr, Ti, Na and Al. Cr, Ti and Al enter into the 6-coordinate site of the pyroxene crystalline structure while Si and Al are found in the 4-coordinate site. Thus the clinopyroxene of basalts erupted on recent spreading ridge axes such as those from the FAMOUS area on the Mid-Atlantic Ridge and those from the East Pacific Rise near 21°N are depleted in Ti (<0.04 atoms/formula unit) and in $Al^{IV+VI}$ (<0.2 atom/formula unit) while alkali basalts from other

TABLE 1-11

Electron microprobe analyses of clinopyroxene from Mid-Oceanic Ridge basalts

| | East Pacific Rise | | | | Mid-Atlantic Ridge | | | | Mid-Indian Oceanic Ridge |
|---|---|---|---|---|---|---|---|---|---|
| | Leg 9 82-7-1/60 (ave. of 3) | E21-16 (ave. of 2) | Leg 54 423-8-1/51 (ave. of 2) | Leg 54 429A-3-1/130 (ave. of 3) | CYP78-10-16 (ave. of 3) | CH4-DR-9T (ave. of 3) | 521-1 (ave. of 3) | V25-DR01-91 (ave. of 2) | Leg 26 251 (ave. of 3) |
| $SiO_2$ (wt.%) | 51.50 | 53.49 | 52.09 | 51.65 | 52.58 | 51.35 | 51.57 | 49.00 | 50.43 |
| $Al_2O_3$ | 2.61 | 2.56 | 2.32 | 3.04 | 2.36 | 3.16 | 3.34 | 5.00 | 3.38 |
| $FeO$ | 10.73 | 6.15 | 6.17 | 9.66 | 6.21 | 8.37 | 5.41 | 9.64 | 10.89 |
| $MnO$ | 0.30 | 0.17 | 0.13 | 0.44 | 0.21 | 0.27 | 0.12 | 0.16 | n.d. |
| $MgO$ | 18.31 | 17.61 | 17.11 | 16.46 | 18.09 | 16.53 | 18.39 | 14.59 | 15.63 |
| $CaO$ | 14.46 | 18.74 | 20.04 | 17.69 | 18.78 | 18.27 | 18.31 | 19.15 | 18.76 |
| $Na_2O$ | 0.23 | 0.27 | 0.22 | 0.24 | 0.27 | 0.15 | 0.18 | 0.40 | 0.36 |
| $TiO_2$ | 0.82 | 0.55 | 0.48 | 0.90 | 0.51 | 0.69 | 0.40 | 1.68 | 1.14 |
| $P_2O_5$ | 0.19 | — | — | — | 0.26 | 0.35 | 0.30 | 0.34 | n.d. |
| $Cr_2O_3$ | 0.17 | 0.22 | 0.38 | 0.20 | 0.65 | 0.09 | 0.51 | 0.22 | 0.24 |
| Total | 99.02 | 99.76 | 98.94 | 100.28 | 99.92 | 99.23 | 98.53 | 100.18 | 100.83 |

Samples from Leg 9, Site 82 are from 02°35.48'N, 106°56.52'W (3707 m); E21-16 is from 54°10'S, 119°49'W (2981 m). The location of Leg 54 samples are shown in Fig. 2-32 and Table 2-7. CYP78-10-16 is from 20°50'N, 109°00'W (≈2650 m). CH4 samples are from 52°59.6'N, 34°58.7'W (1875 m); 521-1 is from 36°48'N, 33°16'N; V25-DR01 is from 25°24'N, 45°18'W (3367 m); Leg 26, Site 251 samples are from 36°30'S, 49°29'E (3489 m) (Kempe, 1974).

TABLE 1-12

Microprobe analyses of clinopyroxene from the Nazca Basin (DSDP Leg 34; Kempe, 1974), from the Wharton Basin (Leg 22, Site 211, Indian Ocean) and from the Mathematicians Seamounts area (RC10-241, eastern Pacific)

| | Leg 34, 319A-2-1/111—114 phenocryst (ave. of 2) | Leg 34, 30B-3-1/54—57 matrix (ave. of 3) | Leg 22, 211-12-2/112—115 phenocryst (ave. of 2) | RC10-241 260 cm (ave. of 3) |
|---|---|---|---|---|
| $SiO_2$ (wt.%) | 49.22 | 50.19 | 42.60 | 49.57 |
| $Al_2O_3$ | 5.20 | 5.40 | 8.76 | 2.94 |
| FeO | 9.89 | 9.45 | 11.07 | 14.18 |
| MnO | 0.20 | — | 0.15 | 0.68 |
| MgO | 14.04 | 16.01 | 9.59 | 11.16 |
| CaO | 19.53 | 16.68 | 22.40 | 18.33 |
| $Na_2O$ | 0.32 | 0.29 | 0.63 | 0.42 |
| $K_2O$ | — | — | — | — |
| $TiO_2$ | 1.55 | 1.76 | 5.29 | 1.01 |
| $Cr_2O_3$ | — | — | — | — |
| Total | 99.95 | 99.78 | 100.49 | 98.29 |

regions are enriched in these cations (Fig. 1-14). Oceanic islands and some aseismic ridges consist of alkali basalts with a clinopyroxene enriched in Al and Ti and depleted in Si (Fig. 1-14). In addition, the clinopyroxene crystallizing within an alkali basalt melt has a lower $SiO_2$ (<48%) and higher $Al_2O_3$ (>4%) content with respect to that found in tholeiitic basalts (Fig. 1-14). However, caution has to be exercised when analysing clinopyroxene because of compositional zoning between the center and the outer margin of the crystals. Indeed it is often noticed that the outer rim of the clinopyroxene analysed is enriched in Ti and Al, depleted in Si, and it falls into the alkali basalt field (Fig. 1-14).

*Olivine.* Olivine is usually the first mineral phase to crystallize. The composition of olivine varies according to that of the various types of basaltic rocks in which it is found. A picritic basalt will have more magnesian olivine ($Fo_{86}$—$Fo_{91}$) than an olivine basalt. A pyroxene-rich basalt could have an olivine with a forsterite content as low as $Fo_{73}$. Olivine usually occurs as euhedral crystals with well-defined crystalline outlines. In pyroxene-rich rocks (PPB), the olivine has a tendency to be rounded with an anhedral form. Magnesian olivine crystallizes in a narrow range of pressure/temperature conditions. Resorbed olivine crystals are often found in a matrix which was not in an equilibrium condition with the original crystalline phase. Equilibrium conditions between the melt and the associated olivine are often shown

40

by a systematic change in the composition of the melt-crystal relationship.
The slope of the linear variation of the FeO/MgO ratio of olivine and that
of its associated glass defines the coefficient ($K_D$) representing the partitioning

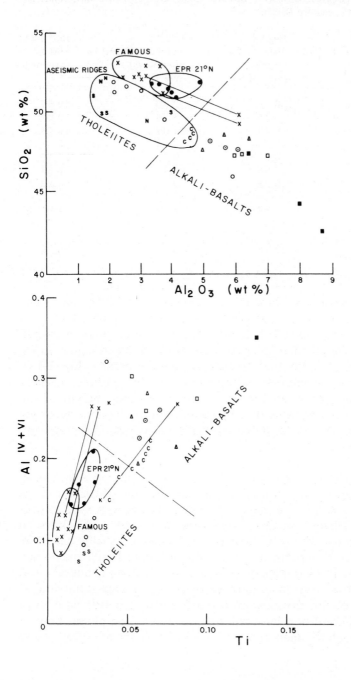

of iron and magnesium between the olivine (ol) and the glass (gl):

$$K_D = \frac{X^{ol}_{FeO}}{X^{gl}_{FeO}} \frac{X^{gl}_{MgO}}{X^{ol}_{MgO}}$$

One example of an olivine-melt relationship is cited using samples from the inner floor of the Rift Valley of the Mid-Atlantic Ridge near 36°50'N, in the FAMOUS area. For the FAMOUS area samples, the value of $K_D$ is 0.28 (excluding xenocrystal olivine). This partition coefficient was based on olivines with a maximum FeO/MgO ratio which shows a linear relationship with the FeO/MgO ratio of the associated glass (Fig. 1-15). The specimens which do not fall along this variation trend of FeO/MgO (glass-olivine) are the plagioclase-rich basalts (HPPB) with an FeO/MgO ratio of their olivine similar to that of an olivine basalt and a picritic basalt (Fig. 1-15). Since the FeO/MgO ratio of the plagioclase-rich basalt glass is 0.7—0.83, which is similar to the plagioclase-olivine-pyroxene basalt (Fig. 1-15), it is likely that the olivine was not in equilibrium with the melt and represents xenocrystic material carried to the surface with the liquid. This interpretation is supported by the small quantity of such olivine (<2 vol.%) and by its rounded and resorbed shape.

*Plagioclase.* The plagioclase of mid-oceanic ridge rocks consists of sodium and calcium end members of solid solutions. No potassic feldspar was found within fresh basaltic rocks associated with recent spreading ridges. However, variability in the range of molecular content of anorthite in the plagioclase occurs both within each individual sample and among various segments of accreting plate boundary regions (the regional variability of the plagioclase will be treated in Chapter 2). The range of composition within each individual sample is related to the order of crystallization of the plagioclase. Early-formed crystals (phenocrysts) will have an anorthite content higher than $An_{75}$. Second-generation plagioclase, such as the microphenocrystal phase, has a composition between $An_{70}$ and $An_{80}$. The last sizeable phase of plagioclase has an anorthite content of less than $An_{75}$.

---

Fig. 1-14. Ti-Al$^{IV+VI}$ (atomic) and SiO$_2$-Al$_2$O$_3$ (wt.%) variation diagrams of clinopyroxene from oceanic basalts collected in different geological settings. $\times$ = basalts from the Rift Valley inner floor near 36°50'N in the FAMOUS area. • = basalts from the East Pacific Rise near 21°N (project RITA). The tie-lines show the difference in composition between zoned pyroxenes (core and margin). The margin of these pyroxenes falls in the alkaline field. ○ = samples W3, WD3-6, WD3-4, WD3-3 and WD3-DR01A from the Walvis Ridge. $N$ = samples (Leg 22, 216-37-4, 214-54-3, and Leg 26, 254) from the Ninety-East Ridge. The alkali basalts are represented by the samples from Christmas Island (Indian Ocean) (△), by those from Annobon Island (⊙) (Atlantic Ocean, Cornen and Maury, 1980), by the rocks from a fracture zone near 43°N (□) on the Mid-Atlantic Ridge (Shibata et al., 1979b) and by DSDP Site 211 in the eastern Indian Ocean (Leg 22, 211-12) (■). $C, S$ = composition of clinopyroxene from the Ceara and the Sierra Leone Rises, respectively. $K$ = a basaltic rock from the Kings Through area in the North Atlantic (Stebbins and Thompson, 1978).

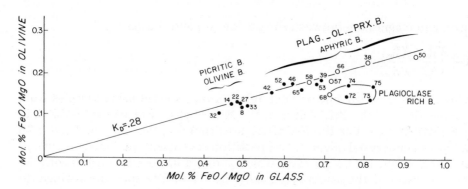

Fig. 1-15. Molecular FeO/MgO ratio of olivine and co-existing basalt glass (total Fe calcu-lated as FeO). Heavy line is visual best fit of plotted points corresponding to $K_D = 0.28$. The empty circles indicate samples without any spinel, filled circles indicate those containing spinel. The sample numbers correspond to those shown on the location map of Fig. 2-5.

The plagioclase found in most oceanic ridge basalts is formed contemporaneously or after the crystallization of olivine. Only in a few cases were large embayed plagioclase crystals (megacrysts) found; they are believed to be formed prior to the associated olivine. Among the trace constituents of the plagioclase, MgO (<1%), FeO (<1.5%) and TiO$_2$ (<1%) are the most prominent compounds. Generally there is a negative correlation between the FeO-MgO pair and the anorthite content of the plagioclase. The decrease of FeO and MgO content with the increase in An is probably due to the lattice replacement of Al by available Mg, Ti, and Fe during crystallization.

*Spinel.* Spinel is often found in small amounts (<4%) in picritic and olivine basalts, but it is rarely seen in plagioclase-rich basalts (HPPB, MPPB). Spinel occurs as euhedral and anhedral brownish-red and light-brown crystals often with a darker rim due to iron enrichment. The types of spinel encountered in the basaltic rocks from the ocean floor are a Mg-chromite (Mg, Cr) and a Cr-spinel (Mg, Al). Chemical analyses of spinel from various oceanic provinces show a variable Cr$_2$O$_3$ content (25—45%) and Al$_2$O$_3$ content (12—30%) (Table 1-13), while both the MgO and the FeO content are around 10—20% (Table 1-13). Often the amount of spinel found in a basalt is proportional to the amount of Mg-olivine in the bulk rock. Spinel is often preserved during the alteration of ocean-floor basalts. An olivine basalt metamorphosed to the green-schist facies mineral assemblage shows fresh spinel, while the olivine has been completely replaced (Table 1-13).

TABLE 1-13

Microprobe analyses of spinel found in basaltic rocks

|  | V25-1-T11 | CH4-D1-41 | Leg 37, 332B-22-1/97—100 | Leg 37, 332B-29-1/69—72 | Leg 22, 212-39-3/149—150 (ave. of 2) |
|---|---|---|---|---|---|
| $SiO_2$ (wt.%) | 0.22 | — | n.d. | n.d. | n.d. |
| $Al_2O_3$ | 27.38 | 27.26 | 29.1 | 28.2 | 26.29 |
| $Fe_2O_3$ | — | — | — | — | — |
| FeO | 17.95 | 18.93 | 13.2 | 11.6 | 15.42 |
| MnO | 0.18 | 0.10 | 0.22 | 0.17 | <0.07 |
| MgO | 15.56 | 15.07 | 15.3 | 16.2 | 15.80 |
| NaO | <0.09 | <0.5 | n.d. | n.d. | n.d. |
| $Na_2O$ | — | <0.78 | n.d. | n.d. | n.d. |
| $K_2O$ | — | — | n.d. | n.d. | n.d. |
| $TiO_2$ | 0.60 | 0.51 | 0.53 | 0.30 | 0.18 |
| $Cr_2O_3$ | 37.26 | 35.82 | 34.5 | 38.0 | 43.19 |

n.d. = not determined.

Sample V25-1-T11 is from the Mid-Atlantic Ridge near 25°24'N, 45°18'W; sample CH4-D1-41 is from the Mid-Atlantic Ridge, 53°0.4'N, 34°59.4'W; Leg 22, Site 212 sample is a fresh spinel from a metamorphosed basalt (Wharton Basin, Indian Ocean; see Chapter 9); Site 332B, Leg 37, is located 30 km west of the Median Valley (Mid-Atlantic Ridge) at 36°52.72'N, 33°38.46'W (Scientific Party, DSDP Leg 37, 1977).

ROCKS OF GRANITIC COMPOSITION

The occurrence of granitic rocks from the ocean floor is limited. Only occasional specimens have been reported from the Pacific, the Atlantic and the Indian Oceans. The term "granitic" is used throughout this paragraph to indicate rocks which are more silicic than basalts and andesites, that is rocks having a $SiO_2$ content higher than 60% (Table 1-14). Rocks of trachytic composition are not included as they have been discussed in the alkali basalt series. The first description of granitic rocks with a composition close to that of rhyolite was mentioned by Murray and Renard (1891) from the South Pacific Ocean at 30—40°S and 94—138°W. These rocks are characterized by a high quartz content (20—45%) and orthoclase content (5—10%) and minor amounts of hornblende, biotite and pyroxene. Peterson and Goldberg (1962) have also reported sediments enriched in anorthoclase, albite, and sanidine on the crest of the East Pacific Rise between 40 and 50°S. However, the most significant samples, because of their location north of the limit of the occurrence of ice-rafted material, are the granitic rocks from a dredge haul recovered near 16°0'S and 77°10'W on the Nazca Ridge. The rocks consist of quartz, microperthite, orthoclase and sanidine as reported by Ruegg (1962).

TABLE 1-14

Silicic rocks from the ocean floor

|  | Diorite 159-35 | Diorite 159-39 | Aplite V25-5 | Aplite 125-16 | Monzonite 125-4 |
|---|---|---|---|---|---|
| $SiO_2$ (wt.%) | 61.97 | 72.47 | 78.39 | 76.37 | 75.07 |
| $Al_2O_3$ | 16.00 | 14.17 | 12.68 | 12.78 | 13.18 |
| $Fe_2O_3$ | 3.22 | 1.85 | 0.38 | 0.39 | 0.76 |
| FeO | 3.57 | 1.19 | 0.41 | 0.46 | 1.15 |
| MnO | 0.09 | 0.08 | 0.01 | 0.02 | 0.03 |
| MgO | 2.43 | 1.39 | 0.54 | 0.87 | 0.23 |
| CaO | 3.24 | 1.48 | 0.55 | 0.84 | 1.10 |
| $Na_2O$ | 5.55 | 5.55 | 6.66 | 7.70 | 4.55 |
| $K_2O$ | 0.75 | 0.24 | 0.06 | 0.07 | 3.27 |
| $TiO_2$ | 0.94 | 0.33 | 0.09 | 0.25 | 0.15 |
| $H_2O$ | 1.28 | 0.90 | 0.41 | 0.28 | 0.28 |
| $P_2O_5$ | 0.22 | 0.06 | 0.01 | 0.02 | 0.12 |
| Total | 99.04 | 99.61 | 100.19 | 100.22 | 99.89 |

The diorites are from the Mid-Atlantic Ridge, 45°42′N, 28°56′W (Aumento, 1969); aplite V25-5 was found in a dredge from the Mid-Atlantic Ridge near 23°31.74′N, 45°07′W (Miyashiro et al., 1970b); aplite 125-16 and quartz monzonite 125-4 were taken near 13°34′S, 66°26′E on the Mid-Indian Ocean Ridge (C.G. Engel and Fisher, 1975).

Diorites were reported from near 45°N on the walls of the Rift Valley of the Mid-Atlantic Ridge, associated with fault breccia made up of serpentinites and gabbros (Aumento, 1968). The anorthite content of the plagioclase is $An_5$—$A_{40}$, and hornblende, plagioclase and clinopyroxene are the major constituents of these diorites. Quartz monzonite veins (a few centimeters thick) were found cutting through a granophyric diabase near the Argo Fracture Zone in the Indian Ocean (C.G. Engel and Fisher, 1975). The granitic rocks dredged from the central Indian Oceanic Ridge near 13°30′S are associated with serpentinized peridotite, gabbros, and diabase (C.G. Engel and Fisher, 1975).

Granitic breccia and quartz arenite fragments found in a piston core (V18-211) from the northern flank of the Broken Ridge in the Indian Ocean near 25°41′S, 99°04′E and identified by J. Didier (personal communication) show the presence of mirmekitic plagioclase, K-feldspars, quartz, chlorite, sericite and biotite. Aplitic rocks composed of quartz, oligoclase-albite, aegirine, apatite, sphene, zircon, and chlorite were also described by C.G. Engel and Fisher (1975) and by Miyashiro et al. (1970b).

ALKALI-RICH VOLCANICS

The division of igneous rocks into subalkalic (tholeiitic) and alkalic groups was first made by Iddings (1892). Within the alkalic rocks various trends of differentiation are observed. As pointed out by Miyashiro (1978), one trend of differentiation of alkalic rocks is characterized by the gradual increase of normative nepheline with the increasing degree of fractionation; another trend could start with hypersthene-normative basalts and then change to a normative quartz type of rock.

The alkali-enriched rock from the ocean floor is made up of a variety of rock types which are primarily identifiable by their increase in $SiO_2$ and $Na_2O + K_2O$ content. Thus alkali-basalts, mugearites, trachybasalts, trachytes and phonolites are among the most common alkali rocks found on the ocean floor. The occurrence of typical minerals such as nepheline and other feld-spathoidal phases facilitates the recognition of alkali-rich volcanics. The fractionated types of alkali-rich volcanics such as trachybasalts, trachytes, and phonolites may be easily recognized by determining their alkalic mineral phases from the nature of their pyroxene (Ti-rich clinopyroxene, aegirine-augite) or from the presence of mica. The recognition of alkali basalts, which are the most primitive among the alkali-rich volcanics, is not always easy. The ocean-floor alkali basalts contain olivine as both an early-formed and late mineral phase; the plagioclase is similar in composition to the other subalkaline basalts seen previously.

Recently, Miyashiro (1978) has summarized various concepts about alkali-rich rocks reported from various subaerial and oceanic environments. From his work it is clear that the best way of classifying the alkali basalts to differentiate them from the subalkalic tholeiitic types found on most accreting plate boundary regions is to use Harker's concept of alkalic rocks. Although the alkali-rich rocks and subalkalic rocks have the same $SiO_2$ content, the alkali basalts have a higher $Na_2O + K_2O$ content than do the subalkalic rocks (tholeiites). The boundary proposed by G.A. Macdonald and Katsura (1964) between the Hawaiian alkalic and tholeiitic rocks ($Na_2O + K_2O$ versus $SiO_2$ diagram) is also valid for dividing ocean-floor alkali-rich basalts. $TiO_2$ and $K_2O$ contents are also good indicators of different basaltic suites. Most alkali-rich volcanics follow a higher $TiO_2$ variation trend than do the tholeiitic suites, as seen in the Hawaiian flows (Fig. 1-16). Subalkalic oceanic basalts from accreting plate boundary regions have a $K_2O$ content less than 0.6% (Fig. 1-16).

*Alkali basalts.* This type of rock consists mainly of plagioclase and olivine. Sometimes it is found that the pyroxene is titaniferous (titanaugite) and has a pinkish color; there is no free silica, and interstitial nepheline may be detected. However, the easiest way of identifying the difference between an alkali basalt and a tholeiite is to calculate the relative amount of total alkali content, which for the alkali basalts is actually higher than 4%, and the $P_2O_5$

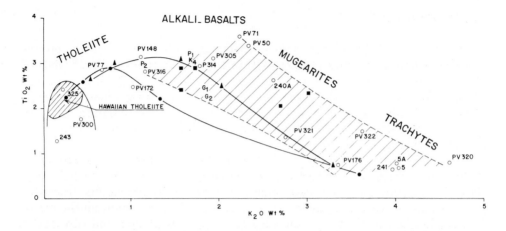

Fig. 1-16. TiO$_2$-K$_2$O (wt.%) binary diagram of various oceanic basalts and their differentiated products. The field of the ocean tholeiites is indicated with that of the Hawaiian tholeiite. The field of the alkali-basalts is represented by rocks dredged from seamounts. The black triangles indicate the fractionated material (basalt-alkali basalt-mugearite-trachyte) from Hawaii. The black dots indicate the trend of tholeiite-granophyre-rhyodacite. Prefixes PV and P indicate samples from the eastern Pacific seamounts (A.E.J. Engel and Engel, 1964; 1966). Prefix K indicates samples from the Kodiak Seamount (Forbes et al., 1969). 240A, 241, 5A and 5 are rocks from the Mathematicians Seamounts. Sample 243 and 325 are basalts from the East Pacific Rise. Samples G1 and G2 are from the Giacomini Seamount (Forbes et al., 1969).

content is close to 1% (by weight). Chemical analyses of the alkali basalts from elevated volcanic edifices are shown in Table 5-4. The normative mineral contents of these rocks show an abundance in olivine and the presence of nepheline.

*Mugearite*. This is a rock type intermediate between the alkali basalts and the trachytes found in the ocean floor. The SiO$_2$ content is about 52—54% by weight and the normative mineral constituents are about 13—15% albite, 15% orthoclase, and 4—5% olivine. The total alkali content of mugearite is about 6%. The CaO and MgO contents are lower than the average alkali basalt from the seamounts of the eastern Pacific Ocean.

*Trachytes*. These rocks consist of plagioclase, amphibole and clinopyroxene. The clinopyroxene is usually an aegirine-augite and an augite. The general texture of trachytes is hypohialline and fluidal due to the parallel arrangement of oligoclase (An$_{28}$—An$_{31}$) laths. Sometimes the rocks are abundantly vesiculated because they were erupted at a relatively shallow depth and the vesicles are often elongated in the direction of the flow. The trachytic rocks are either nepheline normative or quartz normative, and in both cases they are rich in

TABLE 1-15

Trachytes collected from the eastern Pacific seamounts

|  | 241 | 5 | 5A | 1 | PV320 |
|---|---|---|---|---|---|
| $SiO_2$ (wt.%) | 63.30 | 64.00 | 63.60 | 58.57 | 56.12 |
| $Al_2O_3$ | 15.80 | 16.30 | 16.20 | 18.43 | 18.55 |
| $Fe_2O_3$ | 2.33 | 1.86 | 2.05 | 5.69 | 4.00 |
| FeO | 3.22 | 3.10 | 3.20 | 0.91 | 0.32 |
| MnO | 0.13 | 0.18 | 0.12 | 0.21 | 0.16 |
| MgO | 0.40 | 0.50 | 0.50 | 0.64 | 0.71 |
| CaO | 3.00 | 3.00 | 3.00 | 2.74 | 5.43 |
| $Na_2O$ | 5.50 | 5.60 | 5.60 | 6.76 | 6.00 |
| $K_2O$ | 4.00 | 4.00 | 4.00 | 3.28 | 4.63 |
| $TiO_2$ | 0.76 | 0.81 | 0.76 | 0.52 | 0.82 |
| $P_2O_5$ | 0.24 | 0.24 | 0.24 | 0.21 | 1.83 |
| $H_2O$ | 1.40 | 0.00 | 1.80 | 2.71 | 1.07 |
| Total | 100.07 | 99.58 | 101.06 | 100.57 | 99.64 |
| *C.I.P.W. norms* | | | | | |
| Quartz | 9.74 | 9.43 | 9.09 | 0.88 | — |
| Nepheline | 0.0 | 0.00 | 0.00 | 0.00 | 3.12 |
| Orthoclase | 23.63 | 23.63 | 23.63 | 19.38 | 13.94 |
| Albite | 46.53 | 47.38 | 47.38 | 57.20 | 45.00 |
| Anorthite | 6.61 | 7.52 | 7.25 | 10.25 | 10.00 |
| Diopside | 5.74 | 4.93 | 5.17 | 1.52 | 3.10 |
| Hypersthene | 1.02 | 1.87 | 1.75 | 0.88 | — |
| Olivine | 0.00 | 0.00 | 0.00 | 0.00 | 0.23 |
| Magnetite | 3.37 | 2.69 | 2.97 | 2.11 | — |
| Ilmenite | 1.44 | 1.53 | 1.44 | 0.98 | 1.01 |
| Apatite | 0.56 | 0.56 | 0.56 | 0.49 | 4.31 |
| Sphene | 0.00 | 0.00 | 0.00 | 0.00 | 0.00 |
| Hematite | 0.00 | 0.00 | 0.00 | 4.23 | 4.00 |

The chemical analyses were made by the wet analyses method using a Perkin-Elmer atomic absorption. Samples 241, 5 and 5A are from an unnamed seamount of the Mathematicians seamount chain (15°23′N, 110°56′W; Subbarao and Hékinian, 1978). Sample 1 is from the west slope of the Kodiak Seamount and sample PV320 is located at 180°45′N, 158°15′W (7-N Seamount; A.E.J. Engel and Engel, 1964).

$SiO_2$ (56—65%) (Table 1-15). It is likely that trachytic rocks are derived from a fractional differentiation of alkali basalt melt, giving rise to a residue of basanite (Subbarao and Hékinian, 1978). The composition of the clino-pyroxene of some trachytes from the Mathematicians Seamounts area show higher Fe and Ti contents than normal mid-oceanic ridge rocks (Fig. 1-13).

Isotopic studies such as the $^{87}Sr/^{86}Sr$ ratios (0.7040—0.7053) of Pacific Ocean floor alkali-rich rocks show a similarity to the island type of alkaline series. Hence, the typical ocean floor alkaline series consists of sequentially

48

more fractionated rock types going from basanite to trachytes as follows:

Basanite → alkali basalt → mugearite → trachyte


OCEANIC ANDESITE

The term oceanic andesite is applied here to rocks that are made up of sodic plagioclase ($Ab_{54}$ $Or_{2.3}$ $An_{44.7}$) and clinopyroxene. These andesites are found on aseismic ridges (Chapter 3) and they differ from continental or island arc andesitic rocks in their lack of modal hypersthene and/or hornblende. The pyroxene of the oceanic andesite found on the Ninety-East Ridge is iron rich and titanium depleted and corresponds to a pigeonite (Wo = 7.2, En = 45.7, Fs = 47.1) (Table 1-16). The andesite rocks from the ocean floor follow a trend of more iron enrichment than those from the Cascade Range, as shown in the ternary AFM diagram. The textural feature is pilotaxitic and trachytic with interstitial light-brown and light-green glass. Needles of apatite are often found in this type of rock. The $SiO_2$ varies between 54 and 61% and the $K_2O$ content is close to and higher than 1% (Table 1-17). The oceanic andesites are quartz-normative rock (quartz $\simeq$ 8—12%) (see Chapter 3). Andesites similar to those found on the ocean floor

TABLE 1-16

Microprobe analyses of pyroxene and plagioclase from oceanic andesite recovered on the Ninety-East Ridge (Indian Ocean). Details on the location and structural setting of the analyzed sample (Leg 22, 214-48-1-7, 83 cm) are given in Chapter 3

|  | Pigeonite | Sub-calcic augite | Plagioclase | |
|---|---|---|---|---|
| $SiO_2$ (wt.%) | 52.90 | 49.09 | 55.54 | 57.13 |
| $Al_2O_3$ | 0.78 | 0.58 | 26.20 | 26.42 |
| FeO | 25.70 | 28.73 | 0.75 | 0.71 |
| MnO | 0.87 | 1.07 | — | — |
| MgO | 16.43 | 14.27 | — | — |
| CaO | 3.48 | 3.29 | 9.48 | 8.95 |
| $Na_2O$ | 0.11 | 0.12 | 6.18 | 6.11 |
| $K_2O$ | tr. | 0.11 | 0.36 | 0.39 |
| $TiO_2$ | 0.49 | 0.61 | 0.19 | 0.12 |
| $Cr_2O_3$ | — | 0.18 | tr. | — |
| P | tr. | 0.10 | 0.19 | 0.24 |
| Total | 100.70 | 98.12 | 98.70 | 100.07 |
| Wo | 7.38 | 7.09 | — | — |
| En | 48.56 | 42.78 | — | — |
| Fs | 44.06 | 50.13 | — | — |

tr. = trace.

TABLE 1-17

Average composition of seven andesites from the
Ninety-East Ridge (Leg 22, Site 214; Hékinian,
1974a; G. Thompson et al., 1978) and one
andesite from St. Paul Island in the Indian Ocean
(G. Thompson et al., 1978)

|  | Average of 7 andesites | St. Paul andesite |
|---|---|---|
| $SiO_2$ (wt.%) | 56.9 | 55.4 |
| $Al_2O_3$ | 15.9 | 12.6 |
| Total Fe as FeO | 9.84 | 12.9 |
| MgO | 2.48 | 4.61 |
| CaO | 5.79 | 7.53 |
| $Na_2O$ | 3.97 | 3.69 |
| $K_2O$ | 1.50 | 0.97 |
| $TiO_2$ | 1.45 | 2.05 |
| $P_2O_5$ | 0.64 | 0.21 |
| Total | 98.47 | 99.96 |
| Cr (ppm) | 5 | 125 |
| Co | 39 | 38 |
| Ni | 5 | 600 |
| Sr | 647 | 340 |
| Ba | 548 | 285 |
| Zr | 252 | 355 |

were described by Carmichael (1964) from the Thingmuli volcano in Iceland
and from St. Paul Island in the Indian Ocean (Girod, 1971). The oceanic
andesites differ from quartz-normative trachytes found on the eastern Pacific
seamounts in their lower $K_2O$ content and higher $TiO_2$ content (Tables 1-16,
1-17).

## CHAPTER 2

THE WORLD'S OCEANIC RIDGE SYSTEM

The discovery of a sizeable portion of the Mid-Atlantic Ridge system was made by F.B. Taylor (1910), who summarized its main features as follows:

> "The Ridge is a submerged mountain range of a different type and origin from any other on the earth. It is apparently a sort of horst ridge . . . a residual ridge along a line of rifting or parting . . . the earth's crust having moved away from it on both sides."

Washington, in 1930, summarized and commented on the origin of the Mid-Atlantic Ridge, known to extend from 50°N to 40°S latitude.

Today we know that oceanic ridges are made up of elevated volcanic mountains and valleys extending for a distance of about 60,000 km and comprising about 33% of the ocean floor, a surface equal to that of all the earth's continents (Fig. 2-1). Some characteristics of these elevated ridges are their relatively high heat flow values, varying between 1 $\mu$cal/cm$^2$ s to about 10 $\mu$cal/cm$^2$ s; also, the ridge system all around the world is seismically active and is the locus of new crust generation under the oceans. It is only during the past twenty-five years that a considerable body of knowledge has been developed about the extent and the structure of the world's oceanic ridge systems. Oceanic ridges occur in all major oceans and seas (Mediterranean Sea, North Polar Sea). Average elevation from the adjacent sea floor is about 2000 m, and the ridges are about 2000 m wide. The system of oceanic ridges is generally restricted to the ocean floor: however, it also appears in the sub-aerial regions in several specific areas of the world. For example, the East Pacific Rise is seen on the surface of the Gulf of California as a system of complicated fracture zones. Recent volcanic activity related to crustal spreading is also found in the Afar and in the Ethiopian rift system, which is the surface expression of the Mid-Indian Oceanic Ridge. In the Gulf of Tadjoura (Red Sea), the ridge system enters the Afar forming the rift system of Ardoukoba. Here it bifurcates, one branch extending north to the Gulf of Suez and Aqaba, the Dead Sea and the Jordan Valley, and the other branch extending south to the great east African rift valleys. Microearthquakes in the Rift Valley of Kenya seem to be confined primarily to the center of the rift and are directly related to faults (Tobin et al., 1969; Molnar and Aggarwal, 1971). Another known area where an oceanic ridge system reaches the surface is Iceland. The Reykjanes Ridge, part of the Mid-Atlantic Ridge system, continues through the neovolcanic zone of Iceland. Einarsson (1967) suggested that the formation of the neovolcanic zone started at the beginning of the Quaternary and that it is genetically related to major events in the history of the Mid-Atlantic Ridge.

**TABLE 2-1**

Chemical analyses of basaltic rocks collected from the Mid-Atlantic Ridge rift valleys between 28°N and 53°N

| | Latitude 28°N | | | | Latitude 45°N | | | | | | 50°N | Latitude 53°N | | | |
|---|---|---|---|---|---|---|---|---|---|---|---|---|---|---|---|
| | 2762 | 1 | 2 | 3 | MT0 | MT1 | MT2 | MT3 | MT | A | 4 | CH4-DI-8T | | CH4-DI-42T | CH4-DI-18 |
| | PPB | MPPB | MPPB | MPPB | PB | OB | OB | | POPB (ave. of 5) | PPB (ave. of 3) | PPB (ave. of 2) | HPPB bulk | glass | PPB | OB glass (ave. of 2) |
| $SiO_2$ (wt.%) | 49.10 | 49.27 | 48.13 | 47.94 | 47.29 | 48.40 | 49.79 | 49.86 | 49.98 | 50.52 | 50.53 | 48.00 | 50.80 | 49.40 | 50.21 |
| $Al_2O_3$ | 14.50 | 15.91 | 17.07 | 17.45 | 12.05 | 16.12 | 15.96 | 16.00 | 15.73 | 15.10 | 15.92 | 17.75 | 15.67 | 14.60 | 15.19 |
| $Fe_2O_3$ | 1.50 | 2.76 | 1.17 | 1.21 | 2.06 | 1.94 | 1.55 | 1.52 | 2.00 | 0.62 | 0.99 | 2.35 | n.d. | 1.47 | n.d. |
| FeO | 9.04 | 7.60 | 8.65 | 8.47 | 7.34 | 6.34 | 6.96 | 7.05 | 6.34 | 8.24 | 7.89 | 5.01 | 7.91* | 7.91 | 9.96* |
| MnO | 0.18 | 0.13 | 0.13 | 0.13 | 0.16 | 0.16 | 0.16 | 0.16 | 0.15 | 0.13 | 0.13 | 0.13 | 0.16 | 0.17 | 0.14 |
| MgO | 7.80 | 8.49 | 10.29 | 10.19 | 18.50 | 9.63 | 8.79 | 8.76 | 9.11 | 8.65 | 8.76 | 6.37 | 9.14 | 7.66 | 9.19 |
| CaO | 11.50 | 11.26 | 11.26 | 11.26 | 8.35 | 11.48 | 11.89 | 11.89 | 11.15 | 11.60 | 11.32 | 14.21 | 12.54 | 12.85 | 10.15 |
| $Na_2O$ | 2.57 | 2.58 | 2.39 | 2.37 | 2.23 | 2.77 | 2.54 | 2.54 | 2.82 | 2.40 | 2.65 | 2.01 | 2.11 | 2.22 | 2.48 |
| $K_2O$ | 0.13 | 0.19 | 0.09 | 0.09 | 0.31 | 0.53 | 0.27 | 0.26 | 0.57 | 0.25 | 0.10 | 0.08 | 0.03 | 0.10 | 0.05 |
| $TiO_2$ | 1.45 | 1.26 | 0.72 | 0.75 | 1.20 | 1.43 | 1.24 | 1.24 | 1.42 | 1.22 | 1.08 | 0.78 | 0.88 | 1.02 | 1.13 |
| $P_2O_5$ | 0.15 | 0.13 | 0.10 | 0.08 | 0.16 | 0.22 | 0.15 | 0.15 | 0.19 | 0.11 | 0.12 | 0.15 | 0.27 | 0.15 | 0.24 |
| $H_2O$ | n.d. | 0.86 | 0.29 | 0.38 | 0.51 | 1.04 | 0.68 | 0.58 | 1.76 | 0.10 | n.d. | 1.98 | n.d. | 1.16 | n.d. |
| Total | 97.91 | 100.44 | 100.27 | 100.32 | 100.30 | 100.16 | 99.98 | 100.01 | 101.22 | 98.94 | 99.49 | 98.81 | 99.50 | 98.70 | 98.74 |

*Norms*

| | | | | | | | | | | | | | |
|---|---|---|---|---|---|---|---|---|---|---|---|---|---|
| Or | 0.76 | 1.12 | 0.53 | 0.53 | 1.83 | 3.13 | 1.59 | 1.53 | 3.39 | 1.43 | 0.59 | 0.88 | 0.59 |
| Ab | 21.74 | 21.83 | 20.22 | 20.05 | 18.86 | 23.43 | 21.49 | 21.49 | 23.86 | 20.30 | 22.42 | 17.00 | 18.78 |
| An | 27.64 | 31.27 | 35.58 | 36.71 | 21.95 | 29.98 | 31.48 | 31.48 | 28.56 | 29.88 | 31.26 | 39.17 | 29.57 |
| Di | 23.23 | 19.18 | 15.92 | 15.10 | 14.63 | 20.45 | 21.41 | 4.56 | 19.75 | 21.91 | 19.45 | 24.39 | 36.41 |
| Hy | 13.55 | 13.63 | 4.76 | 4.51 | 25.09 | 0.15 | 11.03 | 11.22 | 5.24 | 17.12 | 15.05 | 5.16 | 21.10 |
| Ol | 5.70 | 5.84 | 19.67 | 19.66 | 31.15 | 15.81 | 7.33 | 7.37 | 11.43 | 4.62 | 6.82 | — | 0.68 |
| Mt | 2.17 | 4.00 | 1.69 | 1.75 | 2.98 | 2.81 | 2.24 | 2.20 | 2.89 | 1.29 | 1.43 | 3.40 | 3.05 |
| Ilm | 2.75 | 2.39 | 1.36 | 1.42 | 2.27 | 2.71 | 2.35 | 2.35 | 2.70 | 2.31 | 2.05 | 1.48 | 2.33 |
| Op | 0.35 | 0.30 | 0.23 | 0.18 | 0.37 | 0.51 | 0.35 | 0.35 | 0.45 | 0.25 | 0.27 | 0.35 | 0.35 |

n.d. = not determined.

*Total Fe calculated as $Fe_2O_3$.

PPB = plagioclase-pyroxene basalt, MPPB = moderately-phyric plagioclase basalt, PB = picritic basalt, OB = olivine basalt, POPB = plagioclase-olivine-pyroxene basalt, HPPB = highly-phyric plagioclase basalt.

Samples 2762 (Kay et al., 1970), 1, 2 and 3 (G.D. Nicholls et al., 1964) are basaltic rocks dredged from 28°53'N, 43°20'W (2000 fathoms). Samples MT0, MT1, MT3, MT4 and A are from the inner floor of the Rift Valley at 45°44'N, 27°43'W (mean depth, 3370 m; Muir et al., 1964, 1966; Aumento, 1968). Sample 4 (G.D. Nicholls, 1965) is from the inner floor of the Rift Valley at 50°44'N, 29°52'W (1900 fathoms). Samples CH4-DI-8T, DI-42T and DI-18 are from the western flank of the Rift Valley at 53°00.4'N and 34°59'W (1975 m; this work).

Early work on the exposed outcrops of the ocean floor was carried out on scattered zones along the oceanic ridge system. It is only since the mid-1960's that dredging operations have followed bathymetric, seismic, magnetic and photographic surveys. Such types of surveys have permitted the establishment of a better correlation between sampling sites and structural settings. The sampling of rocks along the world's oceanic ridge systems has not been done systematically. Instead, the data available up to about 1977 and plotted on a generalized map of the world indicate that the most intensively sampled area is located in the North Atlantic (Fig. 2-1). The next most sampled regions are the East Pacific Rise north of latitude 8°N, and an area of the Mid-Indian Oceanic Ridge comprising the centrally located branches of the ridge systems (Fig. 2-1). With the Deep Sea Drilling Program (DSDP) starting in the late 1960's, a new dimension was given to the study of the oceanic crust. By drilling, it became possible to define a vertical distribution of the various lithological sequences which might have existed within the crust. Because of the technical limitations encountered, however, it has not been possible up to now to drill deep into the crust. Also, most of the holes are concentrated on a relatively old crust (>10 Ma) because a sediment cover at least 80 m thick is necessary in order to hold the drilling units in place. The majority of the DSDP sites is located in oceanic basin regions, and only a few sites occur on the isolated sediment ponds of mid-oceanic ridges (Fig. 2-2).

## MID-ATLANTIC RIDGE

The Mid-Atlantic Ridge (MAR) north of latitude 5°S has been the most extensively sampled oceanic ridge area. While the northern portion of the MAR has been sampled along almost its entire length, it is still difficult to find in the literature areas where the sampling sites are related to well-defined geological settings so that both structure and composition may be closely related to each other. Only three zones are known which fulfil this latter prerequisite: the ridges at 45°N, those near 36°N, and those near 22—23°N. These three areas of sampling, described in the following paragraphs, are apparently not associated with transform faults.

### Mid-Atlantic Ridge near 45°N

In 1966, a detailed combined geophysical (bathymetric, magnetic and gravimetric measurements) and petrological study of a relatively small portion of the ridge system near 45°N in the Atlantic Ocean was performed (Loncarevic et al., 1966; Muir et al., 1966; Aumento, 1968). The dredge hauls were primarily recovered from the western side of the Rift Valley axis in the Rift Mountain. The sampling was mainly concentrated on the largest seamount, Confederation Peak, with a mean depth for the mountain top at 600—500

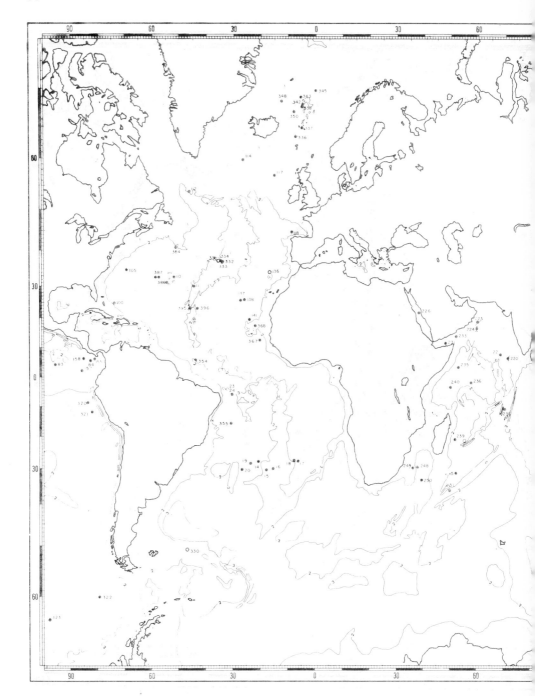

Fig. 2-2. Generalized map of the world showing the distribution of basement rocks recovered from the oceanic crust during deep-sea drilling operations ("Glomar Challenger"). Basement rocks from internal sea regions such as the Mediterranean Sea and the Caribbean region are not reported.

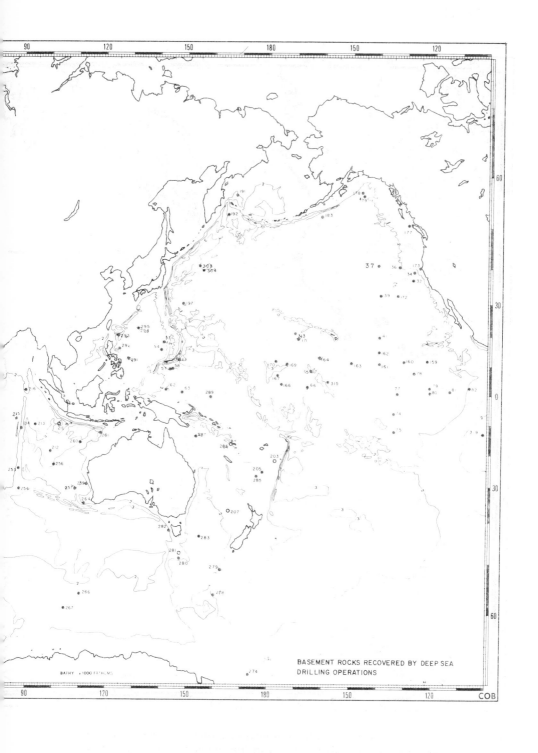

BASEMENT ROCKS RECOVERED BY DEEP SEA
DRILLING OPERATIONS

BATHY 1000 FATHOMS

COB

fathoms (Aumento, 1968). Three more dredges were also taken on another isolated seamount, called Bald Mountain, located 40 miles west of the Rift Valley axis.

Fission track and ferromanganese chronology of the basaltic samples collected at 45°N show a systematic outward increase in age from the axial zone of the ridge. The spreading rate calculated falls into two distinct groups: a faster rate on the Rift Mountain region (average 3 cm/yr) and a slower rate on the High Fractured Plateau (1 cm/yr). The age varies from 23,000 years at 6 km from the Rift Valley axis to about 8.4 Ma at 96 km west of the axial zone. These ages were determined by fission track on glass, and by K-Ar methods on hornblende minerals.

The basalts from Confederation Peak at 45°N are aphyric with very few phenocrysts, and porphyritic containing large crystals (6—15% by volume) of plagioclase. The porphyritic plagioclase-rich basalts contain a few crystals of olivine (as large as 1 mm) and a few rounded augite crystals. Similar rock types from the area sampled at 36°N were labelled moderately-phyric plagioclase basalts (MPPB; see Chapter 1). Highly-phyric plagioclase basalts (HPPB) with about 33% by volume of large plagioclase crystals (up to 6 mm in length) were found on the flank of a seamount in the Rift Mountain region. The plagioclase composition is $An_{86}$—$An_{88}$. These HPPB contain rare olivine crystals ($Fo_{86}$—$Fo_{87}$) and have an $Al_2O_3$ content of about 20%. The most alkaline types of basalt found in the Confederation Peak area are from station 48 (Aumento, 1968) and consist of basalts with a high $K_2O$ (0.9—1.9%) content. However, the occurrence of zeolite, serpentinite and chlorite as vesicle fillings and in minerals might have affected some of the whole-rock analyses and therefore enhanced the degree of alkalinity of the samples.

A great diversity of basalt types was encountered on the Rift Valley walls in the survey area at 45°N. A sequence from transitional tholeiites (olivine normative) to alkali olivine basalts was found (Table 2-1). There is both a lateral and vertical compositional variation of the basalt types. An appreciable correlation of oxide variation with depth exists. A gradual increase in the total alkali content is noticed above 1000 fathoms where the concentration becomes similar to that of islands (Aumento, 1968). The walls and the Rift Mountain regions are made up of volcanic edifices which occur in pairs on opposite flanks of the Median Rift Valley. It was suggested (Aumento, 1968) that as the volcanoes develop, they extrude a progressively more alkali-rich lava on top of the tholeiites. Hence a large volcano ends its eruptive cycle with a capping of alkali lavas. However, the occurrence of large alkali volcanoes on the Rift Mountain area was not observed during recent DSDP drilling operations (Leg 37), and the existence of large volcanoes the size of the Rift Valley is doubtful according to other more recent investigations of the Rift Valley inner floor.

*Mid-Atlantic Ridge near 22—23°N*

The data obtained near 22—23°N were mainly made available as a result of the 1961, 1964 and 1965 expeditions of the R.V. "Chain" and the R.V. "Washington". The dredged rocks recovered during these cruises came from the eastern and western walls (at different levels) of the Rift Valley (Van Andel and Bowin, 1968). Other dredge hauls were taken from the Rift Mountain regions at about 100 km from the Rift Valley axis. The inner floor of the Rift Valley near 22°N has a maximum depth of 4210 m. The east flank is the steepest (15—45°). In the northern part, the valley is crossed by a hill more than 1000 m high above the adjacent valley floor (Van Andel and Bowin, 1968). The Rift Mountain area consists of narrow and linear ridges and valleys with relief of 200—400 m (Van Andel and Bowin, 1968). The water depth in the dredged area of the Rift Valley varies between 2800 and 4200 m (Fig. 2-3).

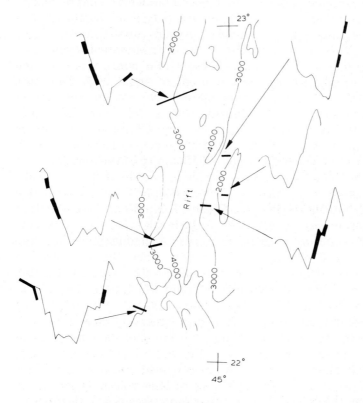

Fig. 2-3. Distribution of dredged samples in the Rift Valley near 22°N in the Atlantic Ocean after Van Andel and Bowin (1968). On the sides of the bathymetric map are the topographic profiles (vertical exaggeration about 12×) in corrected meters.

The basalts found on both walls of the Median Valley and on the Rift Mountain area are of a tholeiitic nature. There is, however, a difference in their degree of weathering. The basalts from the Rift Mountain area (DR10; Van Andel and Bowin, 1968) have a thicker coat of palagonite and manganese (30—40 mm thick) than those from the Median Valley. A basalt taken at about 150 km from the Rift Valley axis, encrusted by palagonite and consolidated foraminifera ooze, was dated as Upper Tertiary (Cifelli et al., 1966). The basalts dredged in the Rift Mountain area are enriched in plagioclase (about 10% phenocrysts) and are classified by G.D. Nicholls et al. (1964) as high-alumina basalts. The phenocrysts of plagioclase are crystals of bytownite ($An_{82}$). Olivine is rare, and plagioclase as both early-formed crystals and in the matrix is the most abundant constituent. Other dredge hauls (DR5, DR6, DR10) from the Rift Valley mountain near 46°W longitude between 60 and 140 km west of the rift valley axis were reported by Van Andel and Bowin (1968).

The dredge hauls from the Rift Valley's eastern and western walls near 22°N (THV15, THV14, THV11, THV10, THV18, THV17, THV8, THV4, DR8, DR2, DR3) were given a general description by Van Andel and Bowin (1968). Unmetamorphosed basalts are olivine tholeiite in nature and are found in most dredge hauls. Metamorphics were found in only three dredges from the eastern flank of the Rift Valley (THV4, DR2 and DR3) and in one from the west flank (THV11; Van Andel and Bowin, 1968). These rocks, described in detail by Melson and Van Andel (1966), vary in metamorphic grade from the low-grade zeolite facies to the greenschist facies. The effect of shearing has been observed in some of these metamorphics; in most samples, however, the original texture of the basalt is recognizable. On the basis of the depth of the sample locations, the authors (Melson and Van Andel, 1966; Van Andel and Bowin, 1968) concluded that the highest-grade metamorphics of the greenschist facies underlie the lower-grade zeolite facies. Thus Van Andel and Bowin (1968) suggested that the oceanic crust exposed for about 2 km on the Rift Valley walls near 22°N is layered, progressing from an unmetamorphosed basalt on the upper part of the wall and the Rift Mountain region to the zeolite and the greenschist facies rocks.

A dredge haul obtained from the top unmetamorphosed basaltic layer on the western flank of the Rift Valley, however, contained metamorphics of the low-grade facies (THV11). Since this dredge was taken near the top of the eastern wall, it is unlikely that the load pressure was sufficient to alter these rocks and Van Andel and Bowin (1968) have suggested the possibility of hydrothermal alteration.

Tuffaceous material (breccia), including basaltic fragments, occurs in some dredges from the Rift Valley walls (THV14, THV15, and DR2). The groundmass is made up of calcareous material intermixed with foraminifera tests.

The model of a layered crust with deep-seated metamorphosed rocks and unmetamorphosed basalts on the top as suggested by Van Andel and Bowin

(1968) and by Melson and Van Andel (1966) is not fully satisfactory. Fresh basalt and dolerite were also reported from the same dredges that contained the metamorphics, and even if they are the minor constituents of the dredges, there is reason to believe that the load pressure of regional metamorphism is not the only way to explain the alteration of rocks in the oceanic crust. More detail is presented in the chapter concerning the metamorphism of ocean floor basalts (Chapter 9).

Basaltic rocks from the Rift Valley wall were also collected at about 100 km north of the surveyed area near 23°N and described by Van Andel and Bowin (1968). A dredge haul containing olivine tholeiites and plagioclase tholeiites, as described by Shido et al. (1971), was taken at 4336—4815 m depth (V25, DR4). Shido et al. (1971) plotted these rocks on a normative plagioclase-pyroxene-olivine diagram and noticed that two trends of crystallization are characteristic of rocks enriched by early-formed olivine and early-formed plagioclase. The phenocrysts of plagioclase in these plagioclase tholeiites consist of 82—85% anorthite (Shido et al., 1971). The variability in the bulk-rock composition of the plagioclase and olivine basalts suggests that these basalts have undergone different stages of fractional crystallization.

During Legs 45 and 46 of the DSDP, two re-entry holes were drilled on the Rift Mountain region of the MAR near 23°N (Scientific Party, DSDP Legs 45, 46, 1976b, c). One hole (Site 395) was located on the western part of the Rift Valley axis on a crust 7 Ma old. The other hole (Site 396) was drilled opposite the first one on the eastern part of the Rift Valley axis on a crust about 8.7—10 Ma old (from magnetic anomaly data). The oldest sediment reached in hole 396 was Middle Miocene (13.6 ± 1.6 Ma) (Scientific Party, DSDP Leg 46, 1978). Site 395 drilled about 550 m into the basement and recovered mainly basaltic rocks intermixed with ultramafic cobbles in the upper section and plagioclase-olivine-pyroxene basalts in the lower part of the core. One layer of ultramafic rocks occurs at a depth of 180 m within the drilled hole (395) and consists of about 130 cm of foliated harzburgite with enstatite (30—50%), all of which are overlying about 90 cm of lherzolite. The lherzolite with orthopyroxene (≃60%) is not foliated, and the contact with the overlying harzburgite consists of red-brick carbonate-rich serpentinized peridotite (which may be ophicalcite). Gabbroic rocks (about 30 cm thick) are lying on top of the harzburgite, and the whole sequence (gabbro-harzburgite-lherzolite association) is included between thick sections of basaltic rocks. A thin layer (about 50 cm thick) of basaltic rock occurs between the two types of serpentinized peridotite, and these basalts were interpreted as being intrusive bodies in the ultramafics (Scientific Party, DSDP Leg 45, 1976b). Gabbroic and ultramafic pebbles and cobbles were also encountered within the sediment column above the basement and are probably slumped material from surrounding outcrops. At Site 396, about 255 m of oceanic crust was penetrated and sequences of basalt pillow flow were recovered. Plagioclase and olivine are the early mineral phases encountered

in the basalt. Basaltic breccia were also found in the lower part of the drilled hole (Scientific Party, DSDP Leg 46, 1976c).

The major difference between drill holes 395 and 396 is the absence of ultramafic rocks in the latter. Also, the occurrences of a more fractionated type of basalt in hole 395 is noteworthy. The fractionated nature of the basalts is shown by the abundance of clinopyroxene and the higher $TiO_2$ content (up to 1.80%) in some units of hole 395 with respect to those of hole 396.

Using some geochemical parameters such as Th, La, Nb and Ta, Bougault et al. (1979b) have shown that the mantle sources giving rise to the rocks found at both Sites 395 and 396 are the same. In order to explain the petrographic diversities encountered among the basalts from these drilled sites, Rhodes et al. (1979) postulated that magma mixing at shallow depth is an important process involved for diversifying basaltic flows. Rhodes et al. (1979) envisioned that mixing takes place episodically in small magma chambers beneath the spreading axis near 23°N and primitive magma derived by partial melting is repeatedly injected and mixed with a fractionated melt within the magmatic reservoirs.

This model of magma mixing explaining the heterogeneous nature of the extruded flows is very attractive and may complete the model of partial melting from a similar source for both drilled holes located on both sides of the Rift Valley near 23°N.

## Mid-Atlantic Ridge near 36°N: the FAMOUS area

The FAMOUS (French-American Mid-Ocean Undersea Survey) project has provided valuable information about different types of volcanic rocks and their structural settings. The operation included the study of only a small portion of the Rift Valley's inner floor and its adjacent first wall escarpments. In addition to the submersible studies, preliminary surveys prior to the dives were carried out by R.V. "Jean Charcot" and "Knorr" in 1972 through 1974. Dredging operations included the recovery of samples from the adjacent walls of the Rift Valley and the Rift Mountain regions (Fig. 2-4). In addition, the DSDP drilled four holes on the Rift Mountain about 15 miles west of the Rift Valley axis. Thus the MAR near 36°48′—37°N has become one of the locations surveyed in most detail among the world's mid-ocean ridge systems. In the studied area, the American and the African lithospheric plates are diverging at a rate of 2.2 cm/yr. This junction is marked by a rift valley with walls 1.2—1.5 km high. The inner floor of the Rift Valley is 1.5—3 km wide. It is dominated by a discontinuous central ridge along its axis. The high points of this ridge include Mt. Venus, Mt. Pluto and other volcanic hills which rise 100—250 m above the inner floor (Renard et al., 1975) (Figs. 2-5, 2-6).

One of the major discoveries of the study carried out near 36°N was to define the actual boundary of the creation of new oceanic crust between

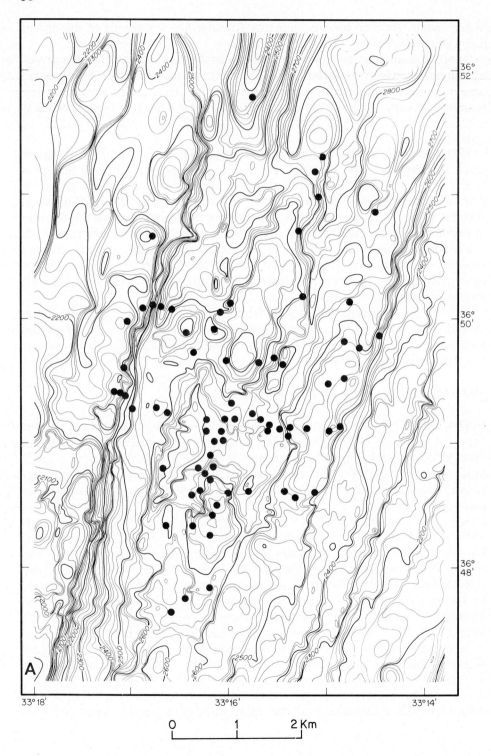

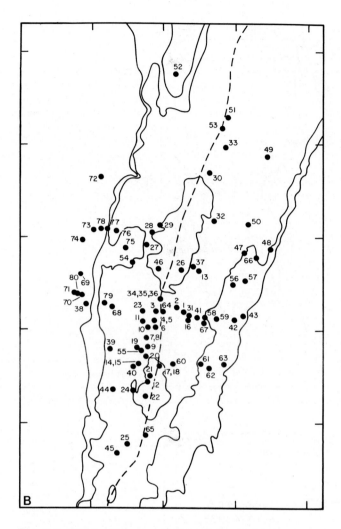

Fig. 2-5. Bathymetric map (A) and sample locations (B) of rocks from the Rift Valley near 36°50′N in the Atlantic Ocean. The solid and the dashed lines represent 20-m and 10-m contour intervals, respectively. These contour lines were taken from the d'Entrecasteaux map (Renard et al., 1975). The numbers in Fig. 2-5B are inventory numbers reported in Hékinian et al. (1976).

the African and the North American plates. The principal investigations in the area focused on the structure (Whitmarsh, 1973; Reid and Macdonald, 1973; Detrick et al., 1973; Fowler and Matthews, 1974; Poehls, 1974; Spindel et al., 1974; Needham and Francheteau, 1974; J.G. Moore et al., 1974; Arcyana, 1975; Ballard et al., 1975), as well as on the magnetic properties of the rocks (Lecaille et al., 1974: Greenewalt and Taylor, 1974) and on the geochronology of the rocks (Störzer and Selo, 1974). Preliminary

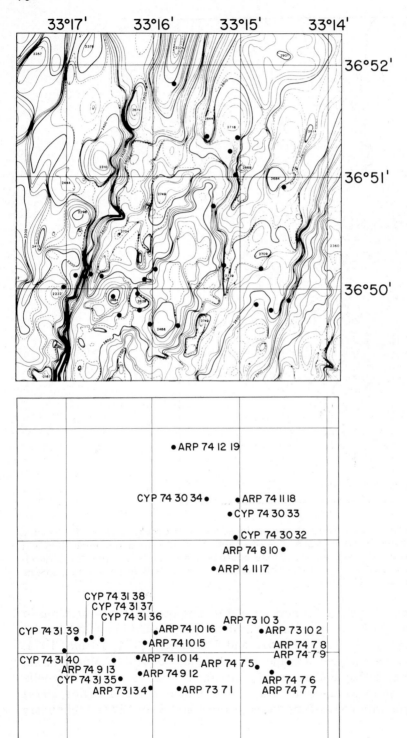

petrological and geochemical data presented by Bougault and Hékinian (1974) were limited to dredge samples on a relatively large segment (about 28 km long) of the Rift Valley near 36°50′N.

Even if the North Atlantic is one of the most dredged oceans in the world, it was difficult prior to Project FAMOUS to assess the relationship between structural settings and volcanicity within the MAR. This is probably due to the difficulties encountered in positioning dredge hauls. Small-scale (on the order of a hundred meters) compositional variation of volcanics is an important aspect for understanding the heterogeneity of the ocean crust and for assessing shallow magmatic processes underneath ridge crests.

The submersible study in the FAMOUS area near 36°50′N (Ballard and Van Andel, 1977) has revealed the presence of small fissures ranging from a few centimeters to as much as 8 m wide (Fig. 2-7). The fissures along the topographic axis of the inner floor have throws of less than 1 m (Ballard and Van Andel, 1977). About 67% of these fissures are elongated parallel to the general direction of the inner floor N20°E, and the majority of the remaining features are within 35° of N20°E (Ballard and Van Andel, 1977). All the fissures observed in the inner floor of the Rift Valley near 36°50′N are attributed to a single stress field due to tension. It was shown by examining the topography of the inner floor structures that flow lines (lobes) exist on the volcanic hills located on the side of the inner floor. This structure shows a down-slope motion of frozen lava tubes often going in all directions. The tops of these hills are probably the crests of small volcanoes (Figs. 2-5, 2-7).

A map of flow lobe patterns established from submersible, photographic, and bathymetric observations (Ballard and Van Andel, 1977) further suggests the existence of isolated and small volcanoes less than 10 km square and less than 300 m high (Fig. 2-7). These volcanoes occur everywhere in the inner floor, either in the axial zone or on the side adjacent to the first wall's escarpment (Fig. 2-7). Using the palagonite thickness as a criterion for determining the relative degree of alteration of the glassy margins, it was noticed that the least weathered rocks are found within the axial part of the Rift Valley within a zone less than 500 m in width.

The detailed study of the Rift Valley near 36°N also permitted the assignment of names to the various topographic provinces encountered in the inner floor. The bathymetric map of Figs. 2-5 and 2-6 shows the centrally-located hills called Mt. Venus and Mt. Pluto, two lateral hills located to the west and called Mt. Jupiter and Mt. Mercury, and a high which seems to have a topo-

Fig. 2-6. Bathymetric map and sample locations from the inner floor of the Rift Valley near 36°50′N on the Mid-Atlantic Ridge. Prefix ARP indicates sample collected by "Archimède"; CYP indicates those collected by "Cyana". The first two digits after the prefix (73 or 74) refer to the year of collection. The next digit (or two digits) refers to the dive number, and the final digit (or two digits) refers to the sample number for that dive.

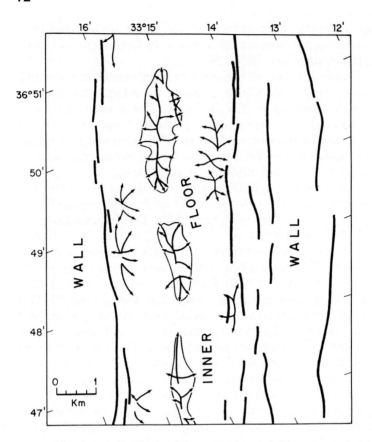

Fig. 2-7. Sketch map of fault pattern (heavy lines) and lava flow direction (arrows) (simplified after Ballard and Van Andel, 1977). The flow lobes depart from the crustal area of volcanic hills located in the inner floor of the Rift Valley in the FAMOUS area near 36°50′N.

graphic continuity from about 36°48′N to about 36°50′N and is called the Eastern Marginal High.

Because of the compositional variability encountered within individual flows for a relatively short distance (tens of meters up to hundreds of meters) it was found useful to classify the rocks by distinct types which reflect the whole-rock petrology (see Chapter 1). This approach will help provide insights into the magmatic and structural evolution of the crust within the dynamic framework of a particular accreting plate boundary.

The different types of basaltic rocks already defined in Chapter 1 are reported on a generalized map of the Rift Valley's inner floor (Fig. 2-8) and in Tables 2-2 and 2-3.

Two kinds of approaches were used to study the rocks from 36°N: (1) a detailed chemical and petrographic study of the glassy margins of individual

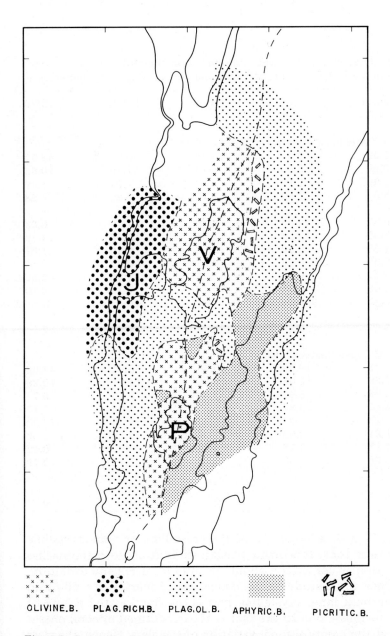

OLIVINE.B.  PLAG.RICH.B.  PLAG.OL.B.  APHYRIC.B.  PICRITIC.B.

Fig. 2-8. Generalized distribution of the different basaltic rocks encountered in the Rift Valley near 36°50′N. *J* = Mt. Jupiter, *V* = Mt. Venus, and *P* = Mt. Pluto. The dashed line represents the topographic axis of the inner floor. The distribution of olivine basalts, picritic basalts, highly-phyric plagioclase basalts (plagioclase-rich basalts) and plagioclase-pyroxene (olivine) basalts (plagioclase-olivine basalts and aphyric basalts) are shown.

TABLE 2-2

Average chemical analyses of basaltic rocks collected from the Mid-Atlantic Ridge rift valley near 36°50'N in the FAMOUS area (bulk-rock analyses made by X-ray fluorescence)

| | Picritic basalt | Olivine basalt | MPPB | POPB | HPPB |
|---|---|---|---|---|---|
| *Submersible samples* ("Archimède", "Cyana", 1973, 1974) | | | | | |
| $SiO_2$ | 47.09 | 50.09 | 49.98 | 50.21 | 48.07 |
| $Al_2O_3$ | 13.29 | 15.09 | 15.75 | 14.49 | 20.62 |
| $Fe_2O_3$ | 0.90 | 1.12 | 1.38 | 1.34 | 1.49 |
| FeO | 8.0 | 7.69 | 7.79 | 9.20 | 4.23 |
| MnO | — | — | — | — | — |
| MgO | 16.97 | 10.05 | 8.18 | 7.99 | 6.63 |
| CaO | 11.02 | 12.14 | 15.04 | 11.78 | 14.89 |
| $Na_2O$ | 1.55 | 2.12 | 2.13 | 2.20 | 1.87 |
| $K_2O$ | 0.09 | 0.16 | 0.17 | 0.27 | 0.15 |
| $TiO_2$ | 0.57 | 0.92 | 1.13 | 1.36 | 0.55 |
| $P_2O_5$ | 0.08 | 0.08 | 0.13 | 0.13 | 0.10 |
| Number of samples | 3 | 9 | 5 | 7 | 2 |
| *Dredged samples* (R.V. "Jean Charcot", 1972) | | | | | |
| $SiO_2$ | 46.91 | 49.89 | 49.56 | 50.01 | 48.92 |
| $Al_2O_3$ | 13.56 | 15.05 | 16.16 | 14.71 | 19.28 |
| FeO | 8.73 | 8.71 | 8.50 | 9.16 | 6.7 |
| MgO | 16.47 | 9.87 | 8.3 | 8.06 | 7.10 |
| CaO | 11.16 | 12.07 | 12.77 | 12.46 | 13.98 |
| $Na_2O$ | 1.61 | 2.09 | 2.03 | 2.14 | 1.85 |
| $K_2O$ | 0.11 | 0.18 | 0.27 | 0.29 | 0.17 |
| $TiO_2$ | 0.68 | 1.03 | 1.10 | 1.14 | 0.79 |
| Number of samples | 7 | 29 | 9 | 23 | 9 |

pillow lavas, and (2) a detailed study of the crystalline or semi-crystalline portion of the basaltic rock. It is obvious that a study from both approaches should result in a similar conclusion about the system of magma evolution. The major difference when studying the chilled glassy margin of a pillow in comparison to the bulk inner crystalline part is the fractionated nature of the glass. This is obviously due to the fact that early-formed mineral phases are not included in the compositional variation of the glass. Hence, the glass undoubtedly represents the fractionated liquid of the bulk-rock composition when early-formed mineral phases are present. Theoretically a glassy margin with no early-formed crystalline phases should have the same composition as the crystalline portion of an aphyric rock. Hence, one must be careful when dealing only with the glass composition of porphyritic rocks, since the glass might not represent the closest parent to the deep-seated magma. Also,

TABLE 2-3

Electron microprobe analyses of Atlantis II dredged glassy basalts from the Mid-Atlantic Ridge rift valley inner floor near 36°30'N, south of the FAMOUS area

|  | Glassy margins* | | |
|---|---|---|---|
|  | AII-77-23 | AII-77-2 28-1 (ave. of 2) | AII-77 76-61 |
| $SiO_2$ (wt.%) | 52.40 | 52.45 | 51.20 |
| $Al_2O_3$ | 13.80 | 13.85 | 15.00 |
| FeO | 11.40 | 10.60 | 7.98 |
| MnO | 0.16 | 0.20 | 0.16 |
| MgO | 6.76 | 7.78 | 9.02 |
| CaO | 10.60 | 11.55 | 13.30 |
| $Na_2O$ | 2.35 | 2.16 | 1.89 |
| $K_2O$ | 0.20 | 0.10 | 0.08 |
| $TiO_2$ | 1.77 | 1.17 | 0.84 |
| $P_2O_5$ | — | — | — |
| $Cr_2O_5$ | — | — | — |
| Total | 99.43 | 99.84 | 99.46 |

*Samples were provided by W. Bryan, Woods Hole Oceanographic Institute; analyst J.M. Morel.

differences in bulk-rock analyses and those of their associated glassy margins are shown by their $K_2O$—$TiO_2$ variation trend of paired samples (Fig. 2-9). In general, the glassy margins have a lower $TiO_2$ and $K_2O$ content than do their more crystallized interior (Fig. 2-9). The $K_2O$ content in the crystalline interior could be up to twice as great as that of the glassy margin. However, the reverse trend of $K_2O$ variability is seen in some picritic basalts (samples 10-3, 30-32) where the $K_2O$ content of the most crystalline portion of the rocks ($K_2O$ = 0.06—0.11%) is lower than that of their glassy margins ($K_2O$ = 0.14—0.16%) (Fig. 2-9).

The chemistry of the glassy portion of the pillow lava shows a systematic chemical variation across and along the Rift Valley's inner floor axis (Hékinian et al., 1976; Bryan and Moore, 1977). The chemical parameters showing changes are represented by the $TiO_2$ content, the FeO/MgO ratio (total Fe calculated as FeO), the Cr content and the Ni content. The lowest values of the FeO/MgO ratio (<0.90) and of the $TiO_2$ content (<1.00%) are found near the axial zone on the summit and flanks of Mt. Venus and Mt. Pluto (Fig. 2-10) (Hékinian et al., 1976), and are classified as olivine and picritic basalts (Table 2-2). A higher $TiO_2$ content (1.00—1.70%) and a higher FeO/MgO ratio (1.00—1.50) are mainly found in the lavas located away from the axial zone of the inner floor near the western and eastern walls (Fig. 2-10). These are the plagioclase-olivine-pyroxene basalts and the aphyric basalts (Table 2-2). The highest values for the FeO/MgO ratio (about 1.7) and the

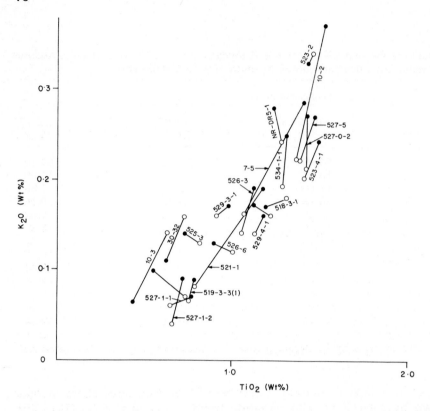

Fig. 2-9. $K_2O$-$TiO_2$ variation diagram of basaltic rocks from the inner floor of the Rift Valley in the FAMOUS area near 36°50′N. The numbers correspond to the submersible ("Archimède", "Cyana" and "Alvin") dives during the 1973—1974 expeditions. NR-DR5-1 is a dredge sample (R.V. "Noroit", 1974) located on the western first escarpment of the Rift Valley wall. The black dots represent the inner most crystalline portion of the sample while the empty circle is the corresponding glassy margin. All these samples are located on the bathymetric map of Fig. 2-5. The localities are described in more detail in Hékinian et al. (1976) and Bryan and Moore (1977).

highest $TiO_2$ content (<1.4%) are found on the eastern part of the inner floor (Fig. 2-10) (Table 2-2).

However, in addition to the lateral variation encountered in the inner floor of the Rift Valley, there is also a change in the $TiO_2$ content and the FeO/MgO ratio along the strike of the inner floor axial zone. $TiO_2$ and FeO/MgO values of about 1.00—1.40 were found to occur on samples located between Mt. Venus and Mt. Pluto and at their northern and southern extensions (Fig. 2-10). Similarly to the inner floor marginal types of volcanics, these specimens consist of plagioclase-olivine-pyroxene basalt and of aphyric basalt (Table 2-2). The Cr and the Ni content decreases continuously from the centrally located hills (from 500 to 200 and 250 to 70 ppm, respectively) to the marginal zones of the Rift Valley's inner floor. Usually the highest values of Cr and Ni content are found on the top of Mt. Venus and Mt. Pluto. This is consistent with the fact that the olivine-enriched basalts (olivine basalt) are found on topographic highs, with the exception of the picritic basalts which also have high values of Cr and Ni and are located at the base of Mt. Venus. Another chemical variable is the $SiO_2$ of the glass, which varies from about 49% to 52% along and across the axial zone of the inner floor. It

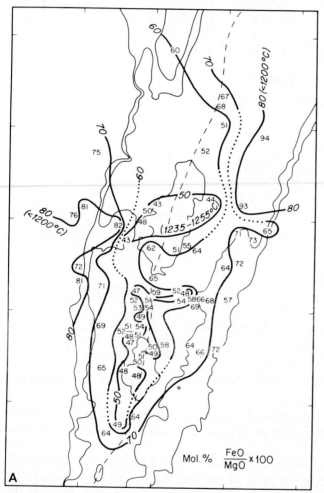

Fig. 2-10. A. Chemical zonation of Rift Valley samples near 36°50′N (FAMOUS area; the area covered is the same as in Fig. 2-5). Each set of numbers represents a sample collected by submersible ("Archimède", "Cyana" and "Alvin"). The contour lines represent the molecular ratio of FeO and MgO of basalt glassy margins (total Fe calculated as FeO) (Hékinian et al., 1976). The temperature values on the sides of the contour lines are inferred from the partitioning of Fe and Mg between olivine and glass in reference to Roeder and Emslie's (1970) experimental work.

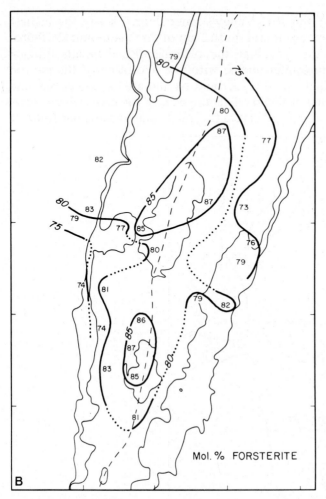

Fig. 2-10 (continued). B. Compositional variation of olivine expressed in molecular percent of forsterite.

is obvious that the type of element variations in the glassy portion of the
rock is typical of a differentiation process affected by crystal liquid fractionation (Table 2-3).

In addition to the chemical zonation there is also a relative age variation in
the volcanics found in the inner floor of the Rift Valley. This is visualized by
measuring the thickness of palagonite of the basaltic glassy margin of the
various types of rocks encountered (Fig. 2-11). The relative age of the inner
floor may be calculated by assuming a manganese accumulation rate of 3
$\mu m/10^3$ yr in relation to the rate of palagonitization. For more details on the
use of these parameters the reader should refer to Chapter 8.

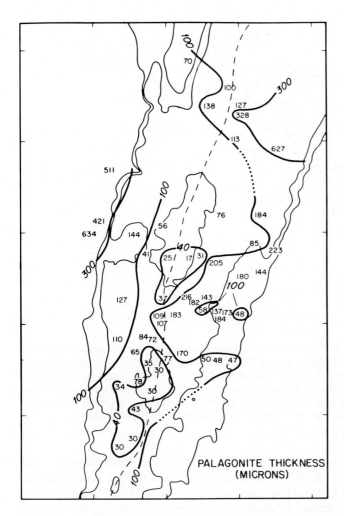

Fig. 2-11. Degree of rock alteration expressed by the relative thickness of palagonite found on pillow flow glassy margins. The area of the map is the same as in Fig. 2-5.

Evidence of crystal settling is shown in the cumulate and glameropor-phyritic texture characterizing the picritic basalt and the highly-phyric plagioclase basalts (HPPB). Differentiation from the fractionation of a residual melt is shown by the presence of reaction rims due to the overgrowth of pyroxene around an olivine host in the plagioclase-olivine-pyroxene basalt (POPB). This type of reaction rim around olivine microphenocrysts and matrix material suggests that the melt, initially enriched in olivine, changed its composition due to a crystal-liquid reaction with the residue. The compo-sition of the groundmass when separated from the picritic basalts expressed in terms of $SiO_2$ (49.50%) and the $Fe_2O_3$ + FeO/MgO ratio (0.8—1) is similar

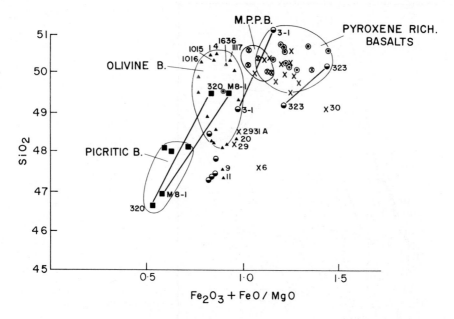

Fig. 2-12. $SiO_2$-($Fe_2O_3$ + FeO/MgO) variation diagram of the different types of basaltic rocks encountered in the Rift Valley and in Transform Fault "A". Samples taken by submersible (ARP73, 74) and by dredging (cruise CH31) are plotted. The tie-line relating individual samples represents the bulk composition (lower $SiO_2$) and the respective groundmass of the rocks. ◕ = highly-phyric plagioclase basalts; ⊗ = moderately-phyric plagioclase basalts; ✕ = pyroxene-rich basalts; ■ = the picritic basalts; ▲ = olivine basalts, ⊙ = plagioclase-pyroxene-olivine basalts.

to the groundmass of the olivine basalts. However, the groundmass of the HPPB is essentially made up of clinopyroxene and plagioclase and is similar to that of the plagioclase-pyroxene basalts (PPB) (Fig. 2-12). The volcanics which do not show any textural evidence of differentiation by crystal settling or by fractionation of a residual melt are some specimens of olivine basalts and some of the moderately-phyric plagioclase basalts (MPPB) found in the Rift Valley. These rocks, too, have intermediate values for their $Fe_2O_3$ + FeO/MgO ratio (averaging 0.9), falling between those of the picritic basalt and the HPPB ($Fe_2O_3$ + FeO/MgO <0.9; C.I. >60) and the pyroxene-enriched basalt (PPB) ($Fe_2O_3$ + FeO/MgO = 1.2; C.I. = 47 on the average). Hence the MPPB and the olivine basalt are both considered to be close in composition to a primitive melt and they were extruded directly from the mantle without accumulating in a magma chamber. These primitive melts, upon differentiation in a shallow magma chamber, have given rise to the other types of rocks. In summary, at least two series of volcanic rocks have erupted in the area of study as follows:

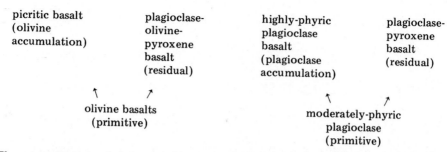

picritic basalt (olivine accumulation)

plagioclase-olivine-pyroxene basalt (residual)

highly-phyric plagioclase basalt (plagioclase accumulation)

plagioclase-pyroxene basalt (residual)

olivine basalts (primitive)

moderately-phyric plagioclase (primitive)

The occurrence of two volcanic series is also evidenced by the distinct structural settings within the inner floor of the Rift Valley near 36°50'N and 36°51'N, where the structural features are best known, and also where the density of sampling is the highest (Figs. 2-5, 2-6). A plagioclase-enriched melt is the main source giving rise to the volcanics west of Mt. Venus (the first step of the western wall), while an olivine-enriched melt built up most of the volcanic edifices of Mt. Venus (axial zone of the Rift Valley) and eastwards (Fig. 2-13). It is believed that the volcanics which are mainly characterized by the crystallization of large plagioclase (HPPB) predate those characterized by the crystallization of olivine (olivine basalt, picritic basalt) (Figs. 2-11, 2-13). Eruptive phases contemporaneous with that of the picritic basalts are seen in the occurrence of the plagioclase-olivine-pyroxene basalts and of the moderately-phyric plagioclase basalts. Compositional variabilities away from and along the strike of the inner floor of the Rift Valley, as well as the con-

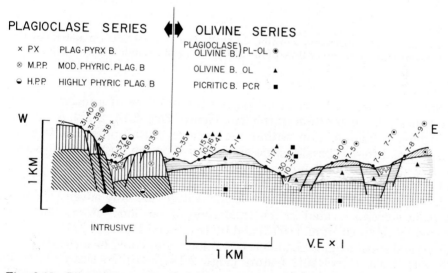

Fig. 2-13. Schematic cross-section of the Rift Valley along a line crossing Mt. Venus (FAMOUS area, Figs. 2-5, 2-9) showing the positions (and projected positions) of sampling sites and the type of rock. The inferred layers are merely to indicate a possible cyclic eruption of different lava types (Arcyana, 1977).

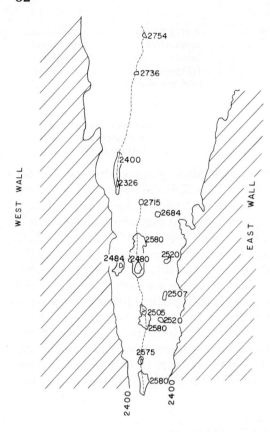

Fig. 2-14. Simplified bathymetry showing the shallowest contour lines (in meters) of volcanic edifices found in the inner floor of the Rift Valley, Mid-Atlantic Ridge, near 36°50′N.

struction of the isolated volcanoes, suggest the presence of a magma source region beneath the inner floor of the Rift Valley (Figs. 2-14, 2-15). Assuming that high-level fractionation within the oceanic crust upper mantle zone occurs before different basaltic lavas are extruded on the surface of the ocean floor, two alternative models of magma sources are possible:

(1) A magma chamber could have the width of the inner floor (~3 km wide), and in this case the isolated volcanic features are only adventive cones of a main volcano located in the axial zone of the inner floor. This was suggested by Hékinian et al. (1976) and by Bryan and Moore (1977).

(2) Several small magma chambers (<1.5 km wide) could be located underneath each isolated volcanic feature (Figs. 2-14, 2-15). In this case, each chamber will act as an independent source of volcanism, and as such it could be a locus for differentiation and cyclic eruption of one or several types of basaltic rock.

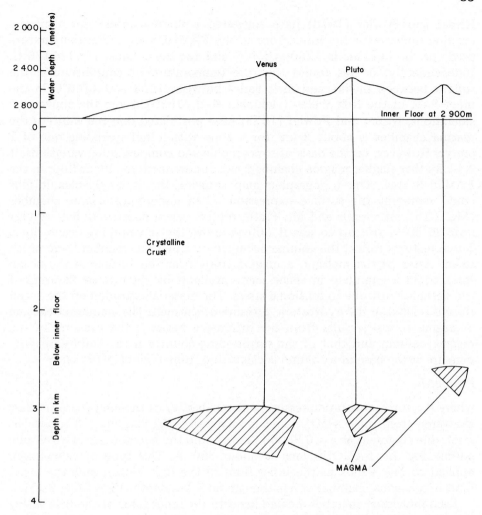

Fig. 2-15. Schematic representation of hypothetical trapped magma pools underneath individual volcanic edifices found in the inner floor of the Rift Valley (near 36°50'N).

The second model is the most appealing and it also seems to be consistent with geophysical data. If a large magma chamber of the size of the inner floor (3—4 km across) occurs, the shear wave velocities (Francis et al., 1977) should either be attenuated or absorbed by the presence of melted material. The steep theoretical temperature gradients (Sleep, 1975) for slow-spreading ridges (half spreading rate of 1 cm/yr) suggest a narrow magma chamber with a maximum width of about 1 km. From Sleep's (1975) thermal model, a temperature of about 1180°C (within the range of basalt melting) could be attained within a depth of less than 5 km. Using the latent heat calculation,

Nisbet and Fowler (1978) have suggested a thermal model for a magma existing underneath the inner floor in the FAMOUS area. The temperature used for the liquidus is 1230–1250°C and for the solidus, 1160–1185°C. Independently, it was found that the temperature of equilibrium at the solidus between olivine and melt varies between 1203 and 1146°C in the inner floor of the Rift Valley (Hékinian et al., 1976). From the above considerations, Nisbet and Fowler (1978) have postulated that the width of the magma chamber is about 2 km for a zone with a half spreading rate of 1 cm/yr. However, on the basis of topographic and compositional variations, it is likely that smaller magma chambers exist underneath the inner floor in the FAMOUS area. The topographic map showing the isolated volcanic hills could represent the surface expression of an underlying magma chamber (Fig. 2-15). Mt. Venus and Mt. Pluto, the two main constructional features near 36°50'N, rise up to about 300 m above the surrounding ocean floor. These features reflect the volume occupied by a magma chamber underneath them. After partial melting, a magma rising near the surface of the ocean floor could accumulate in space made available by the intense fissuring of the oceanic crust due to tensional stress. The material extruded on the ocean floor is related to the hydrostatic pressure of the melt. The main force causing a magma to rise results from the difference between the pressure of the magma column and that of the surrounding country rock. The hydrostatic pressure at the base of a column is calculated from Yoder (1976) as:

$$P = \bar{\rho}g\Delta h$$

where $P$ is the hydrostatic pressure, $\rho$ is the density of the melt (basaltic glass recovered from the FAMOUS area has a density of 2.82 g/cm$^3$). The acceleration due to gravity is $g = 0.980665$ km/s$^2$, and the height ($\Delta h$) of a volcanic edifice for the FAMOUS area is about 300 m. This type of calculation, applied to Mt. Venus in the inner floor of the Rift Valley, puts the lower limit of a magma chamber at less than about 4 km depth (Figs. 2-14, 2-15).

Each individual chamber located beneath the inner floor of the Rift Valley may not solidify all at once; instead, different levels and stages of solidification are likely to occur. The residence time of a magma within the crust is unknown, and depends on factors such as the degree of fissuring and the rate of spreading at a given time. Indirect correlations with structural observations and ocean-floor magnetic stratigraphy suggest that melt could flow within the crust and on the ocean floor a long time after being generated. Within the inner floor of the Rift Valley near 36°N, no change of magnetic polarity was noticed (K.C. Macdonald, 1977). However, deep-tow measurements may only measure the magnetization of fresh surface lava. A 580-m-long section drilled during DSDP Leg 37 on the western flank of the Rift Valley near 36°N shows consistent paleomagnetic inclinations within the single units of basalt flow (Scientific Party, DSDP Leg 37, 1977a). The magnetic stratigraphy of a vertical section within the oceanic crust includes normal, reversed and transitional zones. It is not excluded that each magnetic reversal may represent a distinct cycle of eruption.

## Ultramafics from the Mid-Atlantic Ridge

Ultramafics from the Rift Valley and Rift Mountain region of the MAR not associated with transform faults are found all along the ridge system. The flank of the MAR is formed by faulted blocks which have exposed serpentinized peridotites.

Serpentinized peridotites were found at 30°N at a depth of 1460 m near the top of crestal mountain regions adjacent to the intersection of the Rift Valley with the Atlantis Fracture Zone (Miyashiro et al., 1969a) (Table 2-4). Serpentinized peridotites associated with fault breccia of serpentinized material were reported from drilled holes near 37°N and 22°N, about 30 miles west of the Rift Valley axis (Scientific Party, DSDP Leg 37, 1977a, and Leg 45, 1976b). Serpentinized peridotites and other associated ultramafics were also found on the western fractured plateau region from about 30 to 100 km away from the MAR axis near 45°N (Aumento and Loubat, 1971). These serpentinites occur at all depths: on the top of elevated hills, at the bottom of cliffs, and also on flat areas (Aumento and Loubat, 1971).

## MID-INDIAN OCEANIC RIDGE

The Mid-Indian Oceanic Ridge (MIOR) is composed of a system of seismically active ridges with the shape of the Greek letter lambda: "λ". This ridge system is centrally located and more or less at equal distances from the major continental masses surrounding the Indian Ocean. The north-south extension of the MIOR is called the Central Indian Ridge and extends through the equator down to about 20°S. South of the Rodriguez Fracture Zone (near 20°S) there is the Southeastern Indian Ridge which exhibits the same topography as the Central Indian Ridge. The other extension of the MIOR is the southwestern branch of the ridge, the Southwestern Indian Ridge, almost equidistant from South Africa and Antarctica. The Central Indian Ridge extends northward up to the Owen Fracture Zone and into the Gulf of Aden. The branch of the MIOR between the equator and the Owen Fracture Zone near 9°N is called the Carlsberg Ridge. All these various segments of the MIOR have more or less similar structural features: they all have a rift valley with a central depression or inner floor (Heezen and Tharp, 1966).

Rocks dredged from the various segments of the MIOR, and particularly from the rift valleys, indicate a relative uniformity of fresh glassy rocks. Up to now, it has been extremely rare to find samples collected from the inner floor of the rift valleys in the MIOR. Most basaltic rocks recovered have come from the rift-valley walls and rift mountain regions of the MIOR. Most of the dredge hauls in the Indian Ocean are located on the Carlsberg Ridge, the Central Indian Ridge, the Southwestern Indian Ridge and near the intersection of these three ridge systems (Fig. 2-16). All the freshest material recovered consists of tholeiitic basalt in the wide sense, as described by Yoder and Tilley (1962). All the rocks are porphyritic to various degrees with plagioclase and/or olivine phenocrysts; their $Na_2O$ and $K_2O$ contents are relatively low (2.5 and less than 0.30% by weight, respectively), and their $Al_2O_3$ content varies between 15 and 21%, depending upon the amount of phenocrysts in the sample analysed (Table 2-5).

Compared to the Mid-Atlantic Ridge, very few detailed areas of the MIOR have been studied, with the exception of a small zone on the Carlsberg Ridge. Most of the samples studied are from widely spaced areas along the various segments of the MIOR on the Central and Southwestern Indian Ridge System, and were dredged by the Scripps Institution of Oceanography (S.I.O.) ships R.V. "Dodo", "Circe", and "Antipode" at 36 sites distributed between 8°S and 25°S. Olivine-enriched and plagioclase-enriched basalts are among the most abundant rock types recovered from these segments of the MIOR. Most of the plagioclase-rich basalts are from the Southwestern Indian Ridge (V17-78, Hékinian, 1968) and from the zone of intersection between the Southeastern Indian Ridge and the Central Indian Ridge (D113, D114, D115, C.G. Engel et al., 1965; Circe 109, 110, C.G. Engel and Fisher, 1975). The basalts are all porphyritic and contain phenocrysts of plagioclase; also, they often show early-formed clinopyroxene (generally of augitic composition). The plagioclase basalts show a high $Al_2O_3$ content (16—21%) and a $K_2O$ content of 0.20—0.40 wt.%. Other plagioclase-enriched basalts were collected from the Southeastern Indian Ridge (station Dodo 143 D; C.G. Engel and Fisher, 1975) and from the northern flank of the Rift Valley wall (station 5327; Chernysheva et al., 1975). A common characteristic of the plagioclase-

Fig. 2-16. Locations of dredged and drilled samples from the Indian Ocean. The numbers near each sample are inventory numbers including both dredged and DSDP drilled holes. The samples (84 through 122) from the Gulf of Aden are basaltic rocks dredged by the R.V. "Vema" (1976, courtesy of E. Bonnatti). Sample 83 is a limestone (Laughton, 1966). The samples (78, 79, 80, 124, 76 and 77) from the Carlsberg Ridge are basalts, metabasalts, gabbros and ultramafics (Cann, 1969; Chernysheva and Bezrukov, 1966; Hékinian, 1968). Samples of basalts (82, 132, 133) and gabbro (81) are from Bezrukov et al. (1966), Bunce et al. (1966), and Wiseman (1937). Samples 2, 4, 6, 12, 21, 22, 30, 37 and 40 consist of basalts and sample 5 is an alkali basalt (Hékinian, 1968). Samples 14, 32, 33, 34, 35 and 131 are basaltic rocks from the Mid-Indian Oceanic Ridge and the Wharton Basin (131) (C.G. Engel et al., 1965). Samples 67, 68, 69, 70, 71, 72, 73 are granitic rocks, volcanics and sandstones from the Mascarene Plateau (Bunce et al., 1966; Fisher et al., 1968). Samples 7, 17 and 75 are dunite, harzburgite and basalt, respectively (Udintsev and Chernysheva, 1965); and samples 27, 28 and 29 are basalts (Gladkikh and Chernysheva, 1966). The samples from the central branches of the Mid-Indian Oceanic Ridge (8, 11, 13, 15, 37, 66 and 74) are basalts, gabbros, serpentinized peridotites (C.G. Engel and Fisher, 1975). Squares represent areas with more than one dredge station. Samples 125 through 130 and 134 are DSDP drilled holes (Hékinian, 1974a; Kempe, 1974). Sample 26 consists of small fragments of gabbro, pyroxenite and serpentinite found in a piston core from the Ob Fracture Zone (Hékinian, 1968).

TABLE 2-4

Chemical analyses of serpentinized peridotite from the Mid-Atlantic Ridge near 53°N and from 30°N (Hékinian and Aumento, 1973; Miyashiro et al, 1969a). The data from the Kane and Vema Fracture Zones are from Miyashiro et al. (1969a), and Melson and Thompson (1971), respectively; samples 97 and 93 are lherzolite from the Indian Ocean (C.G. Engel and Fisher, 1969)

| | Atlantic Ocean | | | | | Indian Ocean | |
|---|---|---|---|---|---|---|---|
| | Mid-Atlantic Ridge, 53°N | Mid-Atlantic Ridge, 30°N | Kane Fracture Zone | Vema Fracture Zone | Romanche Fracture Zone | Rodriguez Fracture Zone | Mid-Indian Oceanic Ridge, 12°N |
| | CH4-DR2-1 | A150-6 AM-3 | V25-8-T20 | AII-20-9-1 | CH80-DR10-15 | 97-W | 93-3 |
| $SiO_2$ (wt.%) | 37.80 | 41.76 | 40.68 | 39.71 | 43.00 | 44.50 | 38.29 |
| $Al_2O_3$ | 3.75 | 2.30 | 2.86 | 2.59 | 1.08 | 3.59 | 2.74 |
| $Fe_2O_3$ | 10.73 | 5.66 | 4.69 | 7.31 | 10.29 | 1.60 | 4.65 |
| FeO | — | 3.14 | 3.43 | 1.30 | 1.12 | 5.23 | 2.25 |
| MnO | 0.10 | 0.11 | 0.10 | 0.12 | 0.12 | 0.10 | 0.06 |
| MgO | 32.29 | 33.83 | 33.20 | 33.15 | 31.87 | 34.84 | 35.32 |
| CaO | 1.87 | 1.53 | 3.05 | 1.52 | 0.08 | 5.50 | 3.75 |
| $Na_2O$ | 0.21 | 0.23 | 0.12 | 0.28 | 0.29 | 0.19 | 0.20 |
| $K_2O$ | 0.52 | 0.03 | 0.03 | 0.06 | 0.03 | <0.02 | <0.02 |
| $TiO_2$ | 0.22 | 0.14 | 0.12 | 0.06 | 0.05 | 0.03 | 0.01 |
| $P_2O_5$ | — | 0.02 | 0.01 | 0.02 | 0.08 | — | — |
| $H_2O$ | 12.95 | 11.41 | 12.03 | 13.88 | 12.19 | 4.52 | 11.46 |
| Total | 100.44 | 100.16 | 99.32 | 100.00 | 100.20 | 98.78 | 98.75 |
| Cr (ppm) | n.d. | 3700 | 3600 | 3300 | 2520 | 6400 | 4800 |
| Ni | n.d. | 1500 | 2500 | 2300 | 2238 | 3500 | 3300 |
| Co | n.d. | n.d. | n.d. | n.d. | 144 | 90 | 99 |

enriched basalts is that they are all porphyritic and that their alumina content is related to the amount of plagioclase phenocrysts present.

Olivine-enriched basalts containing a fair amount of early-formed olivine were found on all the various segments of the MIOR. The basalts containing olivine phenocrysts are found on the walls of the rift valleys. However, in the inner floor of the Rift Valley in the Central Indian Ridge near 20°S (Circe 105, C.G. Engel and Fisher, 1975), olivine basalts containing low potash are classified as olivine tholeiites. Most of them have a $K_2O$ content of less than 0.20%. The chromium distribution is generally higher in the olivine basalt (350—450 ppm) than in the plagioclase basalt (150—400 ppm).

TABLE 2-5

Chemical analyses of basaltic rocks from the Mid-Indian Oceanic Ridge

|  | V16-87 90 cm | V16-83 110 cm glass (ave. of 2) | V18-202 glass (ave. of 2) | V16-74 107 cm glass (ave. of 2) | RC8-49 20 cm |
|---|---|---|---|---|---|
| $SiO_2$ (wt. %) | 50.50 | 51.55 | 51.20 | 47.82 | 49.60 |
| $Al_2O_3$ | 15.70 | 15.20 | 15.91 | 15.14 | 16.20 |
| $Fe_2O_3$ | 1.60 | — | — | — | 0.90 |
| FeO | 8.00 | 8.86* | 8.48* | 12.50* | 9.00 |
| MnO | 0.17 | 0.12 | 0.11 | 0.20 | 0.17 |
| MgO | 7.80 | 7.68 | 8.62 | 6.46 | 8.10 |
| CaO | 10.50 | 10.90 | 11.24 | 10.79 | 10.60 |
| $Na_2O$ | 2.60 | 2.68 | 2.96 | 3.78 | 2.50 |
| $K_2O$ | 0.23 | 0.08 | 0.06 | 0.61 | 0.20 |
| $TiO_2$ | 1.40 | 1.36 | 1.34 | 2.17 | 1.50 |
| $P_2O_5$ | 0.16 | — | — | — | 0.18 |
| $H_2O$ | 1.43 | — | — | — | — |
| Total | 100.09 | 98.43 | 99.32 | 99.47 | 98.95 |

*Total Fe of the glassy margin calculated as FeO.

Sample V16-87 is located at 37°38′S, 83°48.5′E at 4316 m; V16-83 is from 29°57.5′S, 78°34′E at 2838 m; V18-202 is from 21°42′S, 70°57′E at 3168 m; V16-74 is from 20°40′S, 57°37′E at 2906 m (southern flank of Maritius Island); RC8-49 is from 51°04′S, 81°33′E at 3908 m.

Deep-sea drilling operations in the Indian Ocean were mainly concentrated on sampling the basin regions, and very little drilling has been done on the accreting plate boundaries. During DSDP Leg 24, one hole (238) penetrated about 80.5 m of basement under 506 m of sediment. This core is located on the Central Indian Ridge at the extreme northeast end of the Argo Fracture

Zone, probably within a buried transform fault (C.G. Engel et al., 1974). The basement is composed of slightly porphyritic plagioclase-pyroxene-bearing basalt with a minor amount of olivine (C.G. Engel et al., 1974). Another hole (251) was drilled during Leg 26 on the southwestern branch of the MIOR and about 10 m of basalt were recovered (Scientific Party, DSDP Leg 26, 1974). The K-Ar age of the crust at this site is 39 Ma. The basalts recovered consist of a plagioclase pyroxene with trace amounts of olivine (Kempe, 1974). The rock has a high (1.4—1.8%) $TiO_2$ content and a low (<7.5%) MgO content; the composition of the plagioclase is $An_{65}$ and the clinopyroxene consists of an augite ($Mg_{44}Fe_{18}Ca_{38}$) (Kempe, 1974).

## Carlsberg Ridge

British geologists have concentrated most of their petrological studies of the Indian Ocean on the Carlsberg Ridge near 1°20′N and 5°20′N. Wiseman (1937) and Cann and Vine (1966) have reported fresh and metamorphosed rocks from a small area of the Carlsberg Ridge. A few small fragments of metamorphosed and of relatively fresh basalts were also found in piston cores

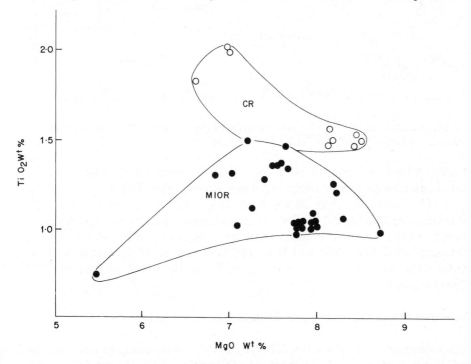

Fig. 2-17. $TiO_2$-MgO variation diagram of basaltic rocks from the Carlsberg Ridge (*CR*) and from both the eastern and western branches of the Mid-Indian Oceanic Ridge (*MIOR*). Data are from C.G. Engel et al. (1965), Hékinian (1968), Cann (1969), Balashov et al. (1970), C.G. Engel and Fisher (1975), and Udintsev (1975).

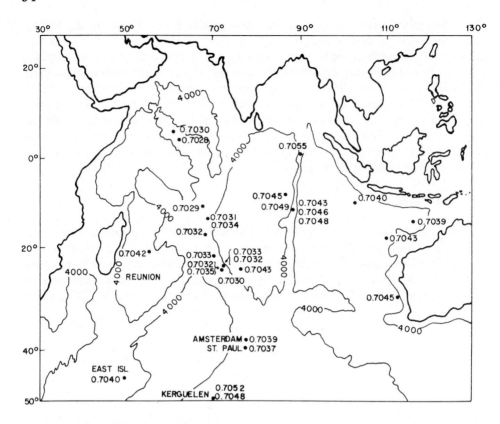

Fig. 2-18. Distribution of $^{87}Sr/^{86}Sr$ isotopic ratios of basaltic rocks from various structural features of the Indian Ocean (Subbarao et al., 1979a).

(V14-99, V19-186) from 6°50′N and 2°50′N, respectively. Generally, the dredged samples from the Carlsberg Ridge show higher $TiO_2$ (1.3—2.5%) and MgO (>8%) contents than those from the other segments of the MIOR (Hékinian, 1968) (Fig. 2-17). This is substantiated by the fact that large-ion lithophile (LIL) elements studied on a few samples from the MIOR show an enrichment of K/Rb (938) and K/Cs (62,000) values of the Carlsberg Ridge samples with respect to the other basalts from the MIOR (Subbarao et al., 1979a) (Fig. 2-18).

*Ultramafics associated with the Mid-Indian Oceanic Ridge*

The ultramafics found on the MIOR system consist mainly of serpentinized peridotite and are found in both transform faults and on the rift valley walls. There is no preferential level of the crust where this type of rock is found. It occurs at various depths on the walls of transform faults and rift valleys. During the second cruise of the "Akademik Kurchatov" and the 36th cruise

of the "Vitiaz" during the period of 1964—1965 (Udintsev, 1975), peridotites showing various degrees of serpentinization were dredged from the
Carlsberg Ridge (station 5319, DR6 and DR30). Other stations from the
western branch of the MIOR (station 5324) also recovered serpentinized
peridotites. More than half the samples from both the Carlsberg Ridge and
from the western MIOR show about 70—80% serpentinization, a smaller
number of rocks show 30—50% serpentinization, and only rare specimens
have undergone as little as 10—20% serpentinization. Transform faults associated with the MIOR are the main types of structure which show outcrops
of serpentinized peridotites and other ultramafic associations, such as dunites
and pyroxenites. The transform faults containing such rock types are the
Vitiaz Fracture Zone near 5°S, the Vema Fracture Zone near 9°S, the Argo
Fracture Zone near 13°S, the Marie Celeste Fracture Zone near 17°S, and the
Melville Fracture Zone on the western MIOR near 30°S and 60°E.

Orthopyroxenite and anorthosite were collected from the southwestern
MIOR in the Melville Transform Fault (C.G. Engel and Fisher, 1975). The
occurrence of pyroxenite is rare in an oceanic environment. Hékinian (1970)
reported eight fragments of clinopyroxenite containing orthopyroxenes from
a piston core collected near the base of the south flank of the Broken Ridge
(Table 1-5). Orthopyroxenites with cumulate textural features are often
found in stratiform bodies exposed on continental regions. The rare occurrence of this type of rock in an oceanic environment may be due to the fact
that pyroxene-enriched bodies are readily altered into talcose and chloritic
types of rocks which are often encountered in association with serpentinized
peridotites.

The ultramafics associated with the Rift Valley walls of the various segments of the MIOR consist primarily of serpentinized peridotites similar to
those found in association with transform faults. Dredging across the Rift
Valley on the Carlsberg Ridge gave rise, at depths varying between 1900 and
4900 m, to fragments of serpentinite and sporadic gabbros associated with
basaltic rocks. This type of material was found at various levels of the Rift
Valley walls, on the western Rift Mountain region, and on the middle slope
at the foot of the rift valley walls (station 5319; Chernysheva et al., 1975).
Abundant serpentinized peridotites were also reported from the Rift Valley
western wall of the southwestern MIOR near 28°S at a depth of 4400 m.
Other ultramafics have been dredged from the lowest east flank (3540—3730
m depth) of the Rift Valley near 12°S (C.G. Engel and Fisher, 1975) (Table
2-5). The most common type of ultramafics from this latter area consists of
lherzolite (ol + en + di + pl) followed by serpentinized harzburgite and minor
orthopyroxene. The freshest specimen (station Circe 93; C.G. Engel and
Fisher, 1975) contains 20% serpentine, hydrogrossular garnet, and picotite.
Generally it is found that the peridotitic material from the MIOR is less
serpentinized than that found in Atlantic fracture zones (C.G. Engel and
Fisher, 1975). Most of the marine geologists who have studied the serpen-

tinized peridotites collected from the MIOR agree that they were emplaced by intrusion into the rift valleys through cracks during the spreading of the lithospheric plates. The intrusive origin of the ultramafics is related to the idea that mantle lherzolite rises into the axial parts of the ridges and is partially melted to produce a liquid phase (harzburgite).

*Gabbroic rocks associated with the Mid-Indian Oceanic Ridge*

The gabbroic rocks found in both transform faults and in the Rift Valley come from widely spaced areas and have various chemical and mineralogical characteristics. For instance, some ferrogabbros enriched in $TiO_2$ (4—8%) were collected from the Rift Valley walls of the Carlsberg Ridge region (station 5319 and DR12-4, DR12-5) (Table 1-4). Plagioclase-gabbros and gabbro-norite were found in the Vema Trench (station 29; Dmitriev and Sharaskin, 1975). Some of the ferro-gabbros have $TiO_2$ contents similar to lunar gabbros (Miyashiro et al., 1970b). The several dozens of gabbroic samples dredged from the Carlsberg Ridge show signs of dynamo-metamorphism including textural features such as slickensides, schistosity, and brittle and plastic deformation effects observed either on the bulk rock or on individual minerals. Recrystallization of broken-up mineral grains is also a characteristic of the gabbros. Granulation is often observed in clinopyroxene mineral grains. The clinopyroxene crystal could be bent and twinned, or it is broken up in mosaic-like fragments. Orthopyroxene is usually more resistant to confined pressure than the clinopyroxene. Olivine could show brittle deformation that reduces its grains into mosaic-like aggregates. The plagioclase also undergoes both plastic and brittle deformation. The plastic deformation is expressed by a bending and stretching of the crystals. Often this plastic deformation is accompanied by a brittle deformation where the larger crystal is broken up into small fragments.

Other effects of rock transformation are shown by the presence of secondary mineral growth such as amphibole, chlorite, prehnite, albite, iddingsite, serpentine, epidote and calcite in gabbroic rocks from the Carlsberg Ridge and on the western branch of the MIOR. The types of amphiboles encountered are uralite-cummingtonite replacing clinopyroxene, and actinolite associated with chlorite and replacing plagioclase and clinopyroxene. Prehnite and calcite occur in veins and veinlets. Such kinds of mineral replacements in many of the gabbroic rocks are probably post-tectonic, since the new minerals do not show any sign of deformation while the remaining bulk rock is affected.

The gabbroic rocks from the Carlsberg Ridge are made up of plagioclase (40—60 vol. %, $An_{65}$—$An_{70}$), diopsidic pyroxene (25—35 vol. %), orthopyroxene of the enstatite-bronzite group (8—12%), and olivine ($Fo_{84}$) occurs as an accessory (<5%). The texture of the gabbros is hypidiomorphic granular (Dmitriev and Sharaskin, 1975).

EAST PACIFIC RISE

The recently active ocean ridge system of the Pacific is not symmetrical with the continental masses bordering this ocean. Most of the ridge is in the southern latitudes and the eastern part of the Pacific Ocean. This ridge system extends approximately from longitude 120°W to longitude 120°E. The eastern branch of the Pacific Ocean ridge system, also called the East Pacific Rise (EPR), runs more or less parallel to the South American and the Central American land masses (Fig. 2-19). However, the EPR is discontinuous between latitudes 30°N and 40°N. The crest of the EPR is not present in the oceanic area south of the Mendocino Fracture Zone. The study of magnetic anomalies over this region led Vacquier et al. (1961) to conclude that south of the Mendocino Fracture Zone the magnetic pattern is offset to the east by about 1600 km with respect to the area to the north of this zone. However, whether the EPR is also offset by this distance is not ascertained. Since continental and oceanic structures are quite different, it is difficult to look for indices of the EPR crestal region under continental masses. North of the Mendocino Fracture Zone there is another region called the Ridge and Trough Provinces, which could be the continuation of the EPR. The faulted topography of this region is comparable to that found in the Basin and Range Province of the North American continent. Also, it has been pointed out (Menard, 1964; Von Herzen, 1964) that both the Basin and Range Province and the Colorado Plateau are higher than the surrounding areas, and have high heat flow, anomalous mantle velocities, and an unusually thin crust. Such properties are characteristic of the axial zone of mid-ocean ridge systems.

The EPR differs from the other mid-ocean ridge systems of the Atlantic and Indian oceans in the lack of a rift valley in the axial zone. Bonatti (1968) suggested that the lack of a rift valley is due to the extrusion of a large volume of flood basalt in the axial zone of the EPR. However, it is also proven that the rate of spreading is much higher on the East Pacific Rise (half rate up to 8—9 cm/yr) than in any other part of the world, and that the absence of a rift valley may be due to the high spreading rate.

In a broad sense, early work on the petrology of the volcanics recovered from the Pacific ridge system has shown that no substantial difference exists between this and other mid-ocean ridges around the world. However, in detail there are striking compositional variabilities among recent volcanics confined to the crestal region of the EPR, the seamount regions, abyssal hills confined to the Pacific basins regions, and the Ridge and Trough Provinces located north of the Mendocino Fracture Zone.

Most of the petrological investigations were concentrated on the eastern branch of the EPR (C.G. Engel and Engel, 1963; A.E.J. Engel and Engel, 1964; Bonatti, 1968; Paster, 1968; Hékinian, 1971a). Diversities of basalt types have been noticed within different regions of the EPR. For instance, the Baja Seamount Province located between the Murray and the Clarion

TABLE 2-6

Chemical analyses of basaltic rocks from the East Pacific Rise

| | Bulk glass analyses (AAA) | | | | | | Bulk rock | Microprobe analyses of glass | | | |
|---|---|---|---|---|---|---|---|---|---|---|---|
| | RC10-243 | V18-325 | RC8-91 | V18-329 | RC9-101 | V18-330 | E21-16 | RC8-41-1 | V18-330 (ave. of 2) | CYP78-2-3 (ave. of 2) | CYP78-10-16 (ave. of 2) |
| $SiO_2$(wt.%) | 51.40 | 51.00 | 49.00 | 51.60 | 46.40 | 51.60 | 49.44 | 50.99 | 50.72 | 50.14 | 50.33 |
| $Al_2O_3$ | 15.50 | 15.50 | 15.50 | 15.90 | 19.10 | 14.60 | 13.24 | 13.68 | 14.58 | 15.79 | 14.02 |
| $Fe_2O_3$ | 0.90 | – | 1.70 | – | – | 1.1 | 3.65 | – | – | – | – |
| FeO | 8.3 | 11.44* | 10.10 | 8.92* | 9.93* | 9.93 | 9.54 | 12.26* | 9.97* | 9.58* | 11.47* |
| MnO | 0.16 | 0.23 | 0.20 | 0.16 | 0.13 | 0.20 | 0.22 | 0.11 | 0.26 | 0.10 | 0.24 |
| MgO | 7.20 | 6.60 | 7.20 | 8.0 | 7.30 | 7.60 | 6.06 | 6.26 | 7.48 | 7.90 | 7.46 |
| CaO | 10.6 | 9.7 | 11.20 | 11.20 | 10.60 | 11.00 | 10.64 | 11.12 | 11.29 | 12.30 | 11.20 |
| $Na_2O$ | 3.40 | 2.90 | 2.40 | 2.70 | 2.60 | 2.70 | 2.98 | 2.89 | 2.65 | 2.88 | 3.10 |
| $K_2O$ | 0.15 | 0.20 | 0.09 | 0.20 | 0.30 | 0.15 | 0.23 | 0.12 | 0.13 | 0.08 | 0.10 |
| $TiO_2$ | 1.30 | 2.20 | 1.90 | 1.40 | 1.90 | 1.60 | 2.12 | 2.16 | 1.55 | 1.34 | 1.70 |
| $P_2O_5$ | 0.28 | 0.26 | 0.23 | 0.18 | 0.22 | 0.18 | 0.29 | n.d. | n.d. | 0.07 | 0.29 |
| $H_2O$ | 0.90 | n.d. | 0.83 | 0.08 | 0.65 | 0.73 | 0.07 | n.d. | n.d. | n.d. | n.d. |
| Total | 100.09 | 100.74 | 99.83 | 100.32 | 99.13 | 100.76 | 98.39 | 99.58 | 98.63 | 100.18 | 99.91 |

*Total Fe calculated as FeO.

Bulk glass analyses performed by atomic absorption method; bulk rock analyses of sample E21-16 were done by spectrometry. Electron microprobe Camebax was used to analyse the glassy margins of the pillow flows (J.M. Morel, analyst).

Locations of the samples are shown in Figs 2-21 and 2-30: sample RC10-243 is from 13°06'N, 104°20'W, 2963 m; V18-325 is from 11°37.5'N, 105°54'W, 5864 m; RC8-91 is from 33°25'S, 111°54'W, 2561 m; V18-329 is from 11°03.5'N, 102°17'W, 3191 m; RC9-101 is from 25°50.5'S, 118°28.5'W, 2981 m; V18-330 is from 11°37.5'N, 101°20'W, 3197 m; E21-16 is from 54°10'S, 119°49'W, 2943 m; CYP78-2-3 and CYP78-10-16 are from near the axis of the EPR.

fracture zones consists mainly of alkali basalt types of rocks. Occasional rounded fragments of peridotite and pyroxenite were also dredged along with volcanic ash and suggest (A.E.J. Engel and Engel, 1964) that explosive or boiling volcanic eruptions caused the fragmentation and ejection of the ultramafics. In the southern region of the EPR, between latitude 40°S and 50°S, granitic types of rock with the composition of rhyolite were dredged by the "Challenger" and described by Murray and Renard (1891) and Bonatti (1968). Sand containing rhyolitic glass, anorthoclase, and quartz was also described from this general area of the EPR by Peterson and Gold-

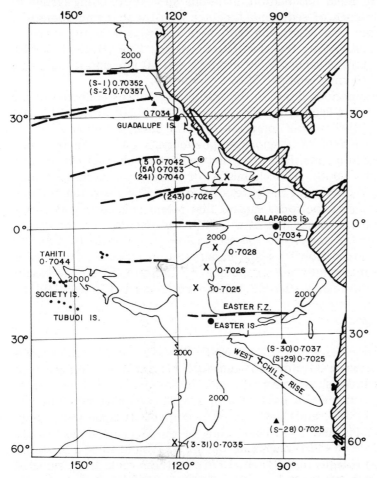

Fig. 2-20. $^{87}$Sr/$^{86}$Sr distribution map of the eastern Pacific Ocean floor basalts. × = average values of $^{87}$Sr/$^{86}$Sr taken from Gast (1965), Tatsumoto et al. (1965), Hedge and Peterman (1970), Subbarao and Hékinian (1978); ▲ = average values taken from Subbarao (1972); ● = isotopic ratios of selected islands (Powell et al., 1965; Bishop and Wollery, 1973; Gass et al., 1973); ⊙ = values from an unnamed seamount of the Mathematicians seamount chain.

berg (1962). The composition of the homogeneous alkali feldspar is rather uniform (Peterson and Goldberg, 1962). The most abundant alkali feldspar has a composition of $Or_{35}(Ab + An)_{65}$.

## East Pacific Rise between 5 and 15°S

This area of the southeastern branch of the EPR is one of the zones of the eastern Pacific awaiting detailed sampling. The dredging operation carried out on board the R.V. "Pillsbury" in 1967 was concentrated in the crestal region of the EPR. Field observation, including an echo-sounding system to determine the presence of sediment, together with the dredging operation, established that the area where the hard rock outcrops is in the shape of relatively narrow hills along the crestal zone of the Rise (Bonatti, 1968). The width of the hills is generally between 40 and 60 km, and in only one case does it reach 80 km (Bonatti, 1968).

K-Ar age determinations done on the freshest specimens collected in the crestal zone of the East Pacific Rise gave values of $0 \pm 1 \times 10^4$ to $4.55 \pm 1 \times 10^4$ years (Bonatti, 1968). The bulk composition of the basalt is similar along the axial zone and consists of plagioclase phenocrysts ($An_{65}$—$An_{80}$), clino-pyroxene, and scarce olivine. However, first-generation phenocrysts are rare or absent (Bonatti, 1968). The chemistry of these rocks permitted their classification as low-potassium tholeiites (Table 2-6). The glassy margins and the relative freshness of the rocks suggest that recent volcanism gave rise to the outcrops in the crestal region of the EPR. Isotopic data on the dredge hauls from the crestal region of the EPR show a low ratio of $^{87}Sr/^{86}Sr$ (0.7026—0.7029) in relation to island basalts (Subbarao et al., 1973) (Fig. 2-20). These isotopic values are comparable to those found on the basalts from the Juan de Fuca Ridge (0.7022, 0.7025; Peterman and Hedge, 1971).

## Juan de Fuca Ridge

The Juan de Fuca Ridge extends in a NNE—SSW direction off the western Canadian coast between latitudes 45° and 49°N. The samples dredged from the Juan de Fuca Ridge show a great diversity of basaltic types. These consist of olivine- and pyroxene-rich basalts, with a wide variability range of FeO (6—16%), $TiO_2$ (1—3%), and MgO (6—10%) contents (Clague and Bunch, 1976). The most Fe- and Ti-enriched ends of the volcanic series are also called ferrobasalts. The most magnesian olivine was found to be a forsterite-rich olivine ($Fo_{86}$) (Clague and Bunch, 1976). In these rocks the plagioclase composition varies between $An_{65}$ and $An_{86}$ and the high $Al_2O_3$ (>18%) content of some specimens suggests the abundance of plagioclase phenocrystal phases. The Clague and Bunch (1976) study indicates that the ferrobasalts from the Juan de Fuca Ridge were formed by a fractionation of up to 74% of the original magma.

TABLE 2-7

Modal analyses of basaltic rocks collected from the East Pacific Rise near 9°N during Leg 54 of the "Glomar Challenger". The mesostasis includes glass, cryptocrystalline material, and Fe-oxide minerals; calcite and zeolite were found in trace amounts

| Minerals (vol.%) | 420-13 cc. 50—57 | 420-14-1 1—8 | 420-15-1 2—22 (ave. of 4) | 421-3-1 28—148 (ave. of 3) | 422-7-1 20—131 (ave. of 3) | 422-9-3 80—95 | 422-9-4 94—97 | 423-8-1 50—52 | 428-6-1 74—79 | 428A-1-1 92—94 | 428A-4-1 50—117 (ave. of 2) | 428A-7-1 15—96 (ave. of 2) | 429-2-1 87—89 | 427-9, 10 sect. 4 (ave. of 2) |
|---|---|---|---|---|---|---|---|---|---|---|---|---|---|---|
| Plagioclase | 18.4 | 46.5 | 12.5 | 47.0 | 41.8 | 52.6 | 44.4 | 10.0 | 48.0 | 50.3 | 50.2 | 49.8 | 35.3 | 55.4 |
| Pyroxene | 4.7 | 27.7 | 3.6 | 32.5 | 46.4 | 34.2 | 42.4 | 14.2 | 28.1 | 27.3 | 28.3 | 28.4 | 43.7 | 32.1 |
| Olivine | — | — | 0.9 | — | tr. | 9.4 | 4.9 | 1.6 | 0.4 | 1.5 | — | 2.5 | 6.7 | — |
| Phyllo-silicates | 2.1 | 0.2 | tr. | 1.73 | n.d. | n.d. | n.d. | n.d. | n.d. | n.d. | n.d. | n.d. | n.d. | 2.1 |
| Mesostasis | 70.8 | 24.1 | 81.8 | 12.4 | 11.3 | 3.5 | 10.0 | 72.5 | 22.0 | 20.2 | 21.3 | 19.4 | 13.4 | 4.8 |
| Vesicles | 3.3 | 1.5 | 1.3 | n.d. | <1 | 1.3 | n.d. | n.d. | n.d. | n.d. | n.d. | n.d. | n.d. | 0.2 |

Site 427 is from the Siqueiros Transform Fault.

TABLE 2-8

Chemical analyses of olivine-bearing basalts collected during Leg 54 of the "Glomar Challenger" on the East Pacific Rise near 9°N

| | 422-10-1 19-25 | 422-7-1 20-25 | 422-7-2 3 | 422-8-5 132-135 | 422-9-1 70-73 | 422-9-3 91-94 | 422-9-5 72-75 | 428-6-2 1-3 | 428A-1-1 21-23 | 428A-1-1 27-29 | 428A-1-2 49-51 |
|---|---|---|---|---|---|---|---|---|---|---|---|
| $SiO_2$ (wt.%) | 50.39 | 50.49 | 50.74 | 50.73 | 50.28 | 48.59 | 49.77 | 50.01 | 50.87 | 50.48 | 50.45 |
| $Al_2O_3$ | 13.92 | 14.17 | 14.28 | 14.32 | 13.99 | 15.50 | 15.83 | 14.17 | 14.61 | 14.53 | 14.28 |
| $Fe_2O_3$ | 3.04 | 3.26 | 2.15 | 2.40 | 2.26 | 2.31 | 2.37 | 3.19 | 2.99 | 3.28 | 3.48 |
| FeO | 7.92 | 6.61 | 7.59 | 7.25 | 7.55 | 6.36 | 6.68 | 6.54 | 6.72 | 6.16 | 6.58 |
| MnO | 0.20 | 0.17 | 0.18 | 0.17 | 0.17 | 0.19 | 0.15 | 0.20 | 0.31 | 0.32 | 0.19 |
| MgO | 7.09 | 7.83 | 8.04 | 8.21 | 7.87 | 8.45 | 8.63 | 7.52 | 7.31 | 7.05 | 7.40 |
| CaO | 11.72 | 11.98 | 12.00 | 11.99 | 11.92 | 11.59 | 11.68 | 11.84 | 11.87 | 12.26 | 12.03 |
| $Na_2O$ | 2.94 | 2.39 | 2.70 | 2.74 | 2.71 | 2.85 | 2.83 | 2.44 | 2.50 | 2.60 | 2.50 |
| $K_2O$ | 0.31 | 0.13 | 0.13 | 0.10 | 0.09 | 0.23 | 0.21 | 0.08 | 0.23 | 0.10 | 0.17 |
| $TiO_2$ | 1.88 | 1.44 | 1.45 | 1.45 | 1.43 | 1.38 | 1.43 | 1.45 | 1.48 | 1.48 | 1.50 |
| $P_2O_5$ | 0.21 | 0.17 | 0.15 | 0.17 | 0.15 | 0.18 | 0.18 | 0.14 | 0.16 | 0.14 | 0.16 |
| $CO_2$ | 0.00 | 0.00 | 0.00 | 0.00 | 0.00 | 0.00 | 0.00 | 0.00 | 0.00 | 0.00 | 0.00 |
| LOI* | 0.00 | 0.76 | 0.21 | 0.14 | 1.04 | 1.33 | 0.33 | 0.36 | 0.00 | 0.50 | 0.33 |
| Total | 99.61 | 99.39 | 99.61 | 99.66 | 99.45 | 98.95 | 100.08 | 97.93 | 99.04 | 98.89 | 99.06 |
| Q | 0.00 | 2.51 | 0.00 | 0.00 | 0.00 | 0.00 | 0.00 | 2.53 | 2.37 | 2.47 | 2.45 |
| Or | 1.83 | 0.76 | 0.76 | 0.59 | 0.53 | 1.35 | 1.24 | 0.47 | 1.35 | 0.59 | 1.00 |
| Ab | 24.87 | 20.22 | 22.84 | 23.18 | 22.93 | 24.11 | 23.94 | 20.64 | 21.15 | 22.00 | 21.15 |
| An | 23.87 | 27.55 | 26.46 | 26.47 | 25.74 | 28.82 | 29.87 | 27.47 | 27.96 | 27.68 | 27.24 |
| Di | 26.67 | 24.73 | 25.98 | 25.75 | 26.25 | 22.09 | 21.62 | 24.41 | 24.10 | 25.91 | 25.24 |
| Hy | 13.59 | 15.00 | 16.05 | 15.94 | 16.07 | 6.04 | 9.24 | 14.33 | 14.57 | 11.85 | 13.38 |
| Ol | 0.30 | 0.00 | 1.07 | 0.95 | 0.55 | 8.81 | 7.26 | 0.00 | 0.00 | 0.00 | 0.00 |
| Mt | 4.40 | 4.72 | 3.11 | 3.47 | 3.27 | 3.34 | 3.43 | 4.62 | 4.33 | 4.75 | 5.04 |
| Ilm | 3.57 | 2.73 | 2.75 | 2.75 | 2.71 | 2.62 | 2.71 | 2.75 | 2.81 | 2.81 | 2.84 |
| Ap | 0.49 | 0.40 | 0.35 | 0.40 | 0.35 | 0.42 | 0.42 | 0.33 | 0.37 | 0.33 | 0.37 |
| C.I.** | 43.24 | 47.24 | 46.27 | 46.69 | 45.36 | 52.18 | 51.86 | 46.70 | 46.41 | 47.94 | 47.12 |

*LOI = loss on ignition.

**Crystallization Index (Poldervaart and Parker, 1964).

*East Pacific Rise near 9°N*

Detailed geophysical studies of this section of the EPR were done by Rosendahl et al. (1976). Multi-channel refraction profiles, magnetism, and seismic reflection studies, together with a bathymetric survey of the area, have permitted the delineation of a horst-like structure about 10 km wide, which is the zone of a diverging plate boundary. There was also speculation about the existence of a low-velocity layer representing a possible partially melted magma reservoir located beneath the accreting plate boundaries. In addition to the geophysical operations conducted in an area of about 200 km$^2$, dredge hauls were taken on various structural features associated with the ridge system near 9°N. These dredges were mainly concentrated on a centrally located east-west-trending feature of the Siqueiros Transform Fault, which offsets the ridge axis near 8°N.

The EPR near 9°N is made up of various types of structural features such as the Siqueiros Transform Fault, and an east-west-trending ridge located on the western flank of the EPR called the OCP (Ocean Crust Panel) Ridge. The OCP Ridge is a discontinuous structural feature which does not offset the EPR axis. Another feature of the EPR at 9°N is its flank fabric which has a small rolling-hills type of structure with a general north-south orientation, parallel to the ridge axis.

Very few dredges have been taken on the flank provinces of the EPR (Batiza et al., 1977). However, a deep sea drilling operation performed in 1977 during DSDP Leg 54 of the "Glomar Challenger" was intended to find out if temporal variation of magmatic events took place across a transect perpendicular to the ridge axis. Another objective of the operation was to determine the chemical, magnetic and physical nature of the transition from EPR fabric to a transverse type of structural feature such as the OCP and the Siqueiros Transform Fault areas. During Leg 54, seven single-bit holes were drilled on the western flank of the ridge. The probable magnetic age of the crust increases from east to west from the axis of the ridge. The basement ages of the drilled sites from east to west are as follows (Scientific Party, DSDP Leg 54, 1980b):

$$Site\ 423 = 1.6\ Ma$$
$$Site\ 422 = 1.7\ Ma$$
$$Site\ 428 = 2.0\ Ma$$
$$Site\ 419 = 2.6\ Ma$$
$$Sites\ 420\ and\ 421 = 3.4\ Ma$$
$$Site\ 429 = 4.6\ Ma$$

From dredging and deep sea drilling operations, three major types of basaltic rocks were found near 8—9°N on the EPR (Tables 2-7, 2-8, 2-9, 2-10):

(1) The ferrobasalts, made up of pyroxene, plagioclase, and titano-magnetite, have a high FeO (total Fe calculated as FeO) content (12—15%), a high TiO$_2$ content (>2%), and a high Zr content (>150 ppm) (Fig. 2-21, Table 2-10).

(2) The plagioclase-olivine-pyroxene basalts, with an olivine content of less

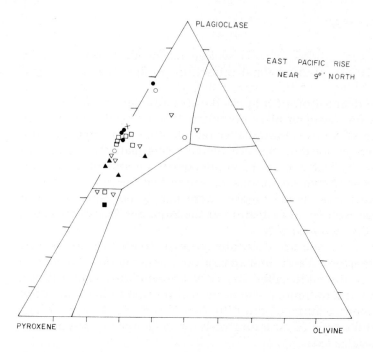

Fig. 2-21. Ternary diagram of modal olivine, plagioclase and clinopyroxene content of basement basaltic rocks drilled during Leg 54 of the DSDP on the flank of the East Pacific Rise near 9°N. The symbols indicate the various sites of drilling: ○ = 421, ● = 420, ▲ = 422, ■ = 423, × = 427, □ = 428 and ▽ = 429. The solid lines define the experimentally derived 1-atm reaction curves of Osborn and Tait (1952).

than 9% by volume, have an FeO content of 10—12%, a TiO$_2$ content of 1.5—2%, and a Zr content of about 100—150 ppm (Fig. 2-21).

(3) The plagioclase-pyroxene basalts which are depleted in olivine content and have an FeO content of less than 11%, a TiO$_2$ content of less than 1.5%, and a Zr content of about 50—100 ppm (Table 2-9) are also similar to the plagioclase-pyroxene basalts found on the Mid-Atlantic Ridge near 36°50′N (FAMOUS area). A smaller amount of basalts with alkaline affinities and picritic basalts were also dredged on east-west-trending features such as the OCP Ridge (J.R. Johnson, 1979) and from another east-west ridge system located in the Siqueiros Transform Fault (Schrader et al., 1980). They differ from the alkali basalts found on eastern Pacific seamounts in their lower Na$_2$O content (2.5—3.5%) and lower K$_2$O content (0.6—0.7%). However, the basalts with alkaline affinities have a higher abundance of heavy rare earth elements and a lower strontium content than do typical island alkali basalts (J.R. Johnson, 1979). The rare earth element (REE) distribution pattern in these latter basalts is similar to that of some aseismic ridge oceanic andesites such as those from the Ninety East Ridge in the Indian Ocean (DSDP Leg 22, Site 214, core 48, section 1) and other plagioclase-olivine-pyroxene basalts from

TABLE 2-9

Chemical analyses of plagioclase-pyroxene basalts collected during Leg 54 of the "Glomar Challenger" on the East Pacific Rise near 9°N (sample 421-2-1/5—7 is a ferrobasalt)

| | 420-14.1 0—8 | 420-17.1 27—30 | 421-2.1 5—7 | 421-3.1 117—120 | 421-4.1 4—7 | 421-BIT 10—13 | 423-5cc 40—42 | 423-6.1 28—34 | 423-8.1 40—42 | 429A-1.1 43—45 | 429A-3.1 #7 |
|---|---|---|---|---|---|---|---|---|---|---|---|
| $SiO_2$ (wt.%) | 50.43 | 50.67 | 50.32 | 51.54 | 51.49 | 50.24 | 50.43 | 50.14 | 50.90 | 49.72 | 50.57 |
| $Al_2O_3$ | 13.77 | 14.28 | 13.11 | 14.14 | 14.04 | 13.85 | 13.59 | 13.40 | 13.55 | 15.87 | 13.89 |
| $Fe_2O_3$ | 2.99 | 4.55 | 5.69 | 4.01 | 2.97 | 5.50 | 4.27 | 4.56 | 4.89 | 3.64 | 3.84 |
| FeO | 8.47 | 5.80 | 7.39 | 6.13 | 6.62 | 6.23 | 6.38 | 7.61 | 6.68 | 4.63 | 7.26 |
| MnO | 0.20 | 0.19 | 0.22 | 0.20 | 0.23 | 0.20 | 0.21 | 0.20 | 0.18 | 0.15 | 0.16 |
| MgO | 7.18 | 6.30 | 6.42 | 6.44 | 6.74 | 7.11 | 6.79 | 6.92 | 6.82 | 7.12 | 7.07 |
| CaO | 11.21 | 11.20 | 9.86 | 11.04 | 11.14 | 10.76 | 11.02 | 10.97 | 10.87 | 12.99 | 11.44 |
| $Na_2O$ | 2.55 | 2.80 | 2.69 | 2.78 | 2.89 | 2.59 | 2.92 | 1.94 | 2.57 | 2.33 | 2.42 |
| $K_2O$ | 0.16 | 0.43 | 0.44 | 0.37 | 0.37 | 0.45 | 0.32 | 0.23 | 0.35 | 0.08 | 0.08 |
| $TiO_2$ | 1.94 | 1.99 | 2.54 | 2.15 | 2.22 | 2.12 | 2.09 | 2.06 | 2.10 | 1.15 | 1.61 |
| $P_2O_5$ | 0.19 | 0.16 | 0.23 | 0.19 | 0.22 | 0.19 | 0.24 | 0.23 | 0.22 | 0.11 | 1.15 |
| $CO_2$ | 0.00 | 0.00 | 0.00 | 0.00 | 0.00 | 0.00 | 0.00 | 0.00 | 0.00 | 0.00 | 0.00 |
| LOI | 0.14 | 1.10 | 0.83 | 0.86 | 0.67 | 0.76 | 1.14 | 0.83 | 0.81 | 0.98 | 0.43 |
| Total | 99.22 | 99.46 | 99.73 | 99.84 | 99.59 | 99.99 | 99.39 | 99.08 | 99.93 | 98.76 | 99.91 |
| Q | 2.33 | 4.34 | 6.02 | 5.22 | 3.41 | 4.51 | 3.32 | 7.09 | 5.52 | 2.67 | 5.67 |
| Or | 0.94 | 2.54 | 2.60 | 2.18 | 2.18 | 2.65 | 1.89 | 1.35 | 2.06 | 0.47 | 0.47 |
| Ab | 21.57 | 23.69 | 22.76 | 23.52 | 24.45 | 21.91 | 24.70 | 16.41 | 21.74 | 19.71 | 20.47 |
| An | 25.65 | 25.12 | 22.39 | 25.01 | 24.24 | 24.83 | 23.02 | 27.17 | 24.40 | 32.60 | 26.80 |
| Di | 23.34 | 23.41 | 20.02 | 22.79 | 23.78 | 21.68 | 24.04 | 20.74 | 22.45 | 24.72 | 18.10 |
| Hy | 16.78 | 8.50 | 11.49 | 9.91 | 11.81 | 11.19 | 10.54 | 14.42 | 11.35 | 9.87 | 16.66 |
| Mt | 4.33 | 6.59 | 8.24 | 5.81 | 4.30 | 7.97 | 6.19 | 6.61 | 7.09 | 5.27 | 5.56 |
| Ilm | 3.68 | 3.77 | 4.82 | 4.08 | 4.21 | 4.02 | 3.96 | 3.91 | 3.98 | 2.18 | 3.05 |
| Ap | 0.44 | 0.37 | 0.54 | 0.44 | 0.51 | 0.44 | 0.56 | 0.54 | 0.51 | 0.5 | 2.71 |
| C.I. | 42.29 | 45.00 | 38.95 | 43.88 | 42.91 | 44.05 | 42.86 | 43.41 | 43.08 | 54.19 | 41.02 |

TABLE 2-10

Chemical analyses of ferrobasalts collected from the Siqueiros Fracture Zone during Leg 54 of the "Glomar Challenger" on the East Pacific Rise near 8°N

|  | 427-9-3/16—19 | 427-10-1/64—67 | 427-10-5/82—84 |
|---|---|---|---|
| $SiO_2$ | 49.78 | 49.41 | 49.88 |
| $Al_2O_3$ | 13.05 | 13.25 | 13.26 |
| $Fe_2O_3$ | 5.10 | 5.18 | 4.85 |
| FeO | 8.21 | 7.89 | 8.00 |
| MnO | 0.20 | 0.22 | 0.19 |
| MgO | 6.59 | 6.78 | 6.80 |
| CaO | 10.15 | 10.21 | 10.31 |
| $Na_2O$ | 2.83 | 2.11 | 3.02 |
| $K_2O$ | 0.07 | 0.08 | 0.06 |
| $TiO_2$ | 2.48 | 2.46 | 2.41 |
| $P_2O_5$ | 0.25 | 0.26 | 0.27 |
| $CO_2$ | 0.00 | 0.00 | 0.00 |
| LOI | 0.29 | 0.67 | 0.10 |
| Total | 98.99 | 98.51 | 99.14 |
| Q | 4.55 | 7.45 | 3.22 |
| Or | 0.41 | 0.47 | 0.35 |
| Ab | 23.94 | 17.85 | 25.55 |
| An | 22.69 | 26.44 | 22.44 |
| Di | 20.98 | 18.10 | 21.70 |
| Hy | 13.42 | 14.73 | 13.53 |
| Mt | 7.39 | 7.51 | 7.03 |
| Ilm | 4.71 | 4.67 | 4.57 |
| Ap | 0.58 | 0.61 | 0.63 |
| C.I. | 38.85 | 41.03 | 39.32 |

the Indian Ocean basins (Leg 26, Sites 250, 256, and Leg 22, Site 215). Also the REE distribution in the plagioclase-olivine-pyroxene basalts dredged from the Mid-Atlantic Ridge associated with the platform of the Azores at 38°—40°N shows a pattern of light-REE enrichment similar to that of the basalts from the OCP Ridge of the EPR (9°N). The origin of the basaltic rocks from the platform of the Azores was attributed (Schilling, 1975b) to the partial melting of a mantle source different from that of other more "normal" types of ocean-ridge basalts. Schilling (1975b) has suggested that a mixing of two magma types originating from two distinct sources occurred; this hypothesis will be further developed in Chapter 12. The OCP basalts with alkaline affinities may also have an origin similar to those from the Mid-Atlantic Ridge as described by Schilling (1975).

*East Pacific Rise near 21°N*

The crestal area of the EPR near 21°N was explored during the 1978 diving program (Cyamex, 1978, 1981) by the submersible "Cyana" and later (1979) by "Alvin".

These programs were part of a project called RITA (named from the initials of the Rivera and Tamayo Fracture Zones) and were designed to conduct a detailed study of a ridge system with an intermediate rate of opening (total rate of about 6 cm/yr).

From structural and morphological studies (Cyamex, 1981; Spiess et al., 1980) three major geological provinces were recognized:

(1) *The axial zone*, also called the extrusion zone, is characterized by con- structional volcanism due to the piling up of recent flows reaching up to about 80 m in height (Cyamex, 1981). The major structural characteristic is the presence of open fissures up to about 1—4 m wide without any vertical displacement. Most of the fissures are oriented parallel to the axis of accretion (about 040°) (Cyamex, 1981). From the degree of freshness of the rocks, it was concluded that this extrusive zone is about 400—1000 m wide. In addition, near the crestal zone, collapsed fossil lava lakes extend for a distance from 30 up to about 1000 m square and they have bordering walls varying from about 5 to 10 m deep. The axial zone is also characterized by active hydrothermal vents which are well identified in the southern area of the dives (Fig. 2-22).

(2) *The horst and graben zone* is the active extensional area characterized by troughs and scarps with reliefs reaching to 70 m in height. This zone shows small amounts of sediment varying from several millimeters to several centi- meters in thickness covering the pillow flows. It is also in this type of terrane that the first appearance of talus piles composed of angular debris occur.

(3) *The active tectonic zone* is located at about 2 km from the axial zone of the area explored. This zone consists of linear tilted blocks several kilo- meters in length and between 25 and 70 m in height. This zone is heavily sedimented and the pillow flows are almost completely buried by sediment. The width of the active tectonic zone seems to be greater than 500 m.

The axial and the horst and graben zones were the sites of extensive sampling (Fig. 2-22). Approximately fifty rock stations were made by the submersibles "Cyana" and "Alvin". In addition, six dredge stations ("Gillis" cruise, 1979) were also studied. Microprobe analyses on the chilled glassy margins and on the early solidified phases indicate the presence of various types of basaltic rocks distributed on a 4-km-wide and 12-km-long band along the ridge axis (Fig. 2-23).

Morphologically, three types of lava flows have been distinguished: (1) common type of bulbous pillow flows which are considered the conducts through which lava has flowed; (2) sheet flows, which are thin flat-lying flows, and (3) the pillar type of extrusives, which are the septa or columns holding the roof of large lava lakes which later collapsed. Among the various types of flows encountered there were no noticeable mineralogical and chemical differences.

Most of the rocks collected are aphyric in nature; however coarse-grained basalts containing plagioclase and olivine were seen in hand specimens. Based on the relative mineral abundance and the early-formed mineral phases in the chilled glassy margin of the basaltic flows, the samples from 21°N have been divided into three major groups (Tables 2-11, 2-12, 2-13):

(1) The plagioclase-rich basalts which correspond to the highly-phyric (HPPB) and the moderately-phyric plagioclase basalts (MPPB) found on other accreting plate boundary regions (Chapter 1). These rock types are characterized by more than 10% by volume of phenocrystal plagioclase set in their glassy matrix. Both plagioclase and olivine seem to be the first to have crystallized; however, the olivine content is very low when compared to

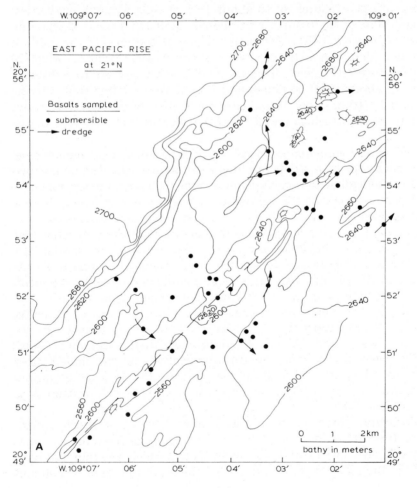

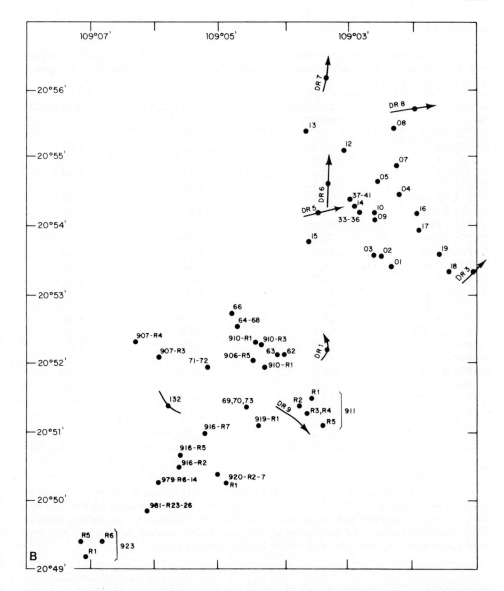

Fig. 2-22. A. Bathymetric map of the East Pacific Rise near 21°N after Spiess et al. (1980). The dashed line indicates the alignment of the sulfide vents and also represents the area where the freshest flows have been seen. The dots indicate the sample stations with corresponding numbers shown in Fig. 2-22B. B. Distribution of the rock samples shown on the bathymetric map (A). Sample numbers having three digits (southern portion of the map) correspond to the "Alvin" dives (ALV979) followed by the station number (ALV979-R7) which is also the sample number. Sample numbers having initially two digits (northern portion of the map) correspond to the "Cyana" 1978 dives (CYP78-02) and are followed by the station number (CYP78-02-02). Sample 132 is a sample from a dredge haul obtained from Scripps Institution of Oceanography.

114

TABLE 2-11

Chemical characteristics of plagioclase-olivine-pyroxene basalts from the East Pacific Rise near 21°N (analyses were performed by electron microprobe (Camebax) on the glassy margin of basaltic flows)

| Sample No. | FeO/MgO × 100 (mol.) | Fo (mol.%) | $TiO_2$ (wt.%) |
|---|---|---|---|
| GL-DR6-1 | 84.1 | 83.1 | 1.67 |
| GL-DR3-1 | 83.3 | 83.3 | 1.50 |
| GL-DR5-2 | 72.6 | 83.9 | 1.53 |
| CYP78-19-70 | 80.0 | n.d. | |
| CYP78-08-14D3 | 78.2 | 85.3 | 1.35 |
| CYP78-03-05 | 77.0 | 83.3 | 1.65 |
| CYP78-19-69 | 76.0 | 83.6 | 1.63 |
| CYP78-03-04 | 73.0 | 83.6 | 1.65 |
| CYP78-10-17 | 73.7 | 84.3 | 1.60 |
| CYP78-10-18 | 71.6 | 83.0 | 1.63 |
| CYP78-04-06 | 73.0 | n.d. | 1.48 |
| ALV907-R4 | 72.1 | 84.1 | 1.60 |
| ALV916-R8 | 69.5 | 84.3 | 1.57 |
| ALV923-R5 | 68.0 | 85.0 | 1.46 |
| CYP78-09-15B | 69.0 | 84.9 | 1.39 |
| CYP78-04-07 | 68.0 | 84.9 | 1.50 |
| ALV910-R1 | 69.0 | 84.2 | 1.50 |
| CYP78-07-12 | 66.3 | 86.2 | 1.60 |
| ALV907-R1 | 66.0 | 85.6 | 1.37 |
| CYP78-19-72D | 63.0 | 86.3 | 1.30 |
| CYP78-12-41A | 64.5 | 85.5 | 1.43 |
| GL-DR9-1 | 66.0 | 84.7 | 1.38 |
| GL-DR1-1 | 65.0 | 85.2 | 1.41 |

n.d. = not determined.

Sample numbers consist of composite letters whose meanings are as follows: GL = R.V. "Gillis"; DR = dredge; ALV = "Alvin"; CYP = "Cyana". The abbreviation ALV is followed by a three-digit number which corresponds to the dive followed by the sample number (R1). The abbreviation CYP is followed by the year (1978), two sets of two-digit numbers which correspond to the dive, and finally by the sample number.

plagioclase. The chemistry of these rocks is variable and depends on the relative amount of olivine phenocrysts with respect to plagioclase. The most fractionated plagioclase-rich basalts composed essentially of plagioclase and pyroxene have a high $TiO_2$ content (1.6—1.9%) and a high FeO/MgO ratio (>1.4). The least fractionated ones, containing olivine phenocrysts (and/or megacrysts) in addition to plagioclase, have a lower $TiO_2$ content (1.4%) (Table 2-13).

(2) The plagioclase-olivine-pyroxene basalts (POPB) consist of early-formed plagioclase, olivine, and pyroxene coexisting together in the glassy margins

TABLE 2-12

Chemical characteristics of olivine basalts from the axial zone of the East Pacific Rise near 21°N* (the analyses were performed on the glassy chilled margin of basaltic flows)

| Sample No. | FeO/MgO × 100 (mol.) | Fo (mol.%) | TiO$_2$ (wt.%) |
|---|---|---|---|
| CYP78-06-09 | 57.6 | 86.3 | 1.30 |
| CYP78-19-71 | 58.1 | 86.0 | 1.35 |
| CYP78-07-13 | 57.6 | 86.2 | 1.29 |
| CYP78-06-10 | 59.5 | 87.0 | 1.23 |
| ALV981-R26 | 52.3 | 88.5 | 1.20 |
| ALV979-R7 | 51.4 | 88.4 | 1.24 |
| ALV916-R7B | 51.8 | 87.5 | 1.16 |
| ALV916-R2 | 53.6 | 87.8 | 1.33 |
| ALV923-R2 | 50.9 | 88.2 | 1.10 |
| GL-DR8-2 | 54.1 | 87.3 | 1.20 |

*The mineral constituents of these rocks are olivine, plagioclase and spinel. Sample numbers are explained in Table 2-11.

TABLE 2-13

Chemical characteristics of moderately-phyric (MPPB) and highly-phyric (HPPB) plagioclase basalts* (chemical analyses performed by electron microprobe on the glassy margin)

| Sample No. | FeO/MgO × 100 (mol.) | Fo (mol.%) | TiO$_2$ (wt.%) |
|---|---|---|---|
| GL-DR5-1 | 74.0 | 84.0 | 1.83 (HPPB) |
| GL-DR7-1 | 87.6 | 82.5 | 1.88 (HPPB) |
| CYP78-02-1 | 79.4 | 83.2 | 1.82 |
| CYP78-02-02 | 59.0 | 86.9 | 1.26 |
| CYP78-02-03 | 57.6 | 86.5 | 1.30 (HPPB) |
| CYP78-18-66B | 68.0 | 85.5 | 1.42 |
| CYP78-10-16 | 86.5 | 82.8 | 1.74 |
| ALV911-R3 | 67.4 | 81.2 | 1.69 |

*The glassy basalts contain more than 10% by volume of plagioclase phenocrysts. Sample numbers are explained in Table 2-11.

of the pillow flows. The texture is often porphyroclastic with all three phases intergrown together. They differ from the plagioclase-rich basalts in their lower content in plagioclase (about 7% or less). These types of basalt show a relatively high TiO$_2$ content (1.4—1.8%), a high FeO/MgO ratio (1.3—1.5) and a forsterite content of their olivine comprised between Fo$_{83}$ and Fo$_{86}$ (Table 2-11).

(3) The olivine basalts are essentially made up of early-formed olivine which is sometimes accompanied by Cr-spinel. The olivine occurs as euhedral and rarely as resorbed crystals. From the relative size of the minerals, two

generations of olivine with no appreciable compositional differences were seen to occur in the glassy matrix of these rocks. The forsterite content of the olivine is $Fo_{86}$—$Fo_{89}$. The chemistry of the glassy chilled margin shows a low $TiO_2$ content (<1.3%) and a low FeO/MgO ratio (0.9—1.1) when compared to the other rock types (Table 2-12).

The most abundant rock type encountered on the EPR near 21°N are the POPB which constitute about 87% by volume of the rock samples recovered. The olivine basalts and the plagioclase-rich basalts comprise about 10% and 3%, respectively, of the total rock sampled.

Both mineralogical and chemical studies seem to point to the heterogeneous nature of the rock types encountered on the accreting plate boundary regions of the EPR near 21°N. Using the FeO/MgO ratio and the forsterite content of olivine, a schematic chemical zonation is shown in Fig. 2-23. This type of compositional zonation observed near 21°N shows a striking similarity

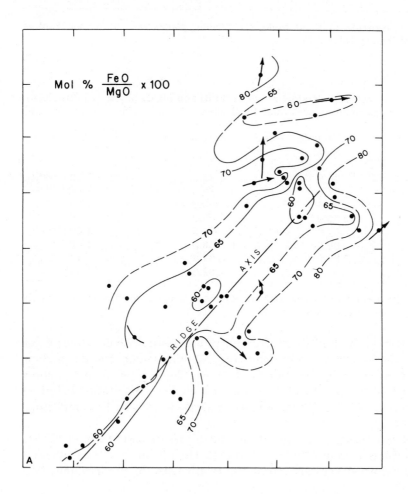

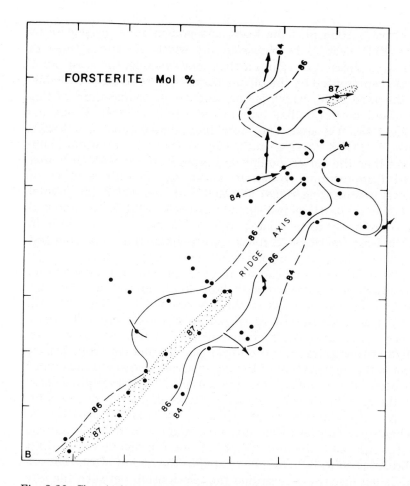

Fig. 2-23. Chemical zonation of East Pacific Rise axial zone near 21°N. The contour lines represent the molecular ratios of FeO and MgO (A) and the forsterite content of olivine analysed from the glassy chilled margins of the various types of basalts (B). Black dots are the sample locations reported in Fig. 2-22.

to that encountered on the inner floor of the Rift Valley in the FAMOUS area (Fig. 2-10). In order to compare these two provinces, similar analytical techniques and sample treatments were used. The FeO/MgO ratio of the basalts from 21°N (EPR), varying between 52 to about 88, differs from that encountered in the FAMOUS area (FeO/MgO = 47—94) and the range of forsterite content is less distinct ($Fo_{83}$—$Fo_{89}$) and less pronounced than for the former zone ($Fo_{79}$—$Fo_{88}$) (Figs. 2-10, 2-23). Also the total width (1—3 km wide) of the petrographic zonation of the EPR is comparable to that of the inner floor of the Rift Valley in the FAMOUS area (Figs. 2-10, 2-23). The size of the inner floor in the FAMOUS area corresponds roughly to that of the axial extrusive zone plus the horst and graben zone, as previously defined for the EPR near 21°N. However, the width of various types of extrusive flows from these two regions differs. For example, the most primitive type of rocks represented by the olivine basalts covers an area (300—500 m wide from the axial zone) which is only about half or one third of that covered by the same corresponding type of rocks found on Mt. Venus and Mt. Pluto in the FAMOUS area. The more fractionated type of volcanics respresented by the POPB cover a width (up to about 2 km wide) about three times larger than the corresponding lava types from the FAMOUS area.

The observed compositional zonations giving rise to various types of basalts on both slow-spreading ridges (FAMOUS area; about 2 cm/yr total spreading rate) and moderately fast-spreading ridges (EPR 21°N; about 6 cm/yr total rate) are likely to be related to fundamental processes. In both ridge systems the least fractionated rocks (olivine basalts) are the youngest flows erupted. The mechanism giving rise to successive compositional changes of the volcanics is believed to depend on the tectonics of spreading and on crystal-liquid fractionation processes at shallow depths underneath the ridge axis. The upwelling magma trapped in shallow reservoirs in the crust will fractionate until it is freed by crustal spreading and volcanism. The early eruptive cycle begins with the most fractionated liquid (plagioclase-, pyroxene-rich melt) which concentrates near the top of the magma reservoir. During the eruptive events the temperature of the liquidus will be lowered and crustal cumulates will occur at the bottom of the reservoir. The last eruptive phase will give rise to the most mafic end member of the liquid left in the reservoir. The eruptive cycle stops until new fissures and cracks are created by the spreading mechanisms. The fact that the most mafic end members of the basaltic rocks (picritic and olivine basalts) are less prominent on the EPR than on the Mid-Atlantic Ridge could be related to the rate of crustal spreading. A more detailed discussion regarding the relationship between the rate of spreading and crustal composition is presented at the end of this chapter.

COMPARISON OF THE VOLCANICS ERUPTED ON VARIOUS ACTIVE SPREADING RIDGES

Comparative studies of crustal and upper mantle material extended to the vast oceanic provinces of the Atlantic, Pacific and Indian Oceans is an ambitious task, especially when one considers the limited data available in proportion to the size of the area. However, in summarizing the oceanic ridge systems it is fair to state that various types of basalts are presently erupted on spreading ridges, i.e. quartz-normative and olivine-normative to nepheline-normative basalts. Major element analyses of basaltic glass from spreading ridges, excluding off-ridge and basin provinces, have been gathered by Melson et al. (1977). These analyses indicate in general that the EPR and the Juan de Fuca Ridge are distinctly higher in FeO (total Fe calculated as FeO) and in $TiO_2$ than the mid-oceanic ridges, both in the North Atlantic and in the Indian Ocean.

Olivine phenocrysts and quenched olivine crystals are conspicuous in both Atlantic and Indian Ocean ridge basalts. The basalts of these mid-oceanic ridges contain early-formed olivine which persists in the groundmass. The volcanics from the EPR also contain microphenocrysts and phenocrysts of olivine in quenched glass but generally they are absent in the groundmass. Spinel is another component which appears mainly in conjunction with the olivine in quenced glass, but generally they are absent in the groundmass. oceanics. Clinopyroxene and plagioclase phases are the most important components of the volcanics from the EPR and from the Juan de Fuca Ridge.

The major compositional difference between the EPR and the other ridge systems in the Atlantic and Indian Oceans is the scarcity of ultramafics recovered from accreting plate boundary regions of the former. In addition there is a slight but constant difference between the volcanic products erupted on the EPR and those from the MAR and the MIOR. These differences are visualized on an AFM diagram (Fig. 2-24), showing the more iron-rich nature and the more pronounced trend of fractionation towards FeO of the EPR basalts with respect to the others. Picritic rocks are most commonly found among volcanics from both the MAR and the MIOR and they fall in the MgO corner of the diagram (Fig. 2-24). However, the presence of the more fractionated types of basalt found on the EPR, when compared to the MAR, does not exclude the fact that an olivine-enriched melt may also exist. Indeed, most of the EPR samples collected are representative of surface flows. The few drilling operations performed along the EPR did not penetrate deeper than 100 m into the basement.

*Titanium and zirconium*

Ti and Zr have also been used as meaningful chemical parameters to differentiate between different types of rocks from both oceanic and island-arc provinces (Pearce and Cann, 1971, 1973; Pearce and Norry, 1979). Using similar parameters, a clear distinction is noticed between the EPR basaltic flows from near 8°N and 9°N and that of the northern MAR and the MIOR. (Fig. 2-25). The observed trends show an increase of Ti and Zr with slopes close to unity (Fig. 2-25). The depletion of Ti and Zr (<1% and 100 ppm, respectively) is probably due to the early crystallization of titanomagnetite and pyroxene or to the depleted nature of the partial melted phase of these elements. The samples from 9°N on the EPR show the highest Ti and Zr values (>1% and >100 ppm, respectively) and this is attributed to the abundance of the clinopyroxene content in these rocks (Fig. 2-24). The basaltic rocks from 21°N (Juteau et al., 1980) show values of Ti content intermediate between those from the MAR and the basalts from 9°N (EPR) (Fig.

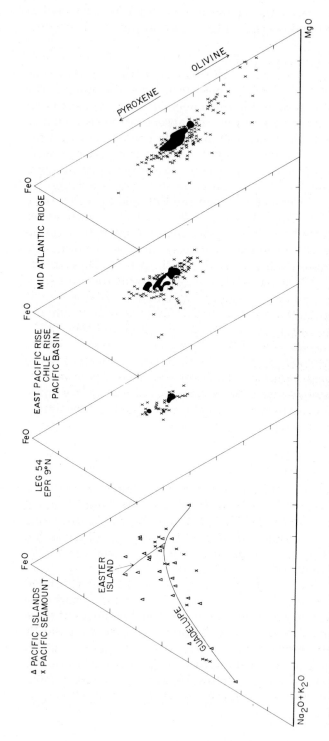

Fig. 2-24. AFM ternary diagrams of rocks collected from various oceanic provinces of the Pacific and Atlantic Ocean. Data from the Pacific seamounts and islands are from C.G. Engel and Engel (1963), Poldervaart and Green (1965), and Kay et al. (1970). The data on the rocks from the Pacific Ocean were taken from Luyendyk and Engel (1969), Hékinian (1971a, b), Yeats et al. (1973), Kempe (1974), G. Thompson et al. (1976), Donaldson et al. (1976), Rhodes et al. (1976), Mazzullo et al. (1976), Bunch and LaBorde (1976) and Hart (1976). The rock samples from the Atlantic Ocean were taken from Muir et al. (1964, 1966), Aumento (1968), Miyashiro et al. (1969a, 1970b), Shido et al. (1971), and Hékinian and Aumento (1973).

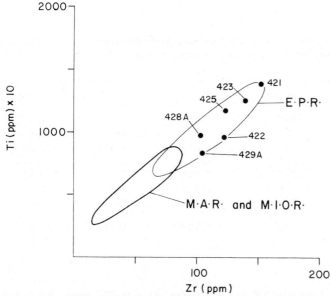

Fig. 2-25. Ti-Zr variation diagram of basaltic rocks from the northern Mid-Atlantic Ridge (including the FAMOUS area; Bougault, 1980) and the Mid-Indian Oceanic Ridge (C.G. Engel and Fisher, 1975). The samples from the boreholes on the East Pacific Rise are from near 9°N (Scientific Party, DSDP Leg 54, 1980b, c; Joron et al., 1980).

2-33). It is also interesting to notice that on average the samples from the MAR have a higher olivine/pyroxene ratio than those from 9°N.

Some other common geochemical criteria for differentiating various types of regional volcanism on accreting plate boundary regions include the distribution of large-ion lithophile (LIL) elements, the rare earth element (REE) content, and the niobium content.

*Large-ion lithophile elements*

In addition to mineralogical and major element compositional variations observed among different mid-oceanic ridge volcanics, there is also a difference in their trace element constituents. LIL constituents of basaltic rocks consist of K, Ba, Rb, Cs, Sr, P, Pb, U, and Th. The LIL elements of mid-oceanic ridge basalts are relatively low with respect to island and seamount basalts (Table 2-14). The spreading-ridge basalts are considered to be LIL element-depleted tholeiites. A compilation of data (Subbarao et al., 1979b) of the LIL elements from the MIOR basalts shows an enrichment in K (1700 ppm), U (0.51 ppm), Rb (2.54 ppm), and $^{87}Sr/^{86}Sr$ ratio (0.7034) when compared to the MAR basalts (K = 1276 ppm; U = 0.28 ppm; Rb = 2.28 ppm, and $^{87}Sr/^{86}Sr$ =

TABLE 2-14

Large-ion lithophile (LIL) element distribution of basaltic rocks from various mid-ocean ridge provinces compared to that of seamount volcanics (data from Subbarao et al., 1973, 1977)

| Province | K/Rb | Rb/Sr | $^{87}Sr/^{86}Sr$ | U (ppm) |
|---|---|---|---|---|
| East Pacific Rise | 1062 | 0.015 | 0.7026 | 0.075 |
| | (27)* | (16) | (23) | (8) |
| Mid-Atlantic Ridge | 869 | 0.016 | 0.7028 | 0.28 |
| | (18) | (16) | (6) | (17) |
| Mid-Indian Oceanic Ridge | 831 | 0.018 | 0.7028 | 0.51 |
| | (6) | (6) | (6) | (4) |
| Seamounts alkali basalt (average of 4) | 947 | 0.032 | 0.7031 | n.d. |
| Seamount trachyte (average of 3) | 349 | 0.388 | 0.7045 | n.d. |

*Number of analyses.

0.7028) and to the EPR basalts (K = 1470 ppm; U = 0.075 ppm, Rb = 1.4 ppm, and $^{87}Sr/^{86}Sr$ = 0.7026) (Table 2-14, Fig. 2-26). This difference suggests that the source material for the MAR and the EPR basalts is depleted in the LIL elements with respect to that of the MIOR. Such a difference in LIL elements could be interpreted in terms of a depletion model for the mantle which occurred during the early differentiation history of the earth. Assuming that isotopic equilibrium was maintained between the various mineral phases in the mantle, the $^{87}Sr/^{86}Sr$ ratio suggests that the mantle underneath the mid-oceanic ridges underwent previous melting processes 1 Ga ago, during which time most of the LIL elements were brought to the surface of the ocean floor.

In making our comparisons among the MAR, EPR, and MIOR, the LIL elements were averaged; such average values, however, of the LIL elements and the major element differences of the volcanics from the various oceanic ridge systems have to be viewed with caution. For instance, within the Atlantic Ocean a value for the $^{87}Sr/^{86}Sr$ ratio (0.7029—0.7036) was found to be as high as that found in the Indian Ocean (Fig. 2-26). The samples giving this ratio were taken from a particular region located near 45°N and between 40°N and 36°N on the MAR. Similarly, the region of the Reykjanes Ridge located south of Iceland is another anomalous zone containing iron- and titanium-rich basalts, and these will be discussed in the next chapter. Another difficulty in considering the LIL elements for comparative purposes is the fact that they are sensitive to moderate degrees of alteration and metamorphism.

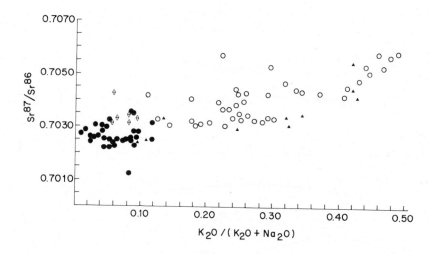

Fig. 2-26. Variation of $K_2O/(K_2O + Na_2O)$ vs. $^{87}Sr/^{86}Sr$ ratios for oceanic rocks from different tectonic environments. ● = Mid-Atlantic Ridge and East Pacific Rise tholeiites (Peterman and Hedge, 1971; Subbarao, 1972; K.V. Subbarao and G.S. Clark, unpublished data); ⊕ = Mid-Indian Oceanic Ridge tholeiites (Subbarao and Hedge, 1973); ○ = oceanic island basalts (Peterman and Hedge, 1971; Hedge et al., 1973); ▲ = seamount basalts (Subbarao et al., 1973; Subbarao and Hékinian, 1978).

## Rare earth elements

Another trace element distribution used for determining magma differences and partial melting processes along mid-oceanic ridges is the rare earth content of the basaltic rocks. The significance of the REE distribution in basaltic rocks is its relationship to various degrees of partial melting (Schilling, 1975a). However, the REE distribution pattern could also depend on the initial abundance, the proportion of phases in the parental material, and the proportion of phases contributing to the formation of the liquid (Yoder, 1976). It is likely that the contribution of the partial melting phases to the distribution of the REE must reflect their eutectic composition. It was also suggested (Bryan et al., 1976) that basalts with high REE abundances have high FeO/MgO values. The trends of these REE variations are the result of a fractional crystallization of Mg-olivine and calcic plagioclase (Kay et al., 1970).

The distribution pattern of the REE is similar to that of the volcanics analysed from the various spreading ridges of the Atlantic, Pacific and Indian oceanic ridge systems (Fig. 2-27). However, as with LIL elements, the REE also could vary from one location to another along the same spreading ridge. Bryan et al. (1976) have shown that basalts from near 22°N on the MAR vary by more than a factor of 2 in REE abundance (Fig. 2-27). Similarly, the Carlsberg Ridge basaltic rocks show a light-REE-depleted pattern, which is similar to other oceanic-ridge basalts, whereas the other MIOR basalts display

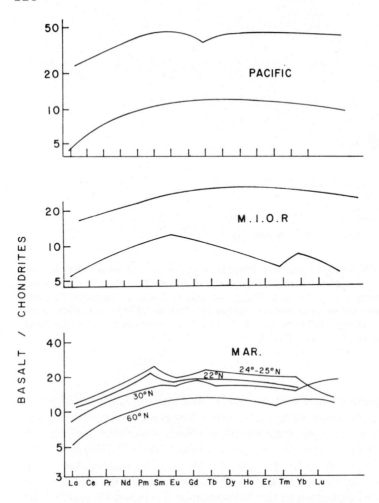

Fig. 2-27. Rare earth element abundances compared to chondritic averages for basaltic rocks from the Mid-Atlantic Ridge (*MAR*), Mid-Indian Oceanic Ridge (*MIOR*) and East Pacific Rise (after Bryan et al., 1976; Subbarao et al., 1979a).

an almost flat relative REE distribution pattern with only a slight light-REE enrichment.

*Niobium distribution*

Niobium is a dispersed element in basaltic rocks and probably occurs as a minor constituent in Fe-Ti-bearing minerals such as titanomagnetite and ilmenite. The most comprehensive studies and summary of data presented on Nb were done by Rankama (1948) and by Parker and Fleischer (1968). Nb has a relatively small radius and high ionic potential and it is not easily

125

removed from crystalline structures. Hence, it seems to be unaffected by even the severe secondary processes of alteration (Cann, 1970a). Metamorphic rocks from the Carlsberg Ridge contain the same amount of niobium as the fresh basalt from spreading ridges, while other elements such as K and Rb have changed considerably (Cann, 1970a). The Nb content of spreading oceanic ridges is usually less than 30 ppm (Fig. 2-28). Cann (1970a) and G.D. Nicholls and Islam (1971) have shown values of Nb varying between 2 and 19 ppm for basaltic rocks collected from the major oceanic ridge systems of the world. The increase in Ti with that of Nb in basaltic rocks from spreading ridges suggests a fractional crystallization of the basaltic melt itself. The concentration of Nb depends on both its initial amount prior to partial melting of the source material and on the degree of fractionation that the magma has undergone during its ascent to the surface.

The various mineralogical and geochemical criteria used to differentiate among the basaltic rocks erupted on different spreading ridges do not provide any simple answer to the origin of these volcanics in relation to their structural setting. The compositional variability observed could be due to the

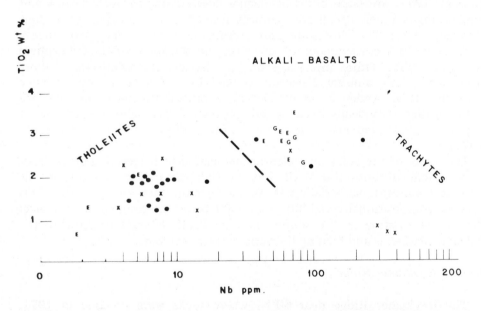

Fig. 2-28. Niobium (Nb, ppm) vs. TiO$_2$ (wt.%) variation diagram of basalts from East Pacific Rise and seamount provinces (data from D. Gottfried, personal communication, and A.T. Anderson and Gottfried, 1971). Dashed line indicates the upper limit of Nb content for mid-oceanic ridge basalts. Crosses and dots are samples from the northern and the southern portions of the East Pacific Rise located between 20°N and 37°S. $E$ = samples described by A.E.J. Engel and Engel (1964); $G$ = samples from Guadalupe Island (A.E.J. Engel and Engel, 1964). The other samples of alkali basalts and trachytic rocks are from the Mathematicians Seamounts.

heterogeneous nature of the source material, or to the degree of its partial melting, or to magma mixing, or to high-level crystal fractionation, or to any combination of these factors. Further aspects on the different magmatic processes will be discussed in Chapter 12.

## REGIONAL VARIATIONS OF VOLCANICS ALONG RECENT OCEANIC RIDGE SEGMENTS

In addition to the petrological variations observed between various oceanic ridges of the Atlantic and Pacific, volcanics erupted along the strike of accreting plate boundary regions also show important compositional changes which appear to correlate with topographic and structural features character- izing particular ridge segments. The surface expressions of many topographic highs representing volcanic platforms, such as those comprising the islands of the Azores and the Galapagos, were defined as "hot spot" locations where the outpouring of abundant mantle material (mantle plumes) has built important edifices on mid-oceanic ridges. The "hot spots" (Wilson, 1965; Morgan, 1971) are large blobs of magma slowly rising through narrow and long-lasting channels which continuously drain the source region of its melt (Morgan, 1981). The "hot spots" are considered to have a variety of surface expressions as a consequence of the changing thickness of the lithosphere (Morgan, 1981). Other inferences on the physical characteristics of these regions are their unusual elevation marked by a free-air positive gravity anomaly (R.N. Anderson et al., 1973), a crustal thickness intermediate between that of typically oceanic and continental areas, and the presence of high geothermal gradients, at least for the first 100 km depth (Talwani et al., 1971).

The study of the volcanic platforms associated with accreting plate boun- dary regions of both the MAR and the Galapagos Ridge systems reveals striking compositional variabilities among their products of volcanism. This type of compositional variability is thought to be the result of petrological differences existing in the source material itself, located underneath the volcanic platforms and feeding the various ridge segments.

### Iceland-Reykjanes Ridge

The Reykjanes Ridge near 60°N, where rocks were dredged in 1971, shows a well-developed structure with regular magnetic lineaments (Talwani et al., 1971) and it has no major offset. The crustal thickness under Iceland varies from 8 to 16 km (Palmason, 1970) and it is about 3.5—4 km under the Reykjanes Ridge (Talwani et al., 1971). Such crustal variation may imply various degrees of magma production. Under Iceland, where the crust is thicker, we expect to have a larger amount of magmatic material, about 2—4 times greater than across the Reykjanes Ridge near 60°N during the last 50—60 Ma (Schilling, 1973).

During the 1971 cruise of the R.V. "Trident", a 9000-pound suite of relatively fresh basalt was dredged along the Reykjanes Ridge. Petrochemically the basalts are classified as tholeiites according to the Yoder and Tilley (1962) normative scheme (Schilling, 1973).

It is primarily through the work of Tatsumoto et al. (1965) and R. Hart (1973) that tholeiitic basalts from such areas were found to be richer in LIL elements and radiogenic strontium and lead than the volcanics derived from other normal ridge segments. Schilling (1973) and J.G. Moore and Schilling (1973) found that there is a change in the chemistry and petrology of lavas exposed at moderate depths (1200 m) when compared to those exposed in subaerial environments along the Reykjanes Ridge south of Iceland. Geochemical studies of basalts erupted along the Reykjanes Ridge and its extension over the median neovolcanic zone of Iceland have revealed the presence of a gradient due to variation of the lead isotopic ratio (Sun et al., 1975). The $^{208}Pb/^{204}Pb$ and $^{87}Sr/^{86}Sr$ ratios vary continuously along the Reykjanes Ridge from about latitude 65°N to 53°N near the Charlie Gibbs fracture (Fig. 2-29) (White et al., 1975; Sun et al., 1975). The compositional changes of volcanics along the Reykjanes Ridge were attributed to two mantle source origins: (1) a mantle plume magma rising beneath Iceland, and (2) a depleted low-velocity layer source beneath the ridge itself. The mixing of these two magma types was assumed (Sun et al., 1975) to be the cause of the gradient observed in the lead isotopic ratios and especially in the lanthanium and samarium (La/Sm) ratio (Schilling, 1973) (Fig. 2-29). The La/Sm values change from about 5—10 near 65°N (Iceland Plateau) to 1—2 at near 60°N (Schilling, 1973) away from Iceland. The modal composition of the Reykjanes-Iceland basalts shows a progressive increase of pyroxene content with respect to plagioclase phenocrysts as the ridge becomes shallower approaching Iceland (Schilling, 1973). Major element analyses of the basalt suites from the Reykjanes Ridge indicate that there is a noticeable tendency for the Fe content to increase in relation to the Mg content towards Iceland (Schilling, 1973). The FeO/MgO weight percent ratio ranges from 1.2 at 60°N to 1.7 at 63°N latitude and is accompanied by an increase of the total iron from 10.5 to about 13 wt.% (Schilling, 1973). Other chemical element variations of the volcanics along the Reykjanes Ridge are shown by the $TiO_2$, the $P_2O_5$ and $K_2O$ (weight percent values) which decrease (1.4—2%; 0.18—0.27%; 0.2—0.28%, respectively) gradually with distance from the Icelandic Plateau, near 65°N (≃1%; 0.10%; 0.07%, respectively) (Schilling, 1973).

The compositional variation encountered along the strike of the Reykjanes Ridge is accompanied by similar changes in the volcanics found across the ridge itself at increasing distances from the ridge axis. From the study of twelve rocks scattered across the Reykjanes Ridge near 60°N, it was noticed (Schilling, 1975a) that while $K_2O$ and $Fe_2O_3/FeO$ vary across the ridge within

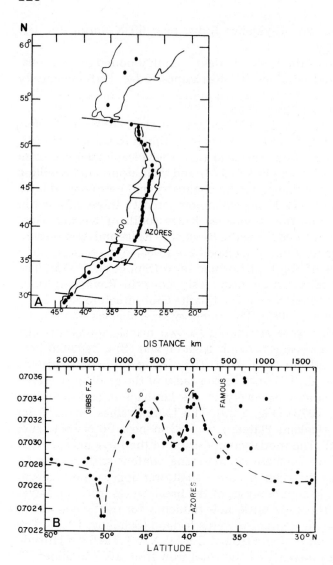

Fig. 2-29. A. Generalized diagram showing the locations of dredged rocks from the North Atlantic Ocean after White et al. (1975). B. $^{87}Sr/^{86}Sr$ variation diagram of the dredged samples shown in A. The distances from the center of the Azores Platform is indicated. Open circles indicate suspected values of isotopic ratios, due probably to secondary alteration processes (White et al., 1975).

a distance of about 40 km from the axis (up to about 3 Ma) due to seawater weathering, the La/Sm ratio (0.4—0.9) stays virtually the same (Schilling, 1975a). The REE best represents the original melt composition and hence reflects the characteristically depleted nature of the asthenosphere beneath the ridge.

Studies of sulfur content and of rock vesicularity (J.G. Moore and Schilling, 1973) have shown that the volcanics erupted along the Reykjanes Ridge characterize the depths at which they were emplaced. Above 200 m depth, the volume percent of vesicles is higher than 40%. This is accompanied by a decrease in sulfur content to less than 850 ppm (J.G. Moore and Schilling, 1973). Below 200 m depth, the sulfur content could reach 1340 ppm and the vesicularity of the rock becomes less. Such a variation in vesicularity and of sulfur content is accompanied, as shown before, by a variation of some other critical oxides, such as $K_2O$, $P_2O_3$ and $TiO_2$, which increase in content with the decrease in sulfur content and an increase in vesicularity.

The systematic changes in LIL elements and of the radio-isotopes of Pb and Sr, together with other oxide variations, led Schilling (1975a) and his co-workers to postulate the existence of a primordial hot mantle plume (PHMP) under Iceland (>250 km depth) which gave rise to a melt which was relatively enriched in LIL elements. In order to explain the variation encountered along the Reykjanes Ridge, Schilling (1973) postulated the presence of another magmatic source, different from that found under Iceland, located under the ridge and depleted in such LIL elements and he called this source the depleted low-velocity layer (DLVL), since such a source is related to pressure and temperature conditions corresponding to a depth of less than 250 km. Mixing of the two source materials, PHMP and DLVL, will give rise to the continuous magmatic variation observed along the Reykjanes Ridge. The mechanical mixing of these two magma sources, one in the mesosphere (>250 km depth), the other in the asthenosphere (<250 km depth), happens in part during the segregation and channelling when both magma sources upwell. During upwelling the rate of partial melting increases because of decompression, thus the melt will accumulate at the base of the crust ready to erupt on the ridge surface during spreading (Schilling, 1973).

According to Schilling, a direct test for this proposed mixing model could be to plot one isotopic ratio against another. The fact that there is a good linear correlation between $^{208}Pb/^{204}Pb$ and $^{207}Pb/^{204}Pb$, respectively, against $^{206}Pb/^{204}Pb$ is an indication that mixing of two basaltic melts is possible. The $^{206}Pb/^{204}Pb$ profile shows high values over Iceland (>18.3—18.8) and low values south of 61°N (<18.3) (Sun et al., 1975). The proposed mixing model as occurring between two types of basaltic melts of very close bulk composition is more acceptable than for very different compositional melts such as rhyolite and basalt. Schilling's (1973) mixing model occurs in elongated magma chambers. Dyke propagation and sills could be present under the Reykjanes Ridge crest and parallel to its trends. It was evidenced that lava in Hawaii had traveled subhorizontally more than 120 km. In order to have mixing it is necessary that the PHMP-derived lava under Iceland rises diagonally along the ridge with a significant component of velocity towards the Reykjanes Ridge. Hence the path of the DLVL- and the PHMP-derived lavas will cross and allow mixing. The PHMP flow will be a more primordial material enriched in LIL elements and volatiles which lie deep in the earth, while the DLVL flow is the shallow depth depleted zone which has been drained of LIL elements and low melting fractions during geological time.

## Azores Platform and other Mid-Atlantic Ridge segments

The Azores Platform consists of the northward and southward extension of the MAR in the vicinities of the Azores where the mantle plume hypothesis of Schilling has been tested. Analyses of basaltic rocks located from 29°N to 60°N in the Atlantic Ocean were studied for LIL elements, such as K, Rb, Cs, Sr, Ba and REE elements (White et al., 1975; Subbarao et al., 1979b). The samples from the Azores islands are alkali basalts and differentiated derivatives. The $^{87}Sr/^{86}Sr$ ratio of the lava from these islands is higher than 0.70332 (0.70332—0.7525). From the radioisotope data two maximum values of $^{87}Sr/^{86}Sr$ are found on the Azores (0.70345) and at 45°N (0.70340), while the transitional zone shows a gentle sloping variation of the isotopic ratio (White et al., 1975) (Fig. 2-30). Ten analyses from the FAMOUS area also show transitional values (0.7028—0.7029) for these isotopic and LIL elements (Fig. 2-30) (White and Bryan, 1977).

The Azores transect is similar to that of the Iceland-Reykjanes Ridge and the Galapagos rift valley zones where geochemical variation along the island platform and the ridge system itself has shown considerable continuity in its chemical variability.

Recent bathymetric work done along the axial zone of the MAR between 10 and 50°N and comprising the platform of the Azores, has revealed the presence of various structural regions as defined by major topographic provinces existing within the inner floor of the Rift Valley (Douaran, 1979). The structural regions were divided according to their average water depth into three categories: (1) a shallow depth region comprised between 2500 and 3500 m and corresponding to the platform of the Azores, (2) a region of intermediate depth between 3500 and 5000 m, and (3) a region of great depth, between 4000 and 5000 m.

In order to verify if the above topographic variability is accompanied by a compositional change, it was necessary to consider a large number of rock samples located on both the axial zone of the Rift Valley and away from it. Thus samples found on a crust younger than 10 Ma were used. Compositional variation across (perpendicular to the ridge axis) the MAR revealed small differences compared to the regional changes along the ridge strike.

The compositional variability observed among the volcanic rocks collected along the MAR is mainly noticed in their LIL element changes which follow the topographic variations.

The changes in FeO*/MgO (total Fe as FeO) and in the $TiO_2$ content of basaltic glasses define two major volcanic provinces in the Atlantic: (1) the region at shallow depths located between 33 and 60°N which consists of

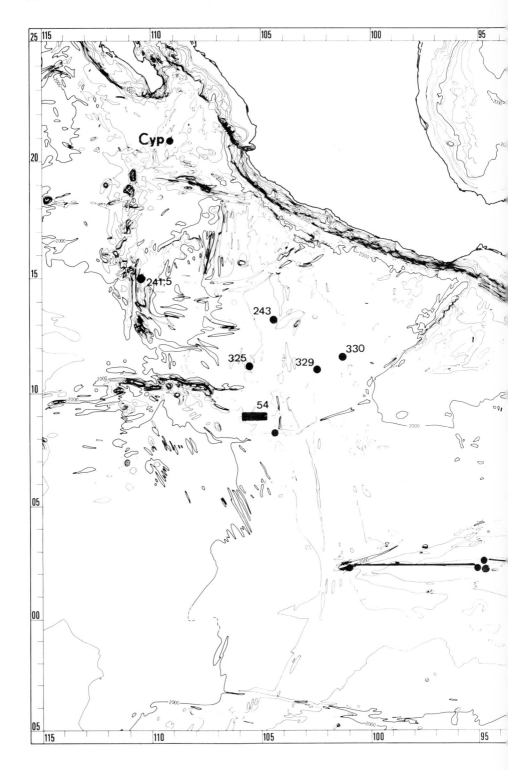

rocks having a low FeO*/MgO ratio (<1.2) and low $TiO_2$ contents (<1.2%), and (2) a region of intermediate and great depth located south of 33°N associated with volcanics having on average a higher FeO*/MgO ratio (1.4) and a higher $TiO_2$ content (1.7%) (Fig. 2-30). Melson and O'Hearn (1979), using the $Na_2O$ and $TiO_2$ contents, have also shown that basalts located near 0—29°N differ from those found at 36—37°N on the MAR.

Another parameter used to show the compositional variability between the two provinces was the degree of saturation with respect to $SiO_2$ of the rocks analysed (Morel and Hékinian, 1980). It appears that the samples from north of about latitude 33° on the MAR show an oversaturated character with respect to those found south of this latitude (Fig. 2-30). This change in the degree of saturation corresponds to changes in the normative content of nepheline, olivine, plagioclase, pyroxene, and quartz. The oversaturated basalts have an average olivine content of 0—13%, a plagioclase content of 54—46%, a pyroxene content of 47—30% and a quartz content of 0—5% (Morel, 1979). The saturated basalts differ in their olivine (0—25%) content, plagioclase (50—59%), and in their depletion in pyroxene (14—43%) and quartz content (0—2%). Among the different variables, such as the composition of the source material and the amount of partial melting, the temperature and pressure conditions might have influenced the basalt variability encountered along the MAR axis. Assuming a garnet peridotite composition as a source material, it is likely that the partial melting at shallow depth of such material will give rise to oversaturated types of basalts whereas at great depth saturated and undersaturated melts will be formed. Although geochemical evidence alone cannot prove the existence of mantle plumes, it is likely that the observed compositional change of the surface lava types is due to a chemical variation in the mantle source area.

*Spreading axis of the Galapagos region*

The accreting plate boundaries of the Galapagos region are located almost at mid-distance between the Cocos and the Carnegie Ridges. These spreading ridges are bounded by a wedge-shaped structure which comprises the Cocos, the Nazca and the Pacific plates which converge through various ridge segments in the Galapagos triple junction located at about 2°N and 102°W (Holden and Dietz, 1972; Hey, 1977). The spreading axes of the Galapagos

Fig. 2-31. Bathymetric map of the north eastern Pacific Ocean showing the Galapagos region. Contour lines are in fathoms and the map was redrawn after the chart of Chase et al. (1970). The heavy lines in the Galapagos region show the approximate trend of the Pacific-Nazca Ridge and the Galapagos Spreading Center. Black dots indicate the locations of dredged samples after Schilling et al. (1976). The analyses of the samples collected from the East Pacific Rise (241, 5, CYP, Leg 54, sites 325, 243, 324, 330) are reported in Tables 2-6, 2-8, 2-9, 2-10, 2-11 and 2-12 of this chapter and in Table 1-16 of Chapter 1.

136

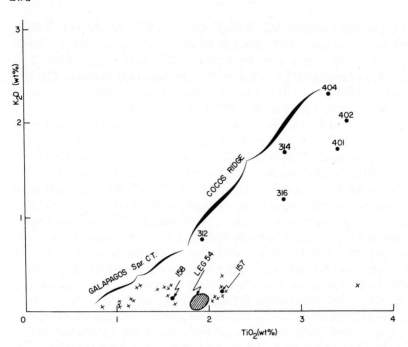

Fig. 2-32. $K_2O$-$TiO_2$ variation diagram of rocks collected on the Galapagos Spreading Center area ($\times$), Leg 54, from the Cocos Ridge (158, 312, 314, 316), Cocos Island (401, 402, 404; Dalrymple and Cox, 1968), the Carnegie Ridge (157). Data from R.N. Anderson et al. (1975), Schilling et al. (1976), C.G. Engel and Chase (1965), and Scientific Party, DSDP Leg 54 (1980c).

region are divided into the Cocos-Nazca Rise, the Galapagos Spreading Center and the Costa Rica Ridge from the triple junction eastward (Fig. 2-31). The Galapagos Spreading Center differs from the EPR by the presence of an intermittent rift valley which appears to be well developed starting from only about 45 km east of the triple junction (Hey, 1977).

Extensive dredging operations undertaken on the Galapagos Spreading Center area between 95 and 85°W have shown a gradient of variability of FeO and $TiO_2$ content away from the Galapagos Islands (Schilling et al., 1976). The decrease in FeO, $TiO_2$ and La/Sm ratio along the ridge, and away from the Galapagos Islands suggested (Schilling et al., 1976) the presence of a mantle plume underneath the Galapagos Platform similar to that encountered for Icelandic volcanic province. Light-REE-depleted basalts are confined away from the Galapagos Platform, east of 87°W and west of 95°W (Schilling et al., 1976). The range of $TiO_2$ content of the volcanics from the Galapagos Spreading Center varies between about 1 and 2.5% (Fig. 2-32). DSDP holes (Site 424, Leg 54) near 85°W also showed the presence of high-Fe and Ti basalt (FeO = 13–15%; $TiO_2$ = 1.9%) (Scientific Party, DSDP Leg 54, 1980c). Both the mineralogy (plagioclase-, pyroxene-rich basalt) and the chemistry of the Galapagos Spreading Center volcanics show similarity in composition to other EPR and adjacent aseismic ridge basalts. However, some volcanics dredged from the Cocos Ridge (C.G. Engel and Chase, 1965) and others from the Cocos Islands (Dalrymple and Cox, 1968) show a higher $TiO_2$ (up to 4%) and higher $K_2O$ content (up to 3%) than the Galapagos Spreading Center basalts (Fig. 2-32).

## COMPOSITIONAL VARIATION RELATED TO THE SPREADING RATE

Several authors (Bass, 1971; Scheidegger, 1973: Melson and O'Hearn, 1979) have suggested that compositional variations along accreting plate boundary regions may depend on tectonic processes enhanced by the spreading rate. Previous attempts to relate spreading rates and observed crustal composition along ridge segments failed to show any meaningful correlations (Schilling, 1975a, b). One major difficulty encountered in correlating spreading rates with crustal composition is in evaluating the amount of stations and the type of material recovered as being representative of the particular ridge segment sampled. Another major difficulty resides in the extrapolation of the spreading which is an average rate evaluated on the scale of millions of years. This generalized spreading rate might not be the same as an instantaneous rate localized in a precise area of sampling.

Morel and Hékinian (1980), using samples collected along various segments of the EPR with crustal age of less than ten million years, have shown the existence of a general correlation between spreading rate and rock composition. The spreading rate in each specific segment of ridge crest sampled was calculated using the method of Minster et al. (1974). Other spreading rates were determined using magmatic anomaly data and sediment age at contact with the basement in deep-sea drilling operations (e.g. DSDP Leg 54 at 8—9°N, Fig. 2-33). The material was treated in a homogeneous fashion, that is, using for all the samples the same instrumental method in order to avoid any analytical discrepancies and to have the highest reproducibility of values. Since the spreading rates in the Pacific Ocean are the most variable among accreting plate boundary regions, particular emphasis was given to samples collected along the EPR. This type of study was possible because of the relatively large number of samples (102) located between 20°S and 20°N on the EPR. The major concentration of data and hence the most meaningful data are located at 21—23°N (total spreading rate of about 6 cm/yr), at 8—9°N (total spreading rate of 12 cm/yr) and at 20—21°S (total spreading rate of 16.5 cm/yr) (Fig. 2-33).

Rock samples from other stations scattered along various segments of the EPR at 54°S (spreading rate of 8.2 cm/yr), at 56°S (9.2 cm/yr), at 33°S (11 cm/yr), at 11°N (11.3 cm/yr), 13°N (10.6 cm/yr) and at 2°N (12.6 cm/ yr) were analysed by Morel and Hékinian (1980) to complement the previously

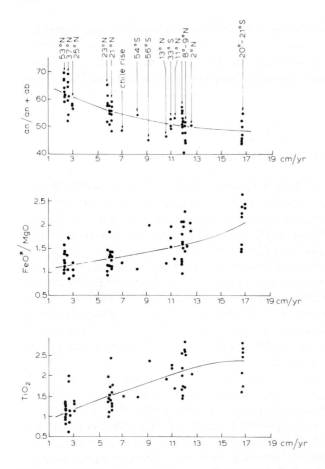

Fig. 2-33. Total spreading rate versus TiO$_2$, FeO*/MgO (total Fe as FeO) and An/An + Ab ×
100 (normative percent) of glassy basalt margin from various segments of the Mid-Atlantic
Ridge and from the East Pacific Rise. The samples from the Mid-Atlantic Ridge (total rate
of spreading of 2—3 cm/yr) are from 53°N (CH4-DR1), 36—37°N (FAMOUS area) and
from 25°N (V25-DR1). The samples from the East Pacific Rise are glassy basalts from
23°N, 21°N (6—7 cm/yr), from 54°S (8 cm/yr), 56°S (9.2 cm/yr), 13°N (10.7 cm/yr), 11°N
(11—12 cm/yr), 8—9°N (12 cm/yr), 2°N (12.7 cm/yr), 13°N (10.5 cm/yr) and from Chili
Rise (7 cm/yr). The samples from 8—9°N (Leg 54, DSDP) are the only aphyric basalts
analysed and are close in composition to their glassy chilled margins. The samples from
20—21°S were collected during the Searise cruise of the R.V. "J. Charcot" from the axial
zone of the East Pacific Rise (1980; J. Francheteau, in preparation).

better-sampled regions mentioned above. In addition, rocks from the MAR
segments at 53°N, 37°N and 25°N (total spreading rate of 2—3 cm/yr) were
also used in order to have a larger spectrum of variabilities of the spreading
rates (Fig. 2-33).

While bulk-rock analyses failed to show any detectable changes along the
ridge axis, both detailed mineralogical and chemical analyses of glass indicate

a compositional variability which could be correlated with the spreading rate. Among the major and minor constituents which are the most sensitive to such changes are the FeO/MgO ratio (total Fe calculated as FeO), the $TiO_2$ and the An/An + Ab normative ratio (Fig. 2-33). Slow-spreading rate (2—3 cm/yr) accreting plate boundary regions have an average FeO/MgO of about 1.1, a $TiO_2$ content of about 1% and An/An + Ab ratio of 65, while on a fast-spreading ridge (12 cm/yr) the same parameters will have values of 1.5, 2% and 55, respectively (Fig. 2-33). Ultrarapid ridge segments (16.5 cm/yr) show volcanics that have reached the highest content of $TiO_2$ and FeO/MgO values (2.5—3% and 1.5—3, respectively) (Fig. 2-33).

Among the early mineral phases to crystallize, plagioclase was used as a parameter for differentiating various oceanic-ridge provinces of the Atlantic and Pacific oceans (Scheidegger, 1973; Morel and Hékinian, 1980). Using the anorthite content of an earlier plagioclase phase found in the basaltic rocks along the EPR, it was possible to show a general change in the composition of the plagioclase with the rate of spreading (Morel, 1979). There is a negative correlation between the rate of spreading and the amount of anorthite in the plagioclase of the basaltic rocks. On ridge segments with half spreading rates of 2—3 cm/yr, the early-formed plagioclase composition averages around $An_{84}$. Volcanics from fast-spreading segments (12 cm/yr) contain plagioclase with an average anorthite content of $An_{70}$. While the basaltic rocks from ultrarapid ridge segments show early-formed plagioclase with the lowest average anorthite content ($An_{60}$). The range of anorthite content of the plagioclase from volcanics erupted along the EPR is $An_{56}$—$An_{88}$. In comparison, the general compositional trend of the plagioclase variation of the MAR basalts is that of a higher anorthite content ($An_{75}$—$An_{92}$) than that of the East Pacific Rise.

CHAPTER 3

ASEISMIC RIDGES

As the name indicates, aseismic ridges are so called because they lack seismic activity. Some will show a few earthquake epicenters but in general they are all considered to be stable oceanic regions. Only a very few aseismic ridges have been sampled, probably because of the thick sediment cover and the difficulty in encountering steep outcrops which would facilitate conventional methods of sampling. Nevertheless, since aseismic ridges comprise about 25% of the world's ocean floor, they must not be overlooked in any study. They are elevated volcanic features which are 2000—4000 m above the surrounding basin floors and form more or less linear features which vary from 250 to 400 km in width and are about 700—5000 km in length. Most aseismic ridges are attached to the continental margin. In the past, these relatively shallow structures could have been important land bridges connecting various islands of the Atlantic Ocean (e.g. Iceland-Faeroe and Walvis Ridges, Rio Grande Rise). The aseismic ridges, like the recently formed mid-oceanic ridges, may be fractured perpendicularly to their axis, such as the Cocos Ridge and the Walvis Ridge, or may form a more straight continuous feature. Steep escarpments along their flanks have been recognized (Walvis Ridge, Francheteau and Le Pichon, 1972; Cocos Ridge, Van Andel et al., 1971) which suggests that they might represent traces of old fracture zones.

All of the aseismic ridges are quite old and represent features which were formed during the early history of the creation of the ocean floor. Their ages, determined from both isotopic methods (K-Ar) and the presence of characteristic fossils, vary between about 50 and 110 Ma. The bathymetry across these ridges and along their crest is generally smooth, in contrast to the more rugged relief of mid-ocean ridges. The aseismic ridges are essentially made up of basaltic and other differentiated volcanics. Sometimes they terminate in an island which generally forms a continuous structural feature with the ridge (e.g. Iceland-Faeroe, Walvis, Cocos and Carnegie Ridges). In the case of the Ninety-East Ridge, Amsterdam and St. Paul Islands are along what seems to be its southern extension. However, these are cut off from the aseismic ridge by the Mid-Indian Oceanic Ridge.

The origin of aseismic ridges is controversial, since it is not known whether they are chains of islands generated from hot "mantle plumes" rising under the lithosphere during the movement of a plate or whether they are just a line of volcanoes bounded by fault zones. Wilson (1965) and Morgan (1971, 1972a, b) have suggested that aseismic ridges throughout the world are the surface expression resulting from plates moving away from fixed hot spots in the mantle. It is obvious that in this case the age of the volcanic crust should decrease as the hot spots move toward the Recent mid-oceanic ridges. This is partially verified if one considers the relatively recent volcanic events in Iceland, in the Galapagos, and on Tristan da Cunha. As far as past eruptive events are concerned, it is interesting to note that on the most systematically sampled aseismic ridge (4 DSDP holes), the Ninety-East Ridge (Indian Ocean), the crustal age is variable along its length.

## ICELAND-FAEROE RIDGE

This ridge system joins Iceland to the Faeroe Islands and separates the Norwegian Basin from the main North Atlantic basin (Fig. 3-1). At the present time it would seem that the Tertiary history of a possible land bridge across the Faeroe-Iceland-Greenland Ridge could be estimated from data obtained

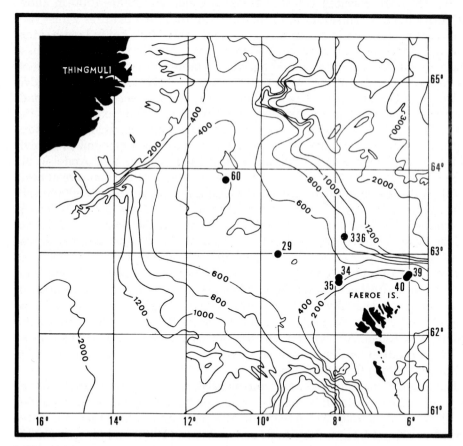

Fig. 3-1. Bathymetric chart of the Iceland-Faeroe Ridge (North Atlantic Ocean) with contour lines in meters (Fleisher, 1971). Basalts and andesite were recovered from Site 60, basalts occur at Sites 29, 34 and 35; andesites occur at Site 39 and basalt at Site 40 (Noe-Nygaard, 1949). DSDP Site 336 (Leg 38) recovered basaltic rocks (Talwani et al., 1976).

from fossil and contemporary biota (Talwani and Udintsev, 1976). The holes drilled by the "Glomar Challenger" during DSDP Leg 38 (Sites 336 and 352) show that the sub-aerial volcanism was covered by late Middle Oligocene sediment (Talwani et al., 1976). The radiometric age of the basement in the drilled area is 41—43 Ma B.P. The nature of the rubble material and the microfauna from the Eocene sediment overlying the basaltic rock suggested submergences at the site of drilling (Scientific Party, DSDP Leg 38, 1976a).

The first samples of basaltic rocks from the Iceland-Faeroe Ridge were taken during a marine biological expedition made by the Danish research ship "Dana" immediately after World War II. Of the ten samples studied, four are made up essentially of plagioclase (34—46%) and clinopyroxene (26—48%), and two samples contain less than 16% olivine (Noe-Nygaard, 1949). Three oceanic andesites were recognized; they consist mainly of andesine (80%) and clinopyroxene (7%) (Noe-Nygaard, 1949; Fig. 3-1). The chemical analogy of these dredged rocks with the adjacent islands of the Faeroes and Iceland is striking. In a comparison between the Thingmuli volcano located on the eastern shore of Iceland and the tholeiitic lava exposed in the Faeroe Islands, it is noticed that the relatively high $TiO_2$ content (1—3%) and the high total Fe content (11—15%) make them quite similar to the basalts from the Iceland-

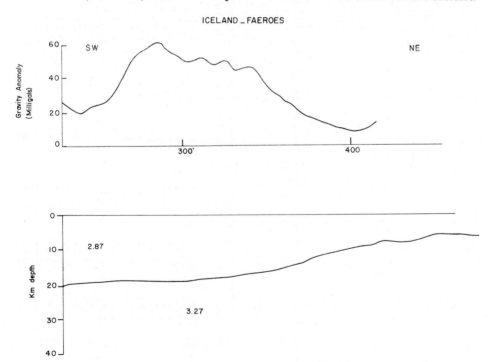

Fig. 3-2. Model of the Iceland-Faeroe Ridge from free-air gravity anomaly data (Fleisher, 1971).

TABLE 3-1

Chemical analyses of rocks from the Iceland-Faeroe aseismic ridge (DSDP Leg 38 basalts are from Kharin, 1974, and the other basalts and andesites are from Noe-Nygaard, 1949)

| | DSDP Leg 38 basalt | | | | | | Basalts | | | Oceanic andesite | | Faeroe Islands basalt |
|---|---|---|---|---|---|---|---|---|---|---|---|---|
| | 336-39-5 40cc | 336-41-1 10 | 336-42-2 82–84 | 336-43-1 123–126 | 336-44-1 94–97 | 336-44-2 105–108 | 6000V | 5839II | 5834I | 6000VII | 5839I | basalt |
| SiO$_2$ (wt.%) | 48.91 | 49.53 | 50.39 | 48.66 | 49.58 | 50.22 | 45.11 | 45.25 | 47.60 | 52.77 | 60.26 | 46.70 |
| Al$_2$O$_3$ | 16.80 | 14.41 | 15.92 | 15.96 | 16.09 | 16.15 | 14.67 | 15.47 | 14.22 | 16.41 | 16.48 | 15.50 |
| Fe$_2$O$_3$ | 7.29 | 5.82 | 3.99 | 3.99 | 6.16 | 7.92 | 5.63 | 7.62 | 3.28 | 6.29 | 2.50 | 3.50 |
| FeO | 5.90 | 7.03 | 6.94 | 6.94 | 5.51 | 4.13 | 5.37 | 4.40 | 10.80 | 2.77 | 3.72 | 8.80 |
| MnO | 0.14 | 0.26 | 0.17 | 0.17 | 0.25 | 0.24 | 0.17 | 0.17 | 0.23 | 0.15 | 0.21 | 0.20 |
| MgO | 5.03 | 6.90 | 6.57 | 6.57 | 6.34 | 5.41 | 7.83 | 8.68 | 5.68 | 3.47 | 1.99 | 9.10 |
| CaO | 7.29 | 10.47 | 9.56 | 9.56 | 9.69 | 8.54 | 11.47 | 11.62 | 9.63 | 5.93 | 4.02 | 12.90 |
| Na$_2$O | 2.86 | 2.38 | 2.86 | 2.86 | 2.70 | 2.76 | 1.90 | 2.24 | 1.97 | 3.96 | 3.39 | 1.90 |
| K$_2$O | 0.22 | 0.18 | 0.16 | 0.16 | 0.25 | 0.38 | 0.83 | 0.63 | 0.78 | 2.71 | 2.62 | 0.30 |
| TiO$_2$ | 1.97 | 1.79 | 1.85 | 1.85 | 1.87 | 1.81 | 1.40 | 1.61 | 3.77 | 2.26 | 2.04 | 1.00 |
| P$_2$O$_5$ | 0.15 | 0.15 | 0.14 | 0.14 | 0.11 | 0.11 | 0.30 | 0.21 | 0.26 | 1.16 | 0.38 | 0.00 |
| LOI | 0.00 | 0.00 | 0.00 | 0.00 | 0.00 | 0.00 | 5.51 | 2.34 | 2.05 | 2.22 | 2.16 | 0.00 |
| Total | 96.55 | 99.01 | 98.54 | 96.85 | 98.54 | 97.66 | 100.18 | 100.23 | 100.26 | 100.09 | 99.76 | 99.89 |

*Norms*

| | | | | | | | | | | | | |
|---|---|---|---|---|---|---|---|---|---|---|---|---|
| Q | 8.52 | 5.01 | 3.74 | 1.98 | 5.44 | 9.05 | 0.00 | 0.00 | 3.60 | 5.37 | 19.19 | 0.00 |
| Or | 1.30 | 1.06 | 0.94 | 0.94 | 1.47 | 2.24 | 4.90 | 3.72 | 4.60 | 16.01 | 15.48 | 1.77 |
| Ab | 24.20 | 20.13 | 24.20 | 24.20 | 22.84 | 23.35 | 16.07 | 18.95 | 16.66 | 33.50 | 28.68 | 16.07 |
| An | 32.35 | 28.37 | 30.12 | 30.23 | 31.04 | 30.55 | 29.04 | 30.29 | 27.65 | 18.99 | 17.46 | 32.87 |
| Di | 2.23 | 18.08 | 13.19 | 13.10 | 12.88 | 8.63 | 20.48 | 20.22 | 15.12 | 2.21 | 0.00 | 25.23 |
| Hy | 13.29 | 14.15 | 16.71 | 16.76 | 12.11 | 9.47 | 11.68 | 6.14 | 18.03 | 7.61 | 6.74 | 2.57 |
| Ol | 0.00 | 0.00 | 0.00 | 0.00 | 0.00 | 0.00 | 0.96 | 4.27 | 0.00 | 0.00 | 0.00 | 14.39 |
| Mt | 10.56 | 8.43 | 5.78 | 5.78 | 8.93 | 8.84 | 8.16 | 10.06 | 4.75 | 2.86 | 3.62 | 5.07 |
| Ilm | 3.74 | 3.39 | 3.51 | 3.51 | 3.55 | 3.43 | 2.65 | 3.05 | 7.15 | 4.29 | 3.87 | 1.89 |
| Ap | 0.35 | 0.35 | 0.33 | 0.33 | 0.25 | 0.25 | 0.70 | 0.49 | 0.61 | 2.73 | 0.89 | 0.00 |
| Hm | 0.00 | 43.52 | 0.00 | 0.00 | 0.00 | 1.81 | 0.00 | 0.67 | 0.00 | 4.31 | 0.00 | 0.00 |
| | | | | | | | | | | | | |
| C.I.* | 35.50 | 43.52 | 41.04 | 41.09 | 43.48 | 40.14 | 48.78 | 55.41 | 37.57 | 21.97 | 17.96 | 59.41 |
| D.I.** | 34.02 | 26.21 | 28.88 | 27.13 | 29.76 | 34.65 | 20.98 | 22.67 | 24.88 | 54.89 | 63.36 | 17.84 |

*Crystallization index (Poldervaart and Parker, 1964).
**Differentiation index (Thornton and Tuttle, 1960).

Faeroe Ridge (Table 3-1). The average mineral composition of ten typical olivine tholeiites from the Faeroe Islands is: plagioclase = 38.3%, clino-pyroxene = 44.7%, olivine = 10.2%, Fe-oxide = 6.6%, alteration products = 0.2% (Noe-Nygaard and Rasmussen, 1968). The percentage of rock types exposed on the Faeroe Plateau (Islands) during the early Tertiary period is: quartz-tholeiite = 70%, olivine-tholeiite = 30%.

The Faeroe Islands lie at the northeastern end of a region of shallow banks incorporating the Rockall Plateau and the Faeroe Plateau. The Faeroe Islands are formed by a sequence, 3000 m thick, of nearly horizontal Early Tertiary lavas about 50—60 Ma old (Noe-Nygaard, 1962; Tarling and Gale, 1968).

Gravity and seismic studies on the Iceland-Faeroe Ridge indicate that the crustal thickness is around 16—18 km with an apparent $P_V$ velocity of 7.84 km/s (Fig. 3-2). The crustal structure is unlike that beneath normal oceanic or continental regions, but in many respects it resembles the crust beneath Iceland (Bott et al., 1971).

## COCOS AND CARNEGIE RIDGES

These two aseismic ridges are located in the central northeastern Pacific Ocean (Fig. 3-3). Both ridges converge at longitude 92°W with the Galapagos Islands. The latter are related with the western extension of the Galapagos Ridge which in turn terminates on the crest of the East Pacific Rise near 102°W, forming a triple junction (near 2°N, 102°W). Both the Cocos and the Carnegie Ridges are bounded by normal faults. The sediment thickness varies between 100 and 400 m. In determining the origin of the Cocos and Carnegie Ridges, it was suggested that both are made up of volcanics extruded on the lithospheric plates drifting across a mantle hot spot (Holden and Dietz, 1972). It was also suggested (Holden and Dietz, 1972) that both these features were separated throughout their histories. It is clear that as far as the crustal com-position is concerned some kind of comparison may be made between the Galapagos rift zone rocks and those of the adjacent aseismic ridges. Many of the Galapagos rift zone volcanics are enriched in total Fe (up to 12%) and in $TiO_2$ (up to about 2%) and the basalt samples from the Galapagos Spreading Center are comparable to those drilled in the crust of the Cocos (Site 158) and the Carnegie Ridges (Site 157). However, other dredge hauls and samples from the Cocos Islands are of an alkali basalt character and are strongly en-riched in $K_2O$ (0.7—2.5%; Table 3-2). Trace element analyses compared with other aseismic ridges are shown in Table 3-3. The composition of the clino-pyroxene from the Cocos Ridge is comparable to that found in recent volcanics from mid-oceanic ridge systems (Table 3-4).

Only a few samples have been described from the Cocos and the Carnegie Ridges. The rocks dredged and drilled are all of basaltic composition. Two samples (314, 316) from a shallow depth were described by C.G. Engel and

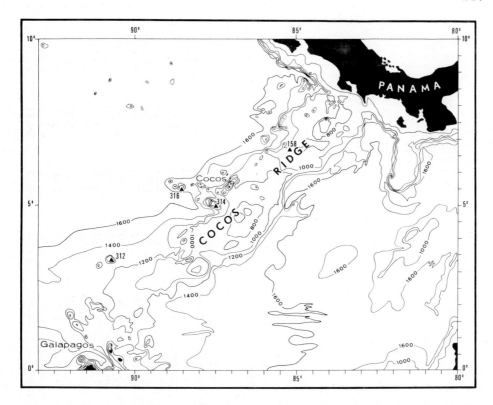

Fig. 3-3. Bathymetric chart of Cocos Ridge in the Pacific Ocean with contour lines in fathoms (Chase et al., 1970). Samples 312, 316 and 314 are plagioclase-rich alkali basalts and DSDP Site 158 (Leg 16) consists of plagioclase-pyroxene basalt.

Chase (1965) as porphyritic alkali basalts containing plagioclase (40—45%), augite (20%) and other opaques and dark mesostasis. One sample (312) is less enriched in alkali (<1% $K_2O$) and consists of 58% plagioclase and 38% augite (C.G. Engel and Chase, 1965). The alkali basalts are highly vesiculated and have a similar composition to those found on the nearby Cocos Islands. The lava flows studied from the Cocos Islands are alkali basalts with porphyritic textural features (Dalrymple and Cox, 1968).

The "Glomar Challenger" boreholes on the Cocos (DSDP Leg 16, Site 158) and on the Carnegie Ridges (Leg 16, Site 157) recovered Fe-enriched tholeiites from both sites. That these rocks are chloritized is shown by their relatively high $H_2O$ content (1.5—3%). The relatively high total Fe content reflects the common occurrence of large-size Fe-oxide minerals (Table 3-2). Samples 157 and 158 are made up essentially of plagioclase (49%), clinopyroxene (26%) and chlorite. The anorthite content of the plagioclase is $An_{58}$—$An_{73}$ (Yeats et al., 1973). A few olivines are present but they are completely altered. The total Fe (11—13%) and $TiO_2$ contents (1.5—2.15%) suggest that these rocks

148

TABLE 3-2

Chemical analyses of the eastern Pacific aseismic ridges

| | Cocos Ridge | | | 158-36cc (1) | 158-36cc (2) | Cocos Island | | | | Carnegie Ridge 157-49-2 146–148 |
|---|---|---|---|---|---|---|---|---|---|---|
| | 312 | 314 | 316 | | | 400 | 401 | 402 | 404 | |
| $SiO_2$ (wt.%) | 48.34 | 44.90 | 42.95 | 45.61 | 49.42 | 58.40 | 48.30 | 48.80 | 48.60 | 50.38 |
| $Al_2O_3$ | 16.36 | 22.80 | 19.10 | 13.30 | 14.02 | 19.00 | 19.00 | 18.60 | 19.10 | 13.01 |
| $Fe_2O_3$ | 2.59 | 1.91 | 4.03 | 6.15 | — | 4.70 | 6.50 | 5.30 | 8.30 | 12.43** |
| FeO | 7.38 | 4.69 | 5.60 | 5.48 | 12.35* | 1.90 | 2.80 | 3.70 | 1.10 | — |
| MnO | 0.17 | 0.12 | 0.16 | — | — | 0.09 | 0.20 | 0.17 | 0.19 | — |
| MgO | 6.23 | 5.09 | 6.65 | 8.71 | 8.60 | 0.69 | 3.70 | 3.80 | 2.60 | 6.48 |
| CaO | 11.97 | 9.16 | 9.37 | 7.86 | 7.44 | 1.70 | 7.00 | 7.20 | 6.20 | 10.84 |
| $Na_2O$ | 2.91 | 4.00 | 3.50 | 2.80 | 3.76 | 5.50 | 3.40 | 3.20 | 4.20 | 2.69 |
| $K_2O$ | 0.79 | 1.76 | 1.15 | 0.17 | 0.14 | 4.50 | 1.70 | 2.00 | 2.30 | 0.20 |
| $TiO_2$ | 1.97 | 2.85 | 2.81 | 1.70 | 1.50 | 1.10 | 3.40 | 3.50 | 3.30 | 2.15 |
| $P_2O_5$ | 0.28 | 1.26 | 0.99 | 0.19 | — | 0.25 | 0.59 | 0.51 | 0.61 | — |
| $CO_2$ | — | — | — | — | — | 0.05 | 0.05 | 0.05 | 0.05 | — |
| LOI | 0.82 | 1.80 | 3.30 | 7.30 | 3.37 | 2.21 | 3.10 | 2.90 | 3.30 | — |
| Total | 99.80 | 100.33 | 99.60 | 99.26 | 100.60 | 100.08 | 99.73 | 99.72 | 99.84 | 98.17 |

*Norms*

| | | | | | | | | | | |
|---|---|---|---|---|---|---|---|---|---|---|
| Q | — | — | — | 1.53 | — | 5.35 | 3.30 | 3.35 | 0.06 | 7.76 |
| Ne | — | 6.55 | 4.59 | — | — | — | — | — | — | — |
| Or | 4.66 | 10.40 | 6.79 | 1.00 | — | 26.59 | 10.04 | 11.81 | 13.59 | 1.18 |
| Ab | 24.62 | 21.73 | 21.13 | 23.69 | — | 46.53 | 28.76 | 27.07 | 35.53 | 22.76 |
| An | 29.24 | 37.21 | 33.00 | 24.22 | — | 6.48 | 30.55 | 30.48 | 26.45 | 22.83 |
| Di | 22.97 | — | 5.53 | 11.42 | — | — | — | 1.23 | — | 18.25 |
| Hy | 0.05 | — | — | 18.50 | — | 1.71 | 9.21 | 8.89 | 6.47 | 7.67 |
| Ol | 9.27 | 10.85 | 11.76 | — | — | — | — | — | — | — |
| Mt | 3.75 | 2.76 | 5.84 | 8.91 | — | 3.22 | — | — | — | — |
| Ilm | 3.74 | 5.41 | 5.33 | 3.22 | — | 2.08 | 6.34 | 2.33 | 2.72 | — |
| Ap | 0.66 | 2.97 | 2.33 | 0.44 | — | 0.58 | 1.39 | 6.64 | 1.43 | — |
| Rt | — | — | — | — | — | — | 0.06 | 1.20 | 1.86 | — |
| Hm | — | — | — | — | — | 2.47 | 6.50 | — | 8.29 | 12.42 |
| Cc | — | — | — | — | — | 2.70 | 0.36 | 3.68 | — | — |
| C.I. | 50.60 | 46.09 | 47.90 | 35.42 | — | 6.65 | 31.48 | 32.61 | 27.10 | 41.86 |
| D.I. | 29.29 | 38.69 | 32.52 | 26.22 | — | 78.48 | 42.11 | 42.25 | 49.19 | 31.70 |

*Total Fe calculated as FeO.

**Total Fe calculated as $Fe_2O_3$.

Samples 312, 314 and 316 are highly-phyric plagioclase basalts (C.G. Engel and Chase, 1965); 158-36cc is a plagioclase-pyroxene basalt from DSDP Leg 16 (Campsie et al., 1973a, Hékinian, 1974b); 400 is a trachyte and 401, 402 and 404 are alkali basalts (Dalrymple and Cox, 1968); 157 is a basalt from DSDP Leg 16 (Campsie et al., 1973a).

TABLE 3-3

Trace element analyses of rocks collected from aseismic ridges around the world

| | Ninety-East Ridge | | | | Cocos Ridge*' | | Carnegie Ridge* | | |
|---|---|---|---|---|---|---|---|---|---|
| | oceanic andesite (ave. of 5) | ferro-basalt 214 (ave. of 6) | ferro-basalt 216 ave. of 7) | basalt 253 | 158-36cc | 158-36-1 146—150 | 157-49-1 130—132 | 157-49-2 10—12 | 157-49-2 65—67 |
| V (ppm) | 39 | 525 | 445 | 210 | 150 | 365 | 385 | 370 | 365 |
| Cr | 5 | 38 | 45 | 345 | 115 | 130 | 175 | 215 | 240 |
| Co | 39 | 65 | 53 | 42 | 56 | 53 | 60 | 57 | 54 |
| Ni | 5 | 50 | 44 | 110 | 55 | 59 | 60 | 60 | 57 |
| Cu | — | — | — | — | 130 | 135 | 165 | 145 | 140 |
| Sr | 647 | 265 | 235 | 77 | 170 | 180 | 160 | 170 | 150 |
| Ba | 578 | 45 | 140 | 14 | 40 | 51 | 34 | 36 | 30 |
| Y | 65 | 26 | 31 | 32 | 37 | 35 | 39 | 43 | 38 |
| Zr | 252 | 120 | 159 | 100 | 135 | 120 | 165 | 165 | 145 |

*DSDP Leg 16.
**DSDP Leg 38.

All the analyses were made by photoelectric emission spectrometry (G. Thompson et al., 1974; Hékinian and Thompson, 1976; Frey et al., 1977) except for Leg 38, Site 336 samples which were analysed by X-ray fluorescence (Ridley et al., 1974).

have followed a different pattern of origin from the normal tholeiites found in the East Pacific Rise. Heath and Van Andel (1971) note that the rare earth element pattern (J. Corliss, personal communication) for these basalts (Sites 157 and 158) is more fractionated than that of other tholeiites. Their total Fe and $TiO_2$ content is also different from other East Pacific Rise tholeiites.

NINETY-EAST RIDGE (INDIAN OCEAN)

The Ninety-East Ridge discovered by Seymour-Sewell (1925) and first shown on a bathymetric map by Stocks (1960) was further delineated by Bezrukov and Kanaev (1963), Heezen and Tharp (1966) and Laughton et al. (1970). The Ninety-East Ridge is remarkably straight and extends some 5000 km along a north-south trend in the eastern Indian Ocean, roughly parallel to the 90° east longitude (Fig. 3-4). Over its entire length it appears to be a single ridge rising generally 1500—2000 m above the surrounding sea floor (Bowin, 1973). Its southern extension is structurally more complex. The 2500-m contour line extends from the Ninety-East Ridge to the adjacent Broken Ridge, also aseismic in nature, and to the seismically active Mid-Indian Oceanic Ridge. Bathymetry (Laughton et al., 1970), seismic refraction (Francis and Raitt, 1967; Bowin, 1973), magnetic and gravity observations (Bowin, 1973) were made on the Ninety-East Ridge. From such geophysical studies, several

| | Walvis Ridge | | | | | | | Iceland-Faeroe Ridge** |
|---|---|---|---|---|---|---|---|---|
| 157-49-2 135—138 | WD 3-3 | WD 3-5 | DR3-33 | WD 4-1 | RD4-A | RD4-B | RD4-6 | 336-41-1 |
| 330 | 550 | 335 | 540 | 445 | 365 | 360 | 465 | — |
| 160 | 47 | 70 | 49 | 38 | 63 | 190 | 55 | — |
| 54 | 96 | 49 | 62 | 65 | 67 | 39 | 60 | — |
| 52 | 60 | 74 | 64 | 31 | 57 | 51 | 48 | — |
| 145 | 140 | 110 | 205 | 40 | 120 | 20 | 135 | — |
| 180 | 290 | 400 | 385 | >400 | 330 | 380 | >400 | 137 |
| 41 | 330 | 320 | 440 | 535 | 245 | 280 | 535 | — |
| 42 | 48 | 51 | 42 | 51 | 40 | 32 | 60 | 30 |
| 165 | 190 | 185 | 205 | 225 | 180 | 175 | 230 | 76 |

hypotheses were formulated to explain its origin: (1) it is a horst-type structure (Francis and Raitt, 1967; Laughton et al., 1971); (2) it is a folded structure due to the overriding of one oceanic crustal plate by another (Le Pichon and Heirtzler, 1968); (3) it represents the trace of a fracture zone (McKenzie and Sclater, 1971); (4) it is an old spreading center (Veevers et al., 1971); or (5) it is the trace left by a hot-spot mantle plume (Morgan, 1972a).

A crustal structure model across the Ninety-East Ridge in its northern position shows the existence of a root of relatively low-density material, 2.94—3.05 $g/cm^3$ (Bowin, 1973) (Fig. 3-5). A density of 2.94 $g/cm^3$ could represent a layer of basaltic rock overlying gabbroic material.

Deep-sea drilling along the Ninety-East Ridge penetrated less than 100 m into the basement (DSDP Leg 22, Sites 214, 216 and Leg 26, Sites 253 and 254). Dredge hauls (Sites 6744, 6742; Kashintsev and Rudnik, 1974) and the DSDP sites recovered four major types of volcanics: (1) oceanic andesite, (2) basalts enriched in iron and titanium (ferrobasalt), (3) olivine basalt, and (4) picritic basalt (Table 3-5).

(1) *Oceanic andesites* are found only in hole 214 (Leg 22) as a differentiated unit about 10 m thick with a fine-grained chilled margin. The contact zones of this flow with the sedimentary units is gradational. There was no evidence of the sharp glassy contact typical of pillow lava. Instead the glassy matrix grades gradually into a more crystalline interior. It is likely that this unit is a thick flow or an intrusive body. These andesites are characterized by the presence of sodic plagioclase and sub-calcic clinopyroxene ($Wo_{7-8}$ $En_{42-48}$

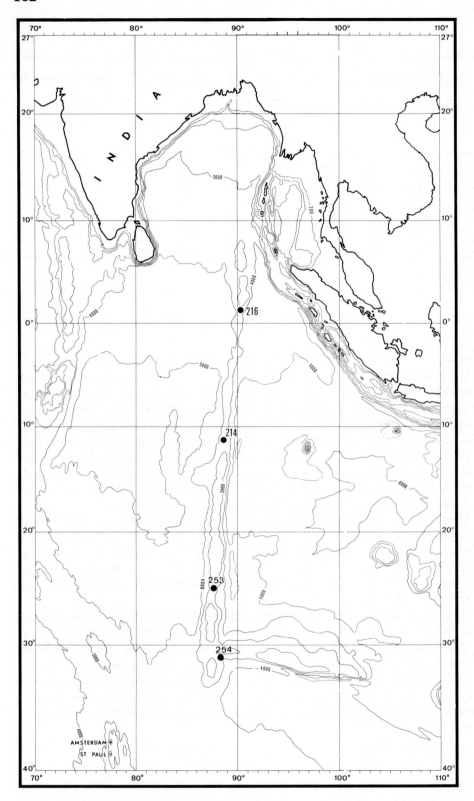

153

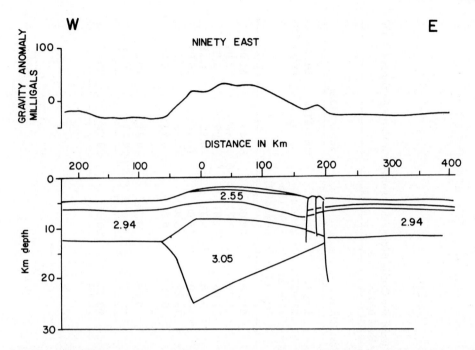

Fig. 3-5. Model across the Ninety-East Ridge at 4°N (Bowin, 1973). The densities of the various layers are shown. The material with densities 3.05 g/cm³ is inferred to be gabbro and serpentinized peridotite (Bowin, 1973).

Fs$_{44-50}$) (Table 1-18). The general texture is a glassy to cryptocrystalline groundmass containing plagioclase laths showing fluidal arrangement. The contents of transitional metals are depleted with respect to the basaltic rocks and the ferrobasalts reported from the Ninety-East Ridge (Tables 1-17 and 3-3). Both the Ba (500—600 ppm) and Se (600—700 ppm) contents are higher than in basaltic rocks (Table 3-3).

(2) *Ferrobasalts* are found in most boreholes (Sites 214, 216, 254). These rocks occur either with characteristic glassy margins similar to that of pillowed structures or as massive and highly vesiculated flow units (Hékinian, 1974a). Some varieties of ferrobasalts are made up of irregular patches (0.5—1 cm in diameter) that are filled by hydrothermal material consisting of vermiculite, smectite, biotite, chlorite and calcite. Dredges from Sites 6742 and 6744 (Kashintsev and Rudnik, 1974) near DSDP Site 214 also consist of Fe-Ti-rich basalts.

The ferrobasalts are enriched in clinopyroxene which exceeds the content

Fig. 3-4. Bathymetric map of the Ninety-East Ridge in the Indian Ocean with contour lines in meters. DSDP Sites 214 and 216 were drilled during Leg 22 of the "Glomar Challenger" and Sites 253 and 254 were drilled during Leg 26 (Kempe, 1974).

TABLE 3-4

Microprobe analyses of clinopyroxene from basaltic rocks recovered from the Cocos Ridge (this work), the Carnegie Ridge (DSDP Leg 16, Site 157) (Yeats et al., 1971), the Walvis Ridge and the Ninety-East Ridge

| | Cocos Ridge | | Carnegie Ridge | Ninety-East Ridge | | | Walvis Ridge | |
|---|---|---|---|---|---|---|---|---|
| | Leg 16 158-36cc (ave. of 2) | Leg 16 157-49-2 146—148 | Leg 16 157-49-2 55—65 (ave. of 2) | Leg 26 254 (ave. of 4) | Leg 22 216-37 109—117 (ave. of 2) | Leg 22 214-54-3 139—135 (ave. of 2) | CH4-WD3 | DR4-4 (ave. of 2) |
| $SiO_2$ (wt.%) | 50.92 | 52.45 | 51.20 | 49.50 | 51.79 | 52.00 | 51.49 | 51.99 |
| $Al_2O_3$ | 4.33 | 2.44 | 4.18 | 3.23 | 1.79 | 1.77 | 2.31 | 2.04 |
| FeO | 6.67 | 7.60 | 7.13 | 10.76 | 11.96 | 11.91 | 8.99 | 9.27 |
| MnO | 0.29 | 0.23 | 0.23 | — | 0.18 | 0.31 | 0.22 | 0.18 |
| MgO | 16.57 | 16.50 | 16.41 | 13.74 | 16.40 | 15.66 | 16.14 | 16.01 |
| CaO | 19.57 | 20.20 | 20.50 | 19.86 | 16.87 | 16.95 | 18.83 | 17.78 |
| $Na_2O$ | 0.26 | tr. | tr. | 0.31 | 0.18 | 0.16 | 0.23 | 0.26 |
| $K_2O$ | — | tr. | tr. | — | — | — | — | — |
| $TiO_2$ | 0.81 | 0.73 | 0.94 | 1.80 | 0.68 | 0.79 | 0.84 | 0.89 |
| $Cr_2O_3$ | 0.47 | — | — | 0.28 | — | — | 0.17 | 0.15 |
| Total | 99.89 | 100.15 | 100.56 | 99.48 | 99.85 | 99.55 | 99.22 | 98.57 |

of plagioclase. The texture is intergranular to subophitic and sometimes pilotaxitic. Titanomagnetite, often crystallizing as microphenocrystal phases, is a common constituent of the rocks. In addition, traces of chalcopyrite and native copper were also found. Chemically these ferrobasalts differ from other basaltic rocks by their higher $TiO_2$ (2—4%) and their total Fe content (>13%) (Table 3-5). They have lower Cr (<60 ppm) and Ni (<50 ppm) contents and higher Sr (>200 ppm) and Ba (>100 ppm) contents than the other basalts recovered from the Ninety-East Ridge (Table 3-5). The depletion of transitional metals and the increase in Sr and Ba contents is due to the alteration of the rocks by hydrothermal circulation. Also the higher water content of these rocks is related to the presence of secondary hydrated minerals such as Fe-nontronite and mixed layer chlorite-smectite.

(3) *Olivine basalts* are made up essentially of olivine and plagioclase and are found overlying picritic basalts in hole 253 (DSDP Leg 26; Kempe, 1974). The olivine basalt differs from the ferrobasalt in its lower total iron content ($FeO + Fe_2O_3$ <11%) and lower $TiO_2$ content (<1%) (Table 3-5).

(4) *Picritic basalt* was found at the base of a scoriaceous olivine basalt unit (Leg 26, Site 253; Kempe, 1974). This is a glassy unit with about 15% modal olivine and an MgO content of about 14%. However, the rock is highly altered and the euhedral crystals of olivine are replaced by serpentine and talc.

Results of geochemical investigations on the Ninety-East volcanics which were carried out on REE and LIL elements (Frey et al., 1977; Subbarao et al., 1979a; Ludden et al., 1980) are reported in Figs. 2-19 and 3-6. The $^{87}Sr/^{86}Sr$ ratios of the ferrobasalts and the olivine and picritic basalts vary between 0.7043 and 0.7048, while the andesites have a higher ratio of 0.7049 (Fig. 2-19). The strontium isotopic data are comparable to those found on the Kerguelen Islands ($^{87}Sr/^{86}Sr$ = 0.7048—0.7052), on St. Paul Rock (0.7037) and on Amsterdam Island (0.7039) (Fig. 2-18). In addition, the different

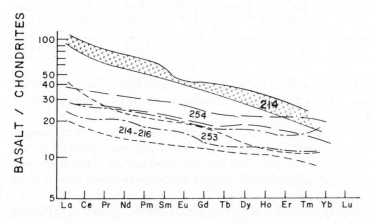

Fig. 3-6. Rare earth element distribution pattern normalized to chondrites of rocks from the Ninety-East Ridge in the Indian Ocean (Subbarao et al., 1979a).

TABLE 3-5

Chemical analyses of rocks collected from the Ninety-East Ridge during Leg 22 (Sites 214, 216; Hékinian, 1974a) and 26 (Sites 253, 254; Kempe, 1974) of the "Glomar Challenger" in the Indian Ocean (analyses made by X-ray fluorescence)

| | Oceanic andesites | | | | | | | |
|---|---|---|---|---|---|---|---|---|
| | 214-48-1 76—83 | 214-48-1 94—100 | 214-48-2 9—13 | 214-48-2 117—123 | 214-49-1 137—144 | 214-49-2 0—7 | 214-50-1 145—150 | 214-51-1 108—114 |
| $SiO_2$ (wt.%) | 56.69 | 54.79 | 55.70 | 54.59 | 57.41 | 58.02 | 55.70 | 51.34 |
| $Al_2O_3$ | 16.10 | 15.82 | 15.40 | 15.67 | 15.81 | 15.95 | 15.30 | 18.73 |
| $Fe_2O_3$ | 2.88 | 3.61 | 3.11 | 2.42 | 2.83 | 2.17 | 2.91 | 3.80 |
| FeO | 7.55 | 6.45 | 7.00 | 7.47 | 7.00 | 7.51 | 7.00 | 2.72 |
| MgO | 2.57 | 2.60 | 2.40 | 2.60 | 2.32 | 2.33 | 2.30 | 1.02 |
| CaO | 5.62 | 5.67 | 5.59 | 5.82 | 5.86 | 5.71 | 5.71 | 7.97 |
| $Na_2O$ | 4.08 | 4.12 | 3.88 | 3.88 | 3.92 | 3.70 | 3.78 | 4.21 |
| $K_2O$ | 1.56 | 1.36 | 1.55 | 1.27 | 1.58 | 1.56 | 1.49 | 1.83 |
| $TiO_2$ | 1.40 | 1.39 | 1.40 | 1.43 | 1.46 | 1.47 | 1.44 | 1.56 |
| $P_2O_5$ | 0.56 | 0.59 | 0.63 | 0.68 | 0.65 | 0.64 | 0.66 | 0.80 |
| $CO_2$ | 0.00 | 0.00 | 0.00 | 0.00 | 0.00 | 0.00 | 0.00 | 1.76 |
| LOI | 1.22 | 1.89 | 1.15 | 1.99 | 1.18 | 0.91 | 1.42 | 5.06 |
| Total | 100.22 | 98.28 | 97.80 | 97.79 | 100.01 | 99.96 | 97.70 | 100.79 |
| *Norms* | | | | | | | | |
| Q | 10.21 | 9.05 | 10.56 | 9.10 | 11.54 | 12.66 | 11.31 | 8.31 |
| Or | 9.21 | 8.03 | 9.15 | 7.50 | 9.33 | 9.21 | 8.80 | 10.81 |
| Ab | 34.52 | 34.86 | 32.83 | 32.83 | 33.16 | 31.30 | 31.98 | 35.62 |
| An | 21.00 | 20.65 | 20.02 | 21.59 | 20.87 | 22.30 | 20.37 | 23.18 |
| Di {Wo | 1.34 | 1.51 | 1.49 | 1.18 | 1.64 | 0.76 | 1.51 | 0.00 |
| Di {En | 0.55 | 0.73 | 0.64 | 0.48 | 0.68 | 0.29 | 0.63 | 0.00 |
| Di {Fs | 0.79 | 0.74 | 0.85 | 0.70 | 0.96 | 0.48 | 0.89 | 0.00 |
| Hy {En | 5.84 | 5.73 | 5.33 | 5.98 | 5.08 | 5.50 | 5.09 | 2.54 |
| Hy {Fs | 8.37 | 5.81 | 7.11 | 8.61 | 7.13 | 9.08 | 7.18 | 0.00 |
| Mt | 4.17 | 5.23 | 4.50 | 3.50 | 4.10 | 3.14 | 4.21 | 4.24 |
| Ilm | 2.65 | 2.63 | 2.65 | 2.71 | 2.77 | 2.79 | 2.73 | 2.96 |
| Ap | 1.32 | 1.39 | 1.48 | 1.60 | 1.53 | 1.50 | 1.55 | 1.88 |
| Hm | 0.00 | 0.00 | 0.00 | 0.00 | 0.00 | 0.00 | 0.00 | 0.87 |
| C | 0.00 | 0.00 | 0.00 | 0.00 | 0.00 | 0.00 | 0.00 | 1.32 |
| Cc | 0.00 | 0.00 | 0.00 | 0.00 | 0.00 | 0.00 | 0.00 | 4.00 |
| Ol | | | | | | | | |
| C.I. | 22.79 | 22.82 | 21.95 | 23.24 | 22.87 | 23.49 | 22.25 | 23.43 |
| D.I. | 53.95 | 51.94 | 52.55 | 49.43 | 54.04 | 53.19 | 52.10 | 54.75 |

types of basalts and the andesites are enriched in light REE (Subbarao et al., 1979a). The highest REE enrichment is encountered in the andesites from Site 214 (Fig. 3-6).

The similar strontium isotopic ratio and REE distribution pattern suggest a genetic relationship between the various types of rocks encountered on the Ninety-East Ridge. Fractional crystallization processes and/or variable degrees of partial melting from a single mantle source could have given rise to these volcanics.

The presence of highly vesicular and amygdalar basalts found at different sub-basement depths suggests subaerial or near-surface volcanic events on the Ninety-East Ridge at Sites 214 and 216. In addition, basal sediment of

| Ferrobasalts | | | | | | Basalt | Picrite | Basalt |
|---|---|---|---|---|---|---|---|---|
| 214-53-1 26—30 | 214-53-1 97—100 | 214-54-3 0—4 | 216-37-2 80 | 216-38-2 79—85 | 216-38 | 253-24-1 84—85 | 253-58-1 2—7 | 254-36-3 95—97 |
| 45.50 | 43.96 | 43.94 | 48.56 | 49.04 | 48.67 | 49.98 | 43.42 | 44.59 |
| 14.70 | 13.51 | 13.65 | 13.67 | 12.97 | 12.63 | 17.39 | 15.61 | 13.93 |
| 6.82 | 5.99 | 7.09 | 5.65 | 5.79 | 6.04 | 3.23 | 5.59 | 5.59 |
| 7.69 | 8.96 | 8.61 | 7.45 | 9.16 | 7.15 | 6.87 | 7.62 | 7.62 |
| 6.22 | 6.59 | 6.07 | 6.52 | 5.33 | 6.55 | 7.46 | 13.85 | 7.65 |
| 8.88 | 8.37 | 6.84 | 8.39 | 9.31 | 9.10 | 9.17 | 7.84 | 9.39 |
| 2.40 | 2.28 | 2.95 | 2.50 | 2.50 | 2.30 | 2.89 | 1.39 | 2.48 |
| 0.21 | 0.30 | 0.26 | 0.44 | 0.83 | 0.90 | 0.30 | 0.37 | 0.15 |
| 2.12 | 2.03 | 2.50 | 2.76 | 2.64 | 2.50 | 0.93 | 0.69 | 2.47 |
| 0.17 | 0.16 | 0.21 | 0.22 | 0.22 | 0.19 | 0.33 | 0.08 | 0.26 |
| 0.00 | 0.00 | 0.00 | 0.00 | 0.00 | 0.00 | 1.62 | 0.38 | 1.21 |
| 5.80 | 7.16 | 6.62 | 4.06 | 0.99 | 3.71 | 4.32 | 7.13 | 4.25 |
| 100.50 | 99.30 | 98.73 | 100.21 | 98.77 | 99.73 | 100.75 | 101.01 | 99.82 |
| 3.71 | 1.71 | 2.03 | 6.55 | 5.52 | 6.13 | 4.14 | — | 1.87 |
| 1.24 | 1.77 | 1.53 | 2.60 | 4.90 | 5.31 | 1.77 | 2.19 | 0.89 |
| 20.30 | 19.29 | 24.96 | 21.15 | 21.15 | 19.46 | 24.45 | 11.76 | 20.98 |
| 28.71 | 25.74 | 23.23 | 24.77 | 21.71 | 21.48 | 33.10 | 35.27 | 26.44 |
| 5.93 | 6.15 | 3.89 | 6.43 | 9.61 | 9.36 | | | |
| 4.12 | 3.85 | 2.60 | 4.59 | 5.77 | 6.81 | — | 0.56 | 8.63 |
| 1.32 | 1.91 | 1.00 | 1.26 | 3.33 | 1.67 | | | |
| 11.36 | 12.55 | 12.51 | 11.63 | 7.50 | 9.49 | 22.90 | 21.39 | 20.56 |
| 3.65 | 6.23 | 4.82 | 3.19 | 4.33 | 2.33 | | | |
| 9.88 | 8.68 | 10.27 | 8.19 | 8.39 | 8.75 | 3.68 | 4.68 | 8.10 |
| 4.02 | 3.85 | 4.74 | 5.24 | 5.01 | 4.74 | 1.77 | 1.31 | 4.69 |
| 0.40 | 0.37 | 0.49 | 0.51 | 0.51 | 0.44 | 0.78 | 0.19 | 0.61 |
| 0.00 | 0.00 | 0.00 | 0.00 | 0.00 | 0.00 | — | — | — |
| 0.00 | 0.00 | 0.00 | 0.00 | 0.00 | 0.00 | — | — | — |
| 0.00 | 0.00 | 0.00 | 0.00 | 0.00 | 0.00 | 3.68 | 0.86 | 2.75 |
| | | | | | | — | 15.64 | — |
| 38.75 | 35.33 | 30.11 | 35.87 | 34.92 | 37.14 | | | |
| 25.26 | 22.77 | 28.53 | 30.31 | 31.58 | 30.91 | | | |

Maestrichtian age (65—67 Ma), composed of glauconitic sediment and indicating a shallow-water environment of deposition, was found (Sclater et al., 1974).

## WALVIS RIDGE

The Walvis Ridge is a non-continuous feature extending from the Mid-Atlantic Ridge near 13°W to the continental shelf of Namibia (South West Africa) over a distance of approximately 3000 km (Fig. 3-7). The origin of the Walvis Ridge has been controversial. Some have suggested a formation

158

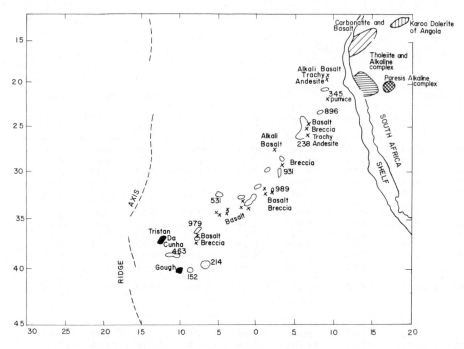

Fig. 3-7. Generalized feature of the Walvis Ridge. The numbers close to the contour lines indicate the shallowest depths recorded along the ridge. The crosses indicate the samples collected from the Walvis Ridge (G. Thompson and S. Humphris, personal communication). Volcanic complexes on the continental area in prolongation to the structure of the Walvis Ridge are also shown.

synchronous with the opening of the South Atlantic and have related its origin to mantle hot spots (Wilson, 1965; Dietz and Holden, 1970; Morgan, 1971, 1972b). Others have noticed that offsets and steep scarps along the Walvis Ridge fit with the transform directions of major South Atlantic fracture zones (Le Pichon and Hayes, 1971; Francheteau and Le Pichon, 1972). The easternmost portion of the Walvis Ridge is a continuous quasi-linear topographic feature which could favor a fracture zone origin. However, its present elevation above the sea floor and its dimensions make it difficult to compare it with oceanic fracture zones (Figs. 3-7, 3-8). Indeed, no remnants of structures as large and as continuous as the Walvis Ridge have been found along any other fracture zones.

Refraction data on the eastern part of the Walvis Ridge yielded a thick cover of sediment and a basement of 5.24—5.49 km/s which are within the range of basaltic layers (Goslin et al., 1973, 1974) (Fig. 3-9). In this area, the Walvis Ridge appears as a double ridge (Goslin et al., 1974) with an internal basin filled with sediment at least 2.5 km thick. A similar double structure was also encountered on the Iceland-Faeroe Ridge.

A DSDP hole on the southwestern part of the Walvis Ridge (Site 359) was

Fig. 3-5. Bathymetric chart of the Walvis Ridge after Simpson (196...); the rock samples are from the "Chacon" cruise (CH14, CH15 and C23) and from the "Atlantis II" cruises (G. Thompson and ..., compilation).

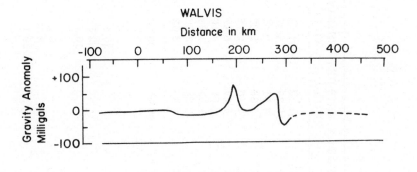

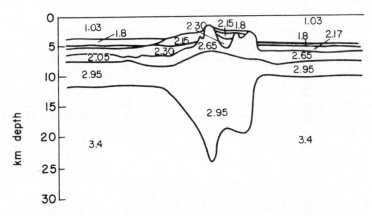

Fig. 3-9. Models of deep structure of the Walvis Ridge near the western coastal margin from gravity data (Goslin and Sibuet, 1975). The densities of the various layers are indicated on the figure. The water density is 1.03 g/cm³ and that of the sediment is 1.8 g/cm³. The densities of layers 2, 3 and 4 are 2.65, 2.95 and 3.4 g/cm³, respectively.

drilled on the flank of a seamount forming part of its structure. Below 57 m of late pelagic carbonate ooze and 30 m of Late Eocene volcanic mud containing pumice, alkalic-tuffaceous material of a probable subaerial origin was encountered (Perch-Nielsen et al., 1975). The continuation of the northern scarp of the Walvis Ridge can be followed under the continental shelf on seismic refraction and reflection records as well as on magnetic anomaly profiles (Goslin and Sibuet, 1975).

Terrestrial volcanism, which could be related at least as far as composition is concerned to the Walvis Ridge, is found in southern Angola and in Namibia (South West Africa) between 12° and 22°S giving rise to post-Triassic or Jurassic continental volcanic ring complexes of mainly alkali basalts (Beetz, 1934; Martin et al., 1960) (Fig. 3-7). Carbonatite and kimberlite are also found in association with doleritic plugs in southern Angola (Lapido-Loureiro, 1968). The Paresis and the Messum complexes of Namibia (South West Africa)

TABLE 3-6

Chemical analyses of alkaline rocks from the Walvis Ridge

| | CH18-DR3 WD 3-5 | CH18-DR3 WD 3-3 | CH18-DR3-A | CH18-DR3-33 | V22-146 | CH19-DR4 WD 4-3 | CH19-DR4-A | CH19-DR4 WD 4-1 | CH19-DR4 WD 4-C |
|---|---|---|---|---|---|---|---|---|---|
| $SiO_2$ (wt.%) | 51.42 | 47.00 | 47.00 | 45.78 | 50.77 | 49.18 | 50.77 | 51.42 | 51.72 |
| $Al_2O_3$ | 14.31 | 16.37 | 16.50 | 16.33 | 15.60 | 16.00 | 15.60 | 16.37 | 15.37 |
| $Fe_2O_3$ | 11.93 | 12.82 | 13.01 | 12.53 | 9.38 | 6.62 | 9.38 | 7.51 | 9.40 |
| FeO | 0.58 | 1.98 | 2.10 | 1.90 | 2.67 | 5.01 | 2.67 | 2.36 | 2.20 |
| MgO | 2.30 | 2.85 | 2.81 | 2.04 | 2.81 | 4.84 | 2.81 | 2.48 | 2.12 |
| CaO | 4.58 | 6.67 | 6.50 | 6.01 | 7.37 | 9.54 | 7.37 | 7.24 | 6.17 |
| $Na_2O$ | 2.71 | 2.76 | 2.86 | 2.80 | 2.74 | 2.60 | 2.74 | 3.00 | 3.10 |
| $K_2O$ | 3.87 | 2.82 | 2.89 | 3.25 | 1.22 | 1.02 | 1.22 | 1.01 | 1.39 |
| $TiO_2$ | 2.31 | 3.41 | 3.31 | 3.29 | 2.90 | 2.21 | 2.90 | 3.28 | 3.22 |
| LOI | 4.27 | 1.88 | 2.19 | 4.81 | 3.02 | 1.83 | 3.02 | 4.43 | 3.62 |
| Total | 98.27 | 98.35 | 99.18 | 98.73 | 98.47 | 98.85 | 98.47 | 99.09 | 98.30 |
| *Modal* (vol.%) | | | | | | 43 | | | |
| Plagioclase | — | 36.6 | 49.0 | — | 49.6 | 43.2 | 58.1 | — | 47.7 |
| Clinopyroxene | — | 0.8 | tr. | — | 8.6 | 3.2 | 6.8 | — | 2.5 |
| Opaques | — | 37.3 | 33.7 | — | 2.3 | 8.9 | 14.9 | — | 10.7 |
| Alteration minerals | — | 4.3 | 0.9 | — | 0.2 | 11.6 | 1.3 | — | 6.1 |
| Quartz | — | — | — | — | — | tr. | 2.4 | — | 3.5 |
| Mesostasis | — | 20.0 | 15.5 | — | 19.3 | 32.7 | 15.9 | — | 29.1 |

*Norms*

| | | | | | | | | | |
|---|---|---|---|---|---|---|---|---|---|
| Q | 9.17 | 3.38 | 3.04 | 2.73 | — | 6.42 | 12.32 | 12.55 | 13.43 |
| Ne | — | | | | 0.77 | | | | |
| Or | 22.86 | 16.66 | 17.07 | 19.20 | 8.39 | 6.02 | 7.20 | 5.96 | 8.21 |
| Ab | 22.93 | 23.35 | 24.20 | 23.69 | 22.68 | 22.00 | 23.18 | 25.38 | 26.23 |
| An | 15.45 | 23.95 | 23.64 | 22.39 | 31.61 | 28.97 | 26.66 | 28.21 | 23.91 |
| Di | 1.14 | 3.83 | 3.99 | 2.58 | 18.38 | 14.29 | 7.70 | 4.21 | 3.11 |
| Hy | 5.19 | 5.31 | 5.31 | 5.14 | — | 5.50 | 3.42 | 4.22 | 3.83 |
| Ol | — | — | — | — | 6.31 | — | | | — |
| Mt | — | — | — | — | 2.75 | — | 0.20 | | — |
| Ilm | 1.22 | 4.18 | 4.43 | 4.01 | 4.42 | 4.98 | 5.50 | 4.98 | 4.64 |
| Sph | 4.08 | 2.96 | 2.44 | 2.88 | — | 1.60 | — | 1.60 | 1.89 |
| Hm | 11.92 | 12.81 | 13.00 | 12.52 | — | 7.50 | 10.54 | 7.50 | 9.39 |
| C.I. | 17.12 | 28.32 | 28.16 | 25/37 | 45.62 | 43.73 | 34.71 | 32.86 | 27.42 |
| D.I. | 54.97 | 43.40 | 44.32 | 45.63 | 31.85 | 34.45 | 42.71 | 43.91 | 47.87 |

Samples WD 4-1 and WD 4-C are trachyandesites.
Analyses were made using a Siemens X-ray fluorescence unit.

consist of alkali basalt, nephelinite and tephrite similar to that found on the islands of Tristan da Cunha and Gough (Fig. 3-7) which are the only islands found on the Walvis Ridge. However, the ridge is relatively shallow in parts and there is indication that subaerial volcanism has occurred along its crest (Perch-Nielsen et al., 1975; Hékinian, 1972). Evidence of shallow-water organisms such as pelecypods, shells and benthic foraminifera was found in dredge hauls (G. Thompson and S. Humphris, personal communication) and in deep-sea drilling cores (DSDP Legs 39 and 40, Sites 359, 362, 363) near elevated structures along the Walvis Ridge. Isolated peaks as shallow as 200 m occur (Fig. 3-7).

The Walvis Ridge, in the area of sampling, consists of tholeiites, alkaline basaltic suites and trachyandesites. Pyroclastic flows and alkali basalts are found to occur in the dredges which consisted mainly of altered rock types. The $Fe_2O_3/FeO$ ratio and the water contents of some specimens are quite abundant (Table 3-6). Also some of the samples were coated with a thick manganese crust. The transitional trace element analyses of the Walvis Ridge samples show values comparable to those of other aseismic ridges in general (Table 3-3). The highly vesiculated nature of several specimens suggest shallow-water volcanic events for their origin.

The interpretation of the gravity anomaly map over the eastern part of the Walvis Ridge (Goslin and Sibuet, 1975) suggest that it is compensated at a depth of 25—30 km. From free-air anomaly data across the Walvis Ridge to the adjacent basins it is suggested that the Walvis Ridge is in isostatic equilibrium and it seems to have been created, together with its compensating root of light material, contemporaneously with the adjacent lithosphere (Fig. 3-9). A model of the deep structure of the Walvis Ridge along two profiles located at the northeastern part suggests a low-density root of 2.95 $g/cm^3$, which corresponds to a basaltic layer and a 3.4 $g/cm^3$ density below 25 km depth (Goslin and Sibuet, 1975).

SIERRA LEONE RISE

The Sierra Leone Rise is located in the eastern central equatorial Atlantic. It consists of various sizeable structural features and forms a discontinuous chain of seamounts as shallow as 2000 m deep (Fig. 3-10). The general trend of the Sierra Leone Rise is oriented towards NE-SW. When considering the 4800-m bathymetric contour lines, the northeastern end of the Sierra Leone Rise is attached to the African continental margin near 9°N (Fig. 3-10). The southwest-trending structure of the Sierra Leone Rise terminates near the Mid-Atlantic spreading ridge at the eastern part of the 4°N fracture zone (Fig. 3-10). Deep-sea drilling of the Ceara Rise penetrated a basaltic basement of the Upper Cretaceous period (Maestrichtian) (Leg 39, Site 354). Similarly, a DSDP hole (Leg 41, Site 366) on the Sierra Leone Rise penetrated a sediment of the Upper Cretaceous period.

A sediment core was taken (5°14.9'N, 20°15.1'W) on the southeastern slope of the Sierra Leone Rise at a depth of 2720 m and in an area which appeared to be the site of significant erosional processes due to bottom currents. The core is about 1.5 m long and consists of about 90 cm of nannoplankton foraminiferal ooze. 60 cm into the core, the sediment is Upper to Middle Eocene in age (about 45 Ma old) and contains a few Paleocene foraminifera (about 60 Ma old). A few rock fragments (<5 cm in diameter) were collected and analysed with the material recovered. They consist of early-formed feldspar set in an altered ferruginous matrix containing pyrite, apatite and tiny laths of parallel-oriented feldspar, indicating a trachytic textural feature. Microprobe analyses of the feldspar are shown on a ternary diagram (Or-Ab-An) in Fig. 3-11. The potassic and the sodic ($Or_{1.8} Ab_{54} An_{44.1}$ —

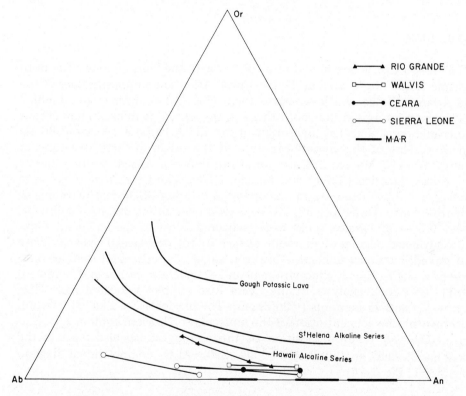

Fig. 3-11. An-Or-Ab ternary diagram of feldspar composition from various localities. The data from Gough Island (Le Maître, 1962), Hawaii (G.A. Macdonald and Katsura, 1964) and from St. Helena (I. Baker, 1969) is compared to that from the Walvis Ridge, the Ceara, the Rio Grande and the Sierra Leone Rises. Open circles indicate phenocrysts of analysed feldspar, filled circles are matrix feldspar found in the corresponding rock samples. The data of the plagioclase from near 37°N in the Atlantic ocean are basalts taken from the inner floor (Hékinian et al., 1976) and those from the Rift Mountain region (Hodges and Papike, 1977).

$Or_{11.5}Ab_{71.7}An_{16.8}$) nature of the feldspathic phases differs considerably from that found on the accreting plate boundary of the Mid-Atlantic Ridge (Fig. 3-11). The molecular percentage of orthoclase in the feldspar from the Mid-Atlantic Ridge basalts is insignificant when compared with that of the Sierra Leone Rise. However, there is a similarity between the composition of the Sierra Leone Rise (Fig. 3-13) feldspar and the alkaline volcanic series from oceanic islands (e.g. Hawaii, Saint Helena and Gough islands) (Fig. 3-11). In addition, the mineral composition and the textural features of the rocks from the Sierra Leone Rise remind us of the trachyandesite type of volcanism encountered on the Walvis Ridge (Hékinian, 1972). The feldspar analysed from the Walvis Ridge alkali basalt (sample WD 3-4) is also of a potassic nature ($Or_{40}Ab_{45.9}An_{50.1}$—$Or_{4}Ab_{28.6}An_{69.0}$) like that from the Sierra Leone Rise (Fig. 3-11).

## CEARA RISE

The Ceara Rise is considered to be paired with the Sierra Leone Rise. Both aseismic ridges are located in the equatorial Atlantic on opposite sides of the Mid-Atlantic Ridge which separates them (Fig. 3-10). Kumar and Embley (1977) have proposed that both these ridges represent thick accumulations of oceanic crust that had their origin at the Mid-Atlantic Ridge about 80 Ma ago. A variant of this hypothesis situated the origin of these structures at about 110—127 Ma ago in a structural gap between Africa, South America and North America (Sibuet and Mascle, 1978). The Ceara Rise rocks were recovered through deep-sea drilling operations during DSDP Leg 39 (Site 354; Scientific Party, DSDP Leg 39, 1977b). After penetrating 886 m of sediment, about 9.5 m of basaltic rocks were recovered. Most of the basement rocks show advanced degrees of alteration as seen by the presence of veins of calcite and mixed layer clay minerals. However, some of the freshest samples have preserved their original clinopyroxene and plagioclase (Fodor and Hékinian, 1981). Bulk-rock analyses of the Ceara Rise samples indicated high $TiO_2$ (3.2 wt.%) and Zr contents ($\simeq 200$ ppm). The plagioclase is alkalic in nature, as shown by the concentration of the molecular orthoclase content ($An_{40-65}$ $Ab_{33-57}Or_{0.8-2}$). The composition of the clinopyroxene also plots in the alkali-basalt field on the $Ti-Al^{IV+VI}$ and $SiO_2$—$Al_2O_3$ compositional diagram (Chapter 1, Fig. 1-14).

## RIO GRANDE RISE (ATLANTIC OCEAN)

The Rio Grande Rise is an irregular feature located in the southwestern Atlantic Ocean between the Mid-Atlantic Ridge and the Brazilian continental shelf. Like the Walvis Ridge, the origin of this rise is of a controversial nature.

The theory that it represents the remainder of a "hot spot" or a leaky transform fault or both is still controversial (Dietz and Holden, 1970; Morgan, 1971; Le Pichon and Hayes, 1971). Deep-sea drilling performed on the Rio Grande Rise failed to penetrate basement (Maxwell et al., 1970). Eocene breccia containing volcanic material which showed affinities to the alkali-basalt type were recovered (Maxwell et al., 1970). The stratigraphic data obtained by the drilling operation (DSDP Leg 39) of the "Glomar Challenger" showed that the rise was a large island during Santonian to Campanian times (85—75 Ma ago; Thiede, 1977). The dredge haul taken by the R.V. "Conrad" at 30°26′S and 36°1′W brought up a variety of volcanics which have undergone various degrees of alteration. The rocks are of an alkali nature and consist of alkali basalt, trachybasalt, trachyandesite and volcanic breccia (Fodor et al., 1977). The alkali basalt suite encountered in the Rio Grande Rise is comparable to that found on the islands of Tristan da Cunha and Gough. In addition, the dredged material from the Walvis Ridge close to the African continent also shows similarities to the Rio Grande Rise samples. Hence, like the Walvis Ridge, the Rio Grande Rise has followed the same structural and compositional regime: they both represent series of islands which have subsided during the early opening of the South Atlantic.

## COMPARISON OF ASEISMIC RIDGES AND OTHER OCEANIC PROVINCES

In addition to petrological data, gravity anomaly and seismic velocity measurements have contributed to our knowledge about the structure of elevated chains of both extinct and active volcanoes. From both gravity and seismic studies it is inferred that aseismic ridges have a thicker crustal structure (15—30 km thick) than that found underneath recent accreting plate boundary regions (average of 7 km thick) (Figs. 3-2, 3-5, 3-9). The type of volcanic activity and structural features encountered on aseismic ridges may be comparable to that of the Hawaiian volcanoes. The structure of the Koolau volcano of the islands of Oahu and Maui in the Hawaiian archipelago suggests a crustal thickness of up to 23 km (Shor and Pollard, 1964). In other areas, such as the western shore of Hawaii, the crustal thickness is about 16 km (Furumoto et al., 1968). Similarly, the crust underneath Iceland varies in thickness from 8 to 15 km, which is comparable to that encountered in the Iceland-Faeroe aseismic ridge.

Thus the geophysical and compositional similarity observed between islands and aseismic ridges suggests that both types of structure might have had comparable processes of formation. Gravity anomalies and seismic studies revealed that aseismic ridges in general are in isostatic equilibrium and that their roots might be formed by layered structures of material differentiated by crystal fractionation. The large structures corresponding to these aseismic ridges suggest that they were built during a period of extensive melting and

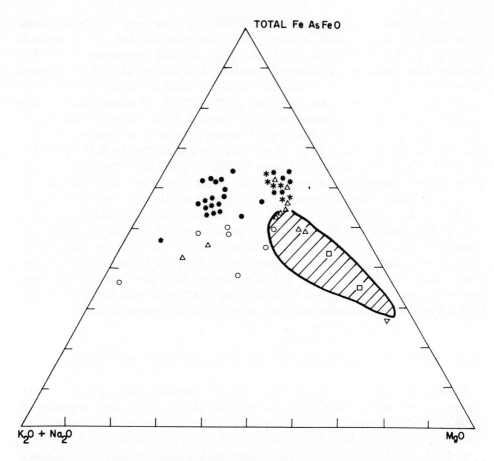

Fig. 3-12. Ternary diagram, MgO-FeO-(Na$_2$O + K$_2$O), of rocks from aseismic ridges. ○ = samples from the Cocos Ridge; △ = samples from the Iceland-Faeroe Ridge; ● = Walvis Ridge. * = Sites 214 and 216 from the Ninety-East Ridge, □ = picritic basalt from Site 254 (Kempe, 1974). The field of mid-oceanic ridge basalt from the Indian and Atlantic Ocean is shown.

uprise of the upper mantle. The mechanism must have been different from that observed on recent spreading ridges. It is likely that the aseismic volcanic edifices were affected by higher heat production during a relatively short period of time when spreading was not disrupted by constructional processes.

Chemically and mineralogically the volcanics collected from the aseismic ridges in general differ from rocks encountered on recent spreading-ridge regions around the world and more closely resemble island and seamount types of assemblages as visualized from the AFM ternary diagrams shown in Figs. 3-12 and 3-13.

The AFM diagram of Fig. 3-12 shows that the bulk composition of aseismic ridge volcanism is generally more fractionated in nature than that found on

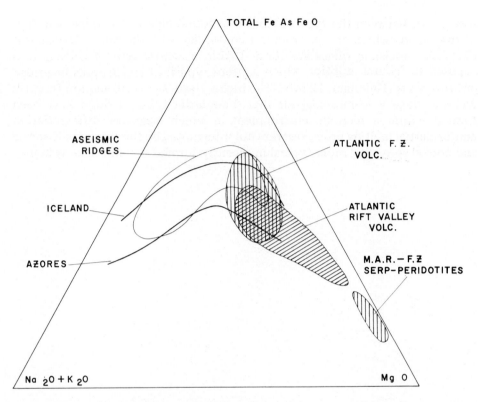

Fig. 3-13. Ternary diagram, MgO-FeO-(Na$_2$O + K$_2$O) showing the fields occupied by rocks of aseismic ridges and other oceanic provinces.

recent accreting plate boundary regions. In addition, trace element studies show that aseismic ridges (e.g. Walvis, Ninety-East Ridges) are enriched in LIL elements and light REE with respect to recent spreading ridge basalts. In general, it is noticed that the REE pattern of the Ninety-East Ridge basalts is comparable to that of oceanic islands (Fig. 3-6). The plagioclase from most aseismic ridges, except for those from the Ninety-East Ridge, is alkaline and differs from that found in recent spreading ridge rocks (Fig. 3-11). Also the clinopyroxene composition of aseismic ridges differs from that of typical oceanic ridge tholeiites (Fig. 1-14). Some critical constituents such as Ti, Al, and Si in clinopyroxene from aseismic ridge basalts show values falling in a field separate from those of clinopyroxene in rocks of other oceanic ridge provinces (Fig. 1-14).

The differences in structure and in composition of the volcanic products from aseismic ridges and from recently active ridges suggest a diverse process of magma generation. Aseismic ridges appear to be lines of islands and/or chains of elevated volcanoes which have subsided during their geological evolution. In previous paragraphs, the existence of a petrological continuity was shown between the type of volcanism found on aseismic ridges and that of the islands which are in structural continuity with the ridges. Hence, it is likely that aseismic ridges are formed from a central type of volcanism as opposed to fissural eruption which is more typical of recent spreading ridge environments (Hékinian, 1974b). The higher degree of fractionation found in aseismic ridge volcanics suggests that these latter types of ridges have been formed within a tectonic environment in which magmatic differentiation and/or partial melting processes were not interrupted by the extensive fissuring and crustal spreading more typical in recently formed oceanic ridge systems.

## CHAPTER 4

# THRUST-FAULTED REGIONS

Some large tectonic features, because of their structural appearance and their locations at the boundaries between plates or in intraplate areas, are tentatively classified as thrust-faulted regions. These features have undergone both faulting and thrusting processes during their geological history; they are recognized as occurring in the southwestern Pacific Ocean (Macquarie Ridge) and in the Atlantic Ocean near 43°N (King's Trough) and near 38°N (Horseshoe Seamounts area). All three regions are unlike the main pattern of normal fracture zones and their relationships with ocean-ridge systems are ill-defined. Thrust-faulted regions are petrographically important features because they may expose deep-seated crustal and upper mantle sections on the ocean floor.

## MACQUARIE RIDGE

This ridge is about 1000 km long and extends from the southern tip of New Zealand down to the Antarctic-Pacific Ridge; it is a curved structure and led Summerhayes (1967) to suggest that it is a continuation of the West Pacific island arc system. To explain the present seismic activity of the Macquarie Ridge, Sykes (1967) and Banghar and Sykes (1969) proposed a combination of thrust faulting and right-lateral motion. It was also suggested that the ridge is different from the spreading type of Recent oceanic ridges (Sykes, 1967).

Because of its location in the southern latitude, the crustal material recovered from the Macquarie Ridge is of a controversial origin. Ice-rafted fragments with characteristic surface striation and rounding were found together with angular rocks considered to be "in situ". These rocks consist of serpentinites, gabbros and basalts quite similar to those recovered from Macquarie Island (Watkins and Gunn, 1971) (Tables 4-1, 4-2).

Macquarie Island located at about the middle of the north-south-trending ridge is aligned along its axis with a fault-bounded structure (Varne and Rubenach, 1972). The extrusives from the island consist essentially of basaltic rocks and breccias; the intrusives are made up of layered gabbros, serpentinized peridotites, and altered dikes. The layered gabbros include troctolites, plagioclase, dunite, anorthosite, norite and hypersthene gabbros (Varne et al., 1969). The dyke swarms are composed of a series of dolerites less than 2 m thick and only attain a maximum thickness of 3 m. The dyke swarms have been metamorphosed at different grades within the amphibolite facies. Mineral assemblages in the pillow-lavas indicate metamorphism under greenschist-facies conditions (Varne and Rubenach, 1972). Based on Ti and Zr abundances, the Macquarie Island rocks show the greatest overlap on the field of ocean-floor basalts (Varne and Rubenach, 1972). It is likely that Macquarie Island represents uplifted oceanic crust.

TABLE 4-1

Locations of rocks collected from thrust-faulted regions

| Sample No. | Latitude | Longitude | Depths (uncorrected meters) | Rock type |
|---|---|---|---|---|
| *Macquarie Ridge (Pacific Ocean)* | | | | |
| E-16-2 | 162°01'E | 51°01'S | 415 | olivine basalt peridotite, greenstone |
| E16-4 | 160°34'E | 52°22'S | 768 | olivine, basalt |
| E16-5 | 159°01'E | 54°31'S | 604 | basalt, scoria |
| E26-2 | 165°17.6'E | 46°43.8'S | 2599 | sandstone |
| | | | | |
| *Horseshoe Seamounts (North Atlantic Ocean)** | | | | |
| RC9-206 | 36°40.5'N | 11°04.4'W | 736 | phonolite, metagabbro |
| RC9-208 | 36°42'N | 11°07.3'W | 104 | phonolite, hornblendite |
| V4-28 | 36°45.8'N | 14°15.7'W | 1000 | alkali-basalt |
| V4-30 | 35°03'N | 12°58'W | 1280 | ankaramite |
| E9 | 36°40.4'N | 14°15'W | 230 | alkali-basalt |
| D7 | 36°40'N | 14°10.6'W | 880—1015 | alkali-basalt |
| D8 | 36°34'N | 11°37'W | 270—600 | serpentinite |
| 120 | 36°42.1'N | 11°22.4'W | 1711 | gabbro |
| | | | | |
| *Palmer Ridge (North Atlantic)*** | | | | |
| 5607 | 42°53.9'N | 20°11.8'W | 3290 | serpentinites |
| 5608 | 42°52.1'N | 20°16.8'W | 4206 | serpentinites |
| 5610 | 42°50.8'N | 20°16.8'W | 4645 | basalt, amphibolite |
| 5619 | 42°53.8'N | 20°16.6'W | 3200 | serpentinites |
| 5968 | 42°50.3'N | 20°12.2'W | 5329 | serpentinites |
| 5975 | 42°54.2'N | 20°12.8'W | 3204 | serpentinites |
| 5976 | 42°53.6'N | 20°15.7'W | 3429 | serpentinites |
| 5977 | 42°54.2'N | 20°15.9'W | 3200 | serpentinites |
| 5978 | 42°54.6'N | 20°11.2'W | 3486 | basalt, serpentinites |
| 5979 | 42°50.7'N | 20°16.2'W | 5300 | gabbros |
| 5981 | 42°51.5'N | 20°16.5'W | 4572 | gabbros |
| 5983 | 42°54.4'N | 20°13.4'W | 3359 | serpentinites |
| 5985 | 42°52.0'N | 20°12.4'W | 4596 | amphibolites, serpentinites |

*Watkins and Gunn (1971).
**Hékinian et al. (1973).
***Cann and Funnell (1967).

TABLE 4-2

Chemical analyses of serpentinites from the Macquarie Ridge, western Pacific Ocean (Watkins and Gunn, 1971) (recalculated anhydrous)

|  | E16-2-7 | E16-4-1 |
|---|---|---|
| $SiO_2$ (wt.%) | 55.10 | 55.30 |
| $Al_2O_3$ | 0.56 | 0.65 |
| Total Fe as $Fe_2O_3$ | 9.54 | 10.06 |
| MnO | 0.05 | 0.05 |
| MgO | 34.49 | 33.60 |
| CoO | 0.07 | 0.12 |
| $Na_2O$ | 0.11 | 0.12 |
| $K_2O$ | 0.01 | 0.01 |
| $TiO_2$ | 0.01 | 0.02 |

KING'S TROUGH

King's Trough is a double ridge system having a central depression located north of the Azores Plateau between 43 and 45°N in the Atlantic Ocean. It is about 100 km wide and about 400 km long, running NW-SE and its structural connection with the Mid-Atlantic Ridge is ill-defined. During the 1972 cruise of the R.V. "Chain", eight dredges recovered significant amounts of alkali-basalts, gabbros and diorite. Various degrees of seawater weathering and alteration by hydrothermalism were recognized in the rocks (Stebbins and Thompson, 1978). Metabasalts containing uralatized pyroxene were also encountered.

The basalts from King's Trough consist of plagioclase-rich and pyroxene-rich rocks. Both minerals occur in phenocrystal phases. The clinopyroxene is usually titaniferous ($TiO_2$ = 1—3%) (Stebbins and Thompson, 1978). Some of the basaltic rocks also contain mica as a secondary alteration product of olivine. Hornblende occurs as reaction rims around clinopyroxene and also as separate subhedral crystals (Stebbins and Thompson, 1978).

PALMER RIDGE

The Palmer Ridge is located at the southeastern end of King's Trough at around 42°N and 20°W. Two deeps, the Peake and Freen Deeps, are found flanked by small ridges to the north and south. They are separated by a 1300-fathom-high ridge called the Palmer Ridge (Cann and Funnell, 1967). Based on the age of the earliest sediments found and on the radiometric data from the freshest amphibolite recovered, the age of the Palmer Ridge is deduced to be around 60 Ma. According to Cann and Funnell (1967) and

TABLE 4-3

Chemical analyses of rocks from the King's Trough region (Palmer Ridge) in the North Atlantic (Cann, 1970a, 1971b)

|  | 5623-1 basalt | 5978-9 dolerite | 5979-10 gabbro | 5607-11 serpentinite | 5983-20 serpentinite |
|---|---|---|---|---|---|
| $SiO_2$ (wt.%) | 46.27 | 46.80 | 47.30 | 38.58 | 38.15 |
| $Al_2O_3$ | 18.43 | 16.75 | 17.82 | 1.53 | 1.36 |
| $Fe_2O_3$ | 8.47 | 5.97 | 4.45 | 8.18 | 7.34 |
| FeO | 3.10 | 2.54 | 2.76 | 0.51 | 0.14 |
| MnO | 0.13 | 0.21 | 0.15 | 0.10 | 0.08 |
| MgO | 3.97 | 7.91 | 8.12 | 36.26 | 36.42 |
| CaO | 11.56 | 12.55 | 14.33 | 0.08 | 0.43 |
| $Na_2O$ | 2.48 | 1.81 | 2.15 | 0.14 | 0.19 |
| $K_2O$ | 0.82 | 0.51 | 0.23 | 0.02 | 0.01 |
| $TiO_2$ | 1.18 | 0.78 | 0.29 | n.d. | n.d. |
| $P_2O_5$ | 0.15 | 0.17 | 0.10 | 0.06 | 0.06 |
| $H_2O^+$ | 1.81 | 1.63 | 1.67 | 11.02 | 11.21 |
| $H_2O^-$ | 1.66 | 2.70 | 0.80 | 2.18 | 3.52 |
| $Cr_2O_3$ | — | 0.05 | 0.09 | 0.43 | 0.43 |
| $CO_2$ | — | 0.16 | 0.23 | 0.20 | 0.15 |
| Total | 100.03 | 100.54 | 100.49 | 100.63 | 99.97 |
| Zr (ppm) | 60 | 52 | — | — | — |
| Ni | 75 | — | — | — | — |
| Cu | 120 | — | — | — | — |
| Zn | 120 | — | — | — | — |
| Rb | 8.5 | 9.5 | — | — | — |
| Sr | 80 | 110 | — | — | — |
| Nb | 2.5 | 7.0 | — | — | — |

Cann (1971b), the origin of the ridge is related to the formation of the Mid-Atlantic Ridge. The Palmer Ridge was created at the crest of the Mid-Atlantic Ridge and together with the rest of the ridge has been moved to its present location (650 km from the Mid-Atlantic Ridge axis) by a process of sea-floor spreading.

However, the association of ultramafics, metamorphics and mafic rocks suggests a more complex tectonic evolution for this ridge than normal ocean-floor spreading as understood for mid-oceanic ridges. The types of rocks encountered in the area of the Palmer Ridge consist of basalt-dolerites with a $K_2O$ content ($>0.5\%$) slightly higher than that encountered on the Mid-Atlantic Ridge, gabbros, serpentinites, and amphibolites (Table 4-3). The serpentinites were the most abundant rock types encountered in the dredge hauls which led Cann and Funnell (1967) to postulate that an intrusion of a serpentinite diapir is an important feature of the Palmer Ridge. However, it is not unlikely that the association of serpentinite and breccia could also be related to thrust faulting. The effect of dynamometamorphism is recognized in some crushed specimens. The readjustment of mineral assemblages from a higher to a lower metamorphic grade (retrogressive metamorphism) is commonly found in folded belts of Alpine and Franciscan (California) amphibolites, which were also the sites of important thrusting and faulting.

HORSESHOE SEAMOUNTS (GORRINGE BANK)

Located in the prolongation of an active seismic zone extending from the Mid-Atlantic Ridge close to the Azores and moving towards the Strait of Gibraltar (Gutenberg and Richter, 1954), the Horseshoe seamount chain marks a major boundary between the Eurasian and African plates in the North Atlantic. In this general area between 10 and 15°W the Gorringe Bank is a dominant structure showing a linear feature rising from about 2600 fathoms to 20 fathoms at its peaks (Purdy, 1975). Two peaks were recognized, the Gettysburg Seamount, and the Ormonde Seamount (Purdy, 1975).

The deep crustal structure was studied by the use of seismic refraction and gravity anomaly interpretations across the Gorringe Bank, and three density layers (2.6, 2.4 and 3.0 $g/cm^3$) were recognized (Purdy, 1975). The density range of 2.4—3.0 $g/cm^3$ may correspond to a crustal composition of basalt, gabbro and peridotite or a mixture of these. The gravity model requires that within the ridge there is a concentration of mass south of the topographic peak. Also the north-facing scarp of the ridge is steeper than the south-facing scarp. These considerations and the presence of a high-density body beneath the ridge led several authors (Le Pichon et al., 1970; Purdy, 1975) to suggest that such an upthrust is due to the overriding of the Gorringe Ridge on the Eurasian plate caused by the slow consumption of the oceanic crust sinking into the asthenosphere. This is similar to the mechanism suggested for the formation of island arcs. However, the major difference between island arcs and the "thrust-faulted" regions is that no trenches and no calc-alkaline volcanoes have been found in the latter areas.

Dredged samples from the Gorringe Bank revealed the presence of abundantly altered serpentinized peridotite (Hékinian et al., 1973) (Table 4-1). A borehole (DSDP; Honnorez and Fox, 1970) recovered about 0.9 m of gabbro (Table 4-1). This shows secondary replacement of amphiboles around fresh cores of clinopyroxenes. The freshest clinopyroxene ($Wo_{44}En_{41.4}Fs_{14.6}$) and the plagioclase ($Ab_{60.4}An_{39.5}Or_{0.05}$) are similar in composition to those found in Mid-Atlantic Ridge volcanics.

Igneous rocks were also recovered by dredging and coring (Gavasci et al., 1970) in the same area (Table 4-1). They consist of gabbros which have undergone various degrees of metamorphism, alkali basalts and phonalites (Gavasci et al., 1970; Hékinian et al., 1973) (Fig. 4-1). Rock samples were also collected from other structural features located in the vicinity of the

TABLE 4-4

Chemical analyses of rocks from the Horseshoe Seamounts area

| | Metagabbro Leg 13 120-8-7 | Serpentinite D8 | Ankaramite V4-30-A | Alkali basalt E9-A |
|---|---|---|---|---|
| $SiO_2$ (wt.%) | 48.61 | 37.90 | 40.10 | 44.20 |
| $Al_2O_3$ | 17.18 | 1.03 | 11.20 | 13.80 |
| $Fe_2O_3$ | 1.19 | 5.96 | 8.79 | 4.42 |
| FeO | 5.61 | 1.39 | 3.00 | 7.43 |
| MnO | 0.14 | 0.10 | 0.15 | 0.19 |
| MgO | 9.94 | 36.18 | 8.42 | 10.82 |
| CaO | 8.65 | 0.77 | 14.41 | 9.64 |
| $Na_2O$ | 3.05 | 0.34 | 2.34 | 2.80 |
| $K_2O$ | 0.20 | 0.06 | 1.36 | 0.96 |
| $TiO_2$ | 0.51 | 0.03 | 2.37 | 2.47 |
| $P_2O_5$ | 0.03 | 0.50 | 0.70 | 0.70 |
| $H_2O^-$ | 0.97 | | | |
| $H_2O^+$ | 3.62 | 13.85 | 6.42 | 1.63 |
| Total | 99.70 | 98.10 | 99.25 | 99.05 |

Sample from Site 120 (DSDP Leg 13) was taken in the Gorringe Bank (36°42′N, 11°23′W, at 1900 m); D8 is from the Gorringe Bank (36°34′N, 11°37′W, at 600—270 m); V4-30-A is from the Ampere Ridge (35°05′N, 12°58′W at 1280 m); E9-A is from the Josephine Bank (36°40′N, 14°15′W, at 230 m).

Gorringe Bank. Some are from the Josephine Bank and are alkali basalts (V4-28, E9; Hékinian et al., 1973); others, from the Ampere Ridge, located south of the Gorringe Ridge, are ankaramites (V4-30), Hékinian et al., 1973) (Fig. 4-1, Table 4-4).

A diving expedition using a submersible ("Cyana") permitted the collection of "in situ" rock samples, and detailed structural observations of the Gettysburg and the Ormonde Seamounts were made (Auzende et al., 1978) (Fig. 4-1). It was found that the Ormonde Seamount is a volcanic feature made up of a basalt-gabbro-phonolite sequence of rocks and that the Gettysburg Seamount consists essentially of serpentinized peridotites (Auzende et al., 1978). Structural observations recognized a faulted-plane zone between the neighboring seamounts. It was postulated (Auzende et al., 1978) that during a first phase of tectonic activity the Ormonde and Gettysburg Seamounts

Fig. 4-1. Topographic map of the Horseshoe Seamounts region showing the location of the rock samples collected during various oceanographic expeditions. The black rectangles show the sites of samples collected by diving (Auzende et al., 1978). 120 DSDP hole (Leg 13, Site 120; Honnorez and Fox, 1970). RC9-206 and RC9-208 represent rock samples recovered by piston coring (R.V. "R. Conrad" from Lamont-Doherty Geological Observatory). V4-28 and V4-30 are dredged samples (C44; R.V. "J. Charcot" from Centre Océanologique de Bretagne).

were subjected to a tilt motion oriented towards the northeast, and during a second phase, the Gettysburg Seamount was uplifted by a significant thrusting phenomenon which also resulted in an uprise (4 km thick) of upper mantle material.

The degree of weathering and the age of some volcanics from the Palmer Ridge and the Gorringe Bank suggest that these were formed on the Mid-Atlantic Ridge. However, it is not excluded that additional intraplate volcanism of a more fractionated nature might have also occurred. It is interesting to notice that the volcanics erupted on these thrust-faulted structures correspond to the type of material exposed in their associated Mid-Oceanic Ridge segments. The continuation of the King's Trough region will intersect the Mid-Atlantic Ridge at 45°N where basalts with alkaline affinities were found (Chapter 2). Also, the Gorringe Bank is on the same east-west tectonic line as the Azores Plateau and both regions are sites of alkali-rich volcanic events.

CHAPTER 5

# THE OCEAN BASINS AND SEAMOUNTS

Compared to the mid-ocean ridge systems and fracture zones, the history of the ocean basins and the structure of their crustal sections are relatively less known at the present time. From studies of plate tectonics it is known that the ocean basins are bounded at their edges by earthquake belts and zones of high heat flow. The edge of an ocean basin may be quite active as far as earthquakes and volcanism are concerned, but it could also be inactive, such as at the foot of continental slopes. Within the ocean basins, recent volcanic activity may also take place in the form of isolated and elevated volcanoes called seamounts, and small volcanic hills which are relatively abundant on the ocean floor of the Pacific. In addition to the conventional dredging of submarine volcanoes, direct sampling of the crust under the ocean basins has been made possible thanks to the Deep Sea Drilling Project which has permitted the recovery of basaltic basement under various thicknesses of sediment cover.

So far, all the major oceans have been sporadically drilled in order to reach basement underneath the sediment and to assess the age of the crust at the various sites (Table 5-1, Fig. 2-2). Underneath the sediment cover, the topography of the oceanic crust is relatively rugged and it is often likely that at least the first hundred meters is made up of talus debris forming breccia complexes which are sometimes difficult to recognize in the drilled sections. The available petrological data from selected boreholes in the ocean basins gives a general idea about the composition of basement rocks and their relationship to recent accreting plate boundary regions.

## INDIAN OCEAN

During Leg 22 of the DSDP, three holes (Sites 211, 212, 213) were successfully drilled in the northeastern Indian Ocean Basin, and basement rock was recovered (Table 5-1, Fig. 2-2). A variety of rock types were found at these three sites. Site 211, close to the Sumatra-Java Trench, consists of alkali basalts and alkali dolerites of an intrusive nature (Hékinian, 1974a). The basalt from this site also has a high content of LIL elements and a highly fractionated REE distribution pattern (Frey et al., 1977). Site 212 recovered samples with a similar age (80—100 Ma old) to rocks recovered at Site 211, but which were entirely made up of metamorphosed basalts. Site 213, drilled on the northwestern side of the Wharton Basin (Fig. 2-2, Table 5-1) on crust about 55 Ma old, recovered weathered basalt with some preserved glassy margins of pillow flows. The content of transitional metals and the low amount of LIL elements suggest that the rocks from this site are similar

TABLE 5-1

Locations of basement rocks drilled during the Deep Sea Drilling Project from oceanic basins around the world (the compilation of the data is up to date to 1976)

| DSDP Leg No. | Site No. | Latitude | Longitude | Province | Rock type | Age of cr (Ma) |
|---|---|---|---|---|---|---|
| 2 | 10 | 29° 56.06′N | 44° 44.80′W | northwest Atlantic Ocean | | 85 (F) |
| 4 | 23 | 6° 08.75′S | 31° 02.60′W | Brazil | basalt | 70—140 |
| | 24 | 6° 16.30′S | 30° 53.53′W | | basalt | — |
| 11 | 105 | 34° 53.72′N | 69° 10.40′W | northwest Atlantic Ocean | basalt | 160 (F) |
| 16 | 160 | 11° 42.21′N | 130° 52.81′W | northeast Pacific Ocean | basalt | 35 (F) |
| | 161 | 10° 50.25′N | 139° 57.21′W | | basalt | 45 (F) |
| | 162 | 14° 52.19′N | 140° 02.61′W | | basalt | 49 |
| | 163 | 11° 14.66′N | 150° 17.52′W | | basalt | 48—57 |
| 18 | 172 | 31° 32.23′N | 133° 22.36′W | northeast Pacific Ocean | basalt | 35—38 |
| | 173 | 56° 57.28′N | 147° 07.86′W | | basalt | 10—20 |
| 20 | 197 | 30° 17.44′N | 147° 40.46′E | West Pacific | basalt | 140 (F) |
| 22 | 211 | 09° 46.53′S | 102° 41.95′E | Wharton Basin | alkali basalt | 72—80 |
| | 212 | 19° 11.34′S | 99° 17.84′E | | metabasalt | 85—110 |
| | 213 | 19° 12.17′S | 93° 53.77′E | | basalt | 55—57 |
| 23 | 215 | 8° 07.30′S | 86° 47.50′E | central Indian Ocean | basalt | 58—60 |
| | 221 | 7° 58.18′N | 68° 24.37′E | | basalt | 46 (F) |
| 24 | 235 | 03° 14.06′N | 52° 41.64′E | Somali Basin | basalt | 70 (F) |
| | 236 | 01° 40.62′S | 57° 38.85′E | | basalt serpentinite | 60 (F) |
| 26 | 250 | 33° 27.74′S | 39° 22.15′E | Mozambique Basin | basalt | 89 ± 6 |
| | 256 | 23° 27.35′S | 100° 46.46′E | Wharton Basin | basalt | 92 ± 4 |
| | 257 | 30° 59.16′S | 108° 20.99′E | | basalt | 177—190 |
| 27 | 259 | 29° 37.05′S | 112° 41.78′E | eastern equatorial | basalt | 100 (F) |
| | 260 | 16° 18.67′S | 110° 17.92′E | Pacific Ocean | basalt | 105 (F) |
| | 261 | 12° 56.83′S | 117° 53.86′E | | basalt | 150 (F) |
| 34 | 319A | 13° 01.04′S | 101° 31.46′W | Nazca plate | basalt | 15—18 |
| | 320 | 9° 00.40′S | 83° 31.80′W | | basalt | 30 (F) |
| | 321 | 12° 01.29′S | 81° 54.24′W | | basalt | 40 (F) |
| 39 | 355 | 15° 42.59′S | 30° 36.03′S | Brazil Basin | basalt | 78 ± 9 |

Water depth is less than 4000 m.
*The crustal age was determined by fission tracks, K-Ar methods on rocks and by fossil-age methods (F) on overlying sediments above basement rocks.

to the Mid-Indian spreading ridge volcanics. Other DSDP holes in the north-eastern Indian Ocean Basin were drilled during Leg 27 and the samples are

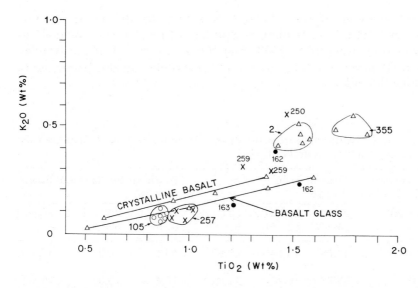

Fig. 5-1. K$_2$O-TiO$_2$ (wt.%) diagram showing the plots of basaltic rocks collected from the oceanic basins in the Atlantic (△), Indian (×), and Pacific (•) Oceans. For purposes of comparison, the glassy and bulk-rock crystalline trends of the FAMOUS area basalts are shown. DSDP Site 2 (Leg 10) and Site 163 (Leg 16) represent analyses of glass (Frey et al., 1974; Yeats et al., 1973). Sites 162 (Leg 16), 250, 257 (Leg 26), 259 (Leg 27), and 355 (Leg 39) are plotted from the data of Yeats et al. (1973), Kempe (1974), Robinson and Whitford (1974), and Fodor (1977).

comparable to rocks recovered from Site 213. Sites 257, 259 and 261, drilled off the continental slope of Australia, recovered rocks which are considered to be typical of a ridge tholeiite type of basalt with a relatively low amount of K$_2$O (generally <0.5%) (Fig. 5-1). Sites 260 and 261 were also drilled into tholeiites with a LIL element abundance and REE distribution intermediate between island tholeiites and LIL-depleted tholeiites. It is also noticed that the ridge type and intermediate types of tholeiites could occur within the same borehole and hence cannot be related to regional magmatic differences.

During Leg 26 of the "Glomar Challenger" in the Indian Ocean, the basement was reached in the Mozambique Basin (Site 250) at a depth of 5119 m, and about 18 m of basalt, composed of olivine (12%), pyroxene (37%), plagioclase (31%), and iron ore (8%), were sampled (Kempe, 1974) (Table 5-2). One analysis of this sample shows a K$_2$O content of 0.57% (Table 5-1). During Leg 24 in the Somali Basin, two holes penetrated the basement at Sites 235 and 236 (C. G. Engel and Fisher, 1975). Rocks from Site 235 consist of a plagioclase-rich basalt with about 40% of plagioclase phenocrysts (C. G. Engel and Fisher, 1975). At Site 236, the basalt is also porphyritic with augite and plagioclase; the augite phenocrysts comprise about 3% of the rock and plagioclase about 10% (C. G. Engel and Fisher, 1975). The trace element analyses of these basalts are similar to Mid-Indian Oceanic Ridge volcanics (Table 5-2). Drilling at Site 221 of Leg 23 penetrated about 10 m of basalts at a depth of 4650 m at the southern end of the Arabian Abyssal Plain (Whitmarsh et al., 1974) and these rocks can also be related to spreading ridge tholeiites.

TABLE 5-2

Chemical analyses of basaltic rocks from the Indian Ocean basins (Leg 26 data are from Kempe, 1974; Leg 28 data are from Ford, 1975)

| | Mozambique Basin (Leg 26) | Wharton Basin (Leg 26) | | | Southeast Indian Ocean Basin (Leg 28) | | |
|---|---|---|---|---|---|---|---|
| | 250A-26-6 45—47 | 256-9-3 45—47 | 257-11-2 82—84 | 257-17-1 91—93 | 267-6-1-1 0—6 | 267-6-1-25 57—61 | 267-6cc |
| $SiO_2$ (wt.%) | 47.72 | 49.59 | 49.70 | 47.97 | 50.1 | 50.8 | 50.5 |
| $Al_2O_3$ | 16.33 | 13.31 | 16.42 | 14.32 | 15.4 | 15.3 | 15.4 |
| $Fe_2O_3$ | 1.70 | 3.06 | 1.71 | 3.70 | 4.3 | 3.8 | 3.3 |
| FeO | 6.43 | 10.21 | 5.02 | 5.16 | 5.3 | 5.2 | 5.7 |
| MnO | 0.14 | 0.23 | 0.28 | 0.15 | — | — | — |
| MgO | 8.76 | 6.06 | 7.97 | 8.08 | 7.1 | 7.4 | 7.4 |
| CaO | 10.86 | 10.43 | 11.76 | 12.41 | 11.7 | 11.6 | 11.9 |
| $Na_2O$ | 2.72 | 2.66 | 2.39 | 2.04 | 2.3 | 2.2 | 2.3 |
| $K_2O$ | 0.57 | 0.24 | 0.12 | 0.45 | 0.54 | 0.40 | 0.36 |
| $TiO_2$ | 1.46 | 2.50 | 1.19 | 0.88 | 1.3 | 1.3 | 1.3 |
| $P_2O_5$ | 0.25 | 0.25 | 0.20 | 0.11 | 0.18 | 0.16 | 0.15 |
| $H_2O$ | 3.04 | 1.65 | 2.88 | 3.23 | 1.40 | 1.50 | 1.40 |
| Ni (ppm) | 110 | 130 | 120 | 150 | — | — | — |
| Cu | 12 | 17 | 17 | 88 | — | — | — |
| Zn | 70 | 130 | 90 | 200 | — | — | — |
| Zr | 200 | 200 | 150 | 75 | — | — | — |

ATLANTIC OCEAN

During DSDP Leg 39 (Site 355), a core of basaltic rocks about 7.5 m long was recovered at a sub-bottom depth of 449 m in the Brazilian Basin (Fig. 2-2). The basalt recovered is generally aphyric, with some agglomerations of plagioclase, olivine, and clinopyroxene. The olivine occurs as relics of altered grains and the general texture of the rock is subophitic. The age of the sediment above the basalt is Late Cretaceous (Perch-Nielsen et al., 1975). Jurassic-age basalts drilled during Leg 11 (Site 105) in the northwestern Atlantic Basin showed a range of compositional variability similar to basalts erupted on the Mid-Atlantic Ridge (Fig. 5-1). During a deep drilling project

TABLE 5-3

Partial chemical analyses of basaltic rocks from the Pacific (Legs 16 and 18; Yeats et al., 1973) and Atlantic Ocean Basins (Leg 2, Frey et al., 1974)

| | Leg 16 161A 15cc | Leg 16 162 17cc | Leg 16 163 (ave. of 6) | Leg 18 172 | Leg 18 178 | Leg 2 10-19 (ave. of 5) glasses | Leg 2 10-20-1 149—150 |
|---|---|---|---|---|---|---|---|
| $SiO_2$ (wt.%) | 50.90 | 49.00 | 49.77 | 51.0 | 45.8 | 50.14 | 47.42 |
| $Al_2O_3$ | 13.87 | 15.73 | 14.85 | 13.3 | 14.9 | 16.14 | 14.90 |
| $Fe_2O_3$ | 17.79 | 9.73 | 10.19 | 3.1 | 7.3 | — | — |
| FeO | — | — | — | 10.95 | 5.5 | 8.75 | 7.74 |
| MgO | 6.74 | 5.43 | 7.64 | 5.4 | 8.0 | 7.71 | 7.31 |
| CaO | 11.17 | 12.16 | 11.60 | 9.3 | 9.0 | 11.98 | 11.65 |
| $Na_2O$ | 2.54 | 2.58 | 2.29 | 2.4 | 2.7 | 2.56 | 2.49 |
| $K_2O$ | 0.24 | 0.40 | 0.12 | 0.35 | 0.37 | 0.46 | 1.13 |
| $TiO_2$ | 1.54 | 1.42 | 1.22 | 2.00 | 1.6 | 1.51 | 1.37 |
| Cr | 0.34 | 0.40 | 0.37 | — | — | — | — |
| $P_2O_5$ | — | — | — | 0.44 | 0.44 | — | — |
| $H_2O$ | 0.55 | 1.37 | 0.82 | 0.21 | 1.8 | — | — |
| Total | 99.81 | 99.00 | 99.94 | 98.65 | 99.71 | 99.25 | 99.72 |

*The samples are crystalline rocks unless otherwise indicated.
— = not determined.

in the northwestern Atlantic Basin (Legs 51, 52 and 53, Sites 417 and 418) Site 418 penetrated 868 m into the crust, of which 548 m were drilled into basement rocks (Scientific Party, DSDP Leg 51, 1980a). The rocks recovered during this exceptional "deep hole" project are Cretaceous basalts which also show a similarity to those from the Mid-Atlantic Ridge in the compositional range of their glassy margins.


PACIFIC OCEAN

The northeastern Pacific Basin has been the site of various deep-sea dril-ling operations (Fig. 2-2). The basaltic rocks recovered consist of typical tholeiite materials which are rather uniform in composition throughout the various basement depths reached (Table 5-3). The bulk-rock analyses of selected basalts from the Pacific Basins, compared to those of the East Pacific Rise, show a similar trend and range of variability. Both the $TiO_2$ contents (1—3%) and the FeO/MgO ratios (1—3) of the basalts collected at various latitudes along the East Pacific Rise are very similar to those from the adjacent basins (Fig. 5-2).

The Nazca plate region located in the southeastern Pacific Ocean is named

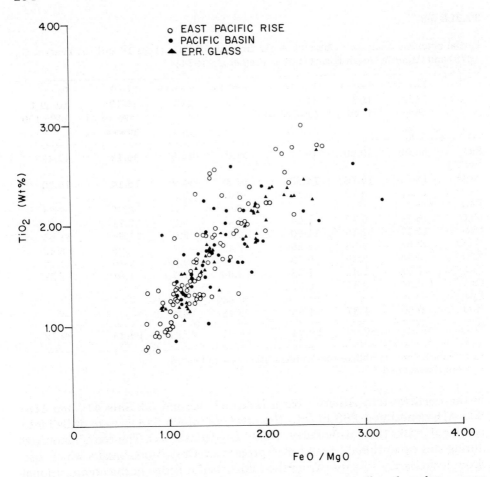

Fig. 5-2. TiO$_2$—FeO/MgO (total Fe as FeO) binary diagram of bulk-rock analyses versus pillow flow glassy margin analyses for rocks recovered from various structural features of the eastern Pacific ocean.

after the Nazca Ridge and is considered to be a rigid plate bounded at the north by the Galapagos Ridge/Carnegie Ridge system, at the west by the East Pacific Rise, and to the south by the Chile Rise. Since most of the samples from this region were collected from small hills in abyssal plains from the Nazca plate, it will be treated here as an ocean basin region. The peculiarity of the Nazca plate region is the discovery of metalliferous deposits. Sediment enriched in Fe and Mn was found in the Bauer Deep, located on the northeastern corner of the plate, and has led many marine geologists to become interested in this particular region.

The tectonic history of the Nazca plate is a complex one since this region comprises most types of the oceanic structures encountered elsewhere in the ocean. For example, there is a fossil ridge (Galapagos Rise) and an aseismic ridge (Nazca Ridge) within the complex. A major tectonic feature, the Easter Island Fracture Zone, divides the Nazca plate into two portions, with a shallower depth to the north and deeper zones to the south of 25°S (Fig. 2-20). The maximum topographic expression of the fracture zone occurs between Easter Island and the intersection of the zone with the Nazca Ridge where no seismic activity was observed (Fig. 2-20).

An active oceanic ridge called the Galapagos Rise was recognized (Menard, 1964); it extends north-south, roughly parallel with the East Pacific Rise. As in other mid-oceanic ridges, it is transected by transform faults. The most intensive petrologic work on the Nazca plate was carried out during the DSDP operation in 1974 (Leg 24) when the oceanic crust was sampled at three different sites. Thick units (15 m) and massive flows of basaltic rocks of diabasic texture were drilled at two sites: one on the flank of the Galapagos Rise (Site 319), and one in a basin area near the Peru-Chile Trench (Site 321). The other site (320) consists of a series of pillow lavas composed of glassy basalts (Scientific Party, DSDP Leg 34, 1974). Another sample of fresh glassy basalt was found in a sediment piston core (RC8-98) collected from the Galapagos Rise at a depth of 3768 m (Hékinian, 1971a). The Leg 34 basalts are remarkable for their paucity or lack of olivine, and the abundance of clinopyroxene microphenocrysts and Fe-oxide minerals. Geochemical studies of the basalts from the Nazca plate show that they consist of tholeiites which are typical of recent mid-oceanic ridge erupted flows in that they are depleted in light REE when compared to chondrites, and have low abundances of K, Sr, Ba, and Zr in relation to island and continental basalts (G. Thompson et al., 1976). Samples from Sites 319 (DSDP) and RC8-98 from the Galapagos Rise are similar in their low $K_2O$ content (<0.2%) and low total Fe content (<10%) to the crustal composition of rocks from the adjacent East Pacific Rise spreading center.

Comparing recent spreading ridge volcanics with basement rocks drilled in intraplate regions associated with the various oceanic basins, it was noticed that a large-scale variation trend exists within the basaltic rocks from oceanic basins. Basalts which range widely in composition from relatively primitive to highly fractionated compositions are found in the oceanic basins as well as in recent spreading ridges. For example, samples from Leg 11, Leg 26 (Site 257) and Leg 27 (Sites 259 and 261) are depleted in light REE and in LIL elements, while others, such as those from Site 215 (Leg 22) located in the central Indian Basin, or those from Site 10 (Leg 2) off the New England seamount chain area are enriched in $K_2O$ (>0.4), in LIL, in the FeO/MgO ratio (>1.10), and in light REE (Bryan et al., 1979) (Table 5-3). These rocks are also characterized by a higher La/Sm ratio (1—3), and a higher La/Yb ratio (1—7) (Sites 10 and 215) (Bryan et al., 1979). However, similar high values for these ratios were found in fractionated basalts from recent spreading ridges in the FAMOUS area near 36°N (basaltic glass; Bryan and Moore,

1977) and at 45°N in the Atlantic Ocean (Bryan et al., 1979). Volcanics from recent spreading ridges and from basins show a wide variability range in FeO/MgO (0.45—2), Ca/Sm (0.68—5), and La/Yb ratios (0.38—7).

Most of the basalts recovered during the DSDP show a similar variability range of $K_2O$ (0.10—0.60%) and $TiO_2$ (0.9—2.0%) contents, when compared to those encountered on recent accreting plate-boundary regions (Fig. 5-1). However, there is still a very limited amount of data for the volcanics from oceanic basins, as far as drilling sites are concerned, which could enable a conclusive comparison between the various structural provinces.

## ABYSSAL HILLS OF THE PACIFIC

The abyssal hills are small-scale isolated edifices, usually less than 500 m high above the floor, and surrounded by a sediment cover of more than 100 m thick. Only a few samples from abyssal hill regions have been described in the literature.

Several kilograms of a dredge haul near 125°W, 32°N were taken at a depth of 4400 m (Luyendyk and Engel, 1969). The structure of the sampled area is that of a 100-m-high linear scarp. Abyssal hill provinces of the southeast Pacific regions were also sampled, either by dredging or coring operations (Hékinian, 1974a). Usually the samples collected from the southeastern Pacific basins are taken at depths greater than 3500 m.

The chemical differences found among the abyssal hill rocks were thought to be due to the weathering action by seawater and hence strictly time-dependent. Increases of the $Fe_2O_3/FeO$ ratio and the $K_2O$ content of up to 1 and 0.8%, respectively, are characteristic factors of an alteration effect (see Fig. 8-5).

Subbarao (1972), however, has shown from his study on strontium isotope ratios that the same abyssal hill basalt could have had a different origin from that of the adjacent recently active East Pacific Rise. The East Pacific Rise basalts have a $^{87}Sr/^{86}Sr$ ratio of 0.7026, while those from the abyssal hill have values in the range 0.7025—0.7037 (Subbarao, 1972) (see Fig. 2-21). These differences were thought to be due to a different source material associated with the two provinces. Subbarao (1973) concluded that the source for the abyssal hill basalt in this region is not different from that of volcanic islands and seamounts in general.

## EASTERN PACIFIC SEAMOUNTS

Elevated and isolated volcanic edifices, called seamounts, are most preponderant in the Pacific Ocean. Petrological investigations of the Pacific seamounts have shown that they are built up by an accumulation of mainly alkali types of basalts and late fractionated material such as mugearites and trachytes (A. E. J. Engel and Engel, 1964). Most of the seamounts studied

Fig. 5-3. A. Bottom photograph showing pillow flow basalt lying near the top of the Explorer Seamount (northeastern Pacific Ocean) at about 630 fathoms. B. The dredged material (RC10-DR3) from the area was taken at about 790 fathoms and consisted of freshly fragmented tholeiitic basalts.

TABLE 5-4

Chemical analyses of alkali basalts and tholeiites from eastern Pacific seamounts

|  | GCM 1 | GCM 2 | PV 50 | PV 71 | RC10-251 | RC10-RD3 | RC10-RD3 glass |
|---|---|---|---|---|---|---|---|
| $SiO_2$ (wt.%) | 41.62 | 47.95 | 45.33 | 44.93 | 50.00 | 47.00 | 46.60 |
| $Al_2O_3$ | 16.42 | 16.56 | 15.50 | 15.99 | 14.00 | 18.40 | 20.60 |
| $Fe_2O_3$ | 9.94 | 10.39 | 1.81 | 4.11 | 5.70 | 1.40 | 0.87 |
| FeO | 3.47 | 3.02 | 10.64 | 6.15 | 4.90 | 7.80 | 8.30 |
| MnO | 0.14 | 0.14 | 0.23 | 0.30 | 0.15 | 0.17 | 0.16 |
| MgO | 1.65 | 1.62 | 6.77 | 7.51 | 5.50 | 7.90 | 7.80 |
| CrO | 8.13 | 7.96 | 8.50 | 9.38 | 11.30 | 11.00 | 10.50 |
| $Na_2O$ | 4.79 | 4.85 | 4.30 | 3.82 | 2.60 | 2.80 | 2.90 |
| $K_2O$ | 1.81 | 1.82 | 2.36 | 2.21 | 0.75 | 0.17 | 0.04 |
| $TiO_2$ | 2.42 | 2.24 | 3.53 | 3.64 | 1.60 | 1.30 | 1.30 |
| $P_2O_5$ | 1.84 | 1.43 | 0.73 | 0.76 | 0.34 | 0.13 | 0.15 |
| $H_2O$ | 1.66 | 1.73 | 0.21 | 1.16 | 2.50 | 0.62 | 0.65 |
| Total | 99.88 | 99.70 | 99.90 | 99.95 | 99.33 | 98.69 | 99.87 |

*Norms*

|  | GCM 1 | GCM 2 | PV 50 | PV 71 | RC10-251 | RC10-RD3 | RC10-RD3 glass |
|---|---|---|---|---|---|---|---|
| Q | 0.37 | — | — | — | 7.05 | — | — |
| Ne | — | — | 12.39 | 8.89 | 4.43 | — | 0.39 |
| Or | 10.69 | 10.75 | 13.94 | 13.05 | 22.00 | 1.00 | 0.24 |
| Ab | 40.53 | 40.93 | 13.50 | 15.90 | 24.31 | 23.69 | 23.82 |
| An | 17.95 | 18.04 | 16.02 | 19.95 | 12.32 | 37.14 | 43.08 |
| Di | 8.05 | 8.70 | 17.51 | 17.04 | 23.30 | 13.48 | 6.56 |
| Ny | 0.37 | — | — | — | 4.64 | 0.65 | — |
| Ol | — | — | 15.29 | 9.30 | — | 17.30 | 21.07 |
| Mt | 4.62 | 3.69 | 2.62 | 5.95 | 8.26 | 2.03 | 1.26 |
| Ilm | 4.59 | 4.25 | 6.70 | 6.91 | 3.03 | 2.47 | 2.47 |
| Ap | 4.34 | 3.37 | 1.72 | 1.79 | 0.80 | 0.30 | 0.35 |
| Hm | 6.74 | 7.83 | — | — | — | — | — |

GCM 1, GCM 2 are samples from Giacomini Seamount at 56° 24'N, 146° 34'W (Forbes et al., 1969); PV 50 and PV 71 are from Hodgkins Bank Pinnacles at 53° 15'N, 136° 46'W and an unnamed seamount at 27° 42'N, 119° 17'W, respectively (C. G. Engel and Engel, 1963); RC10-251, sampled at 05° 02'N, 81° 24'W, and RC10-RD3, at 49° 03.5'N, 130° 57'W, are from the Explorer Seamount.

are located in the Gulf of Alaska south of the Aleutian Trench (Giacomini Seamount) and in the Eastern Pacific Ocean (Baja California province). Major and trace element studies on seamount rocks suggest a similarity to island volcanics. The total alkali content (3—7%), the $SiO_2$ content (47—64%), and the range of variation in the $^{87}Sr/^{86}Sr$ isotopic ratios of the seamount volcanics (0.7028—0.7054) is also similar to that of alkali-island types of volcanism (Subbarao et al., 1973) (Fig. 2-27). Hence, it is most likely that elevated volcanoes found at a distance from recent spreading centers could have tapped material having a different composition than that found on the

East Pacific Rise. G. A. Macdonald and Katsura (1964) used Tilley's (1950) method of plotting total alkalies against silica to distinguish Hawaiian basalts from tholeiites. The same diagram could be applied to differentiate between the seamount basalts and the deeper-seated basalt of the East Pacific Rise.

*Explorer Seamount.* Not all the Pacific seamounts are made up of alkali basalts or late fractionated material; some made up of tholeiitic basalts are also known to exist. The Explorer Seamount located in the northeast Pacific near 49°N contains tholeiitic rocks with a high alumina content (18—20%). The high alumina values of both the glassy margin and the most crystalline portion of a pillow lava are not justified by the presence of plagioclase phenocrysts. Indeed, the early-formed plagioclase does not exceed 2% of the bulk rock. Hence it is likely that we are here dealing with a high-alumina basalt sensu stricto as defined by Kuno (1960) for the Japanese series (Chapter 1). The Explorer Seamount rock sampled consists of relatively fresh pillow lava with characteristic radial jointings and glassy margins (Fig. 5-3). The rocks are vesiculated and generally aphyric in appearance. Their major mineral phases are plagioclase laths, olivine, and very little clinopyroxene.

The high alumina content of these rocks is not associated with a high CaO content (<12%) (Table 5-4). The $TiO_2$ content (1.3%), the MgO content (7.8—8.3%) and the $K_2O$ content (0.04—0.19%) are similar to other tholeiitic rocks recovered from the East Pacific Rise. The rocks from the Explorer Seamount are relatively depleted in light REE. Their Sr content (190 ppm) and K/Rb values (980; Kay et al., 1970) are similar to those of oceanic ridge basalts (K/Rb = 700—1536; Subbarao and Hékinian, 1978).

*Cobb Seamount.* Cobb Seamount is located about 500 km off the coast of Washington. It has 2700 m of relief, is 31 km wide, and rises to within 34 m of the surface. The mineralogy of two samples collected from Cobb Seamount shows that they are pyroxene-rich basalts with an intermediate plagioclase ($An_{65}$). In addition, chlorite and carbonate are present in these rocks, which suggests a certain degree of alteration. K-Ar age dating shows a variable age of 1.7 ± 0.3 Ma (Dymond and Deffeyes, 1968). Buddinger and Enbysk (1967) reported an age of 27 Ma for the same seamount. Dymond and Deffeyes (1968) conclude that an age of less than 3.5 Ma does not conflict with a crustal spreading model for the area. C.G. Engel and Engel (1963) have also described the occurrence of a pyroxene-rich basalt from Cobb Seamount. This basalt consists of about 60% pyroxene, 32.4% plagioclase, and 5.8% olivine. The chemistry of this rock shows a relatively high $TiO_2$ content (2.27%) and a low $K_2O$ content (0.26%).

*Kodiak Seamount.* This seamount lies about 167 km southeast of Kodiak Island in the Gulf of Alaska (Forbes and Hoskin, 1969). Kodiak Seamount was surveyed by Menard and Dietz (1951), who described it as a flat-topped elevated edifice 2017 m high. Its position, located at the break of the continental slope and the continuation of the Aleutian Trench, is of particular interest as it corresponds to a zone of transition between continental and oceanic crust. Dredged samples from the flank and the summit of Kodiak Seamount brought up trachytic rocks intermixed with glacial erratics (Forbes and Hoskin, 1969). These rocks are characterized by a high $SiO_2$ content (56—59%), a high $Na_2O$ content (6—7%), and a high $K_2O$ content (3—5%) (Forbes and Hoskin, 1969).

*Giacomini Seamount.* This elevated structure is also located in the Gulf of Alaska southeast of Kodiak Island. Two angular blocks of alkali basalts were dredged from the summit (790 m) of Giacomini Seamount. However, the majority of the dredge hauls contained erratics of variable rock types such as metamorphic and sedimentary rocks. The alkali basalts consist of plagioclase phenocrysts (calcic andesine), and olivine pseudomorphs. The groundmass is composed of augite, plagioclase, K-feldspar, and opaques (Forbes et al., 1969). The texture is pilotaxitic with a subparallel orientation of the feldspar. The chemistry of these basalts consists of relatively high $TiO_2$ (2—3%), low $K_2O$ (1.8%), and moderately high $Al_2O_3$ (16—17%) contents (Forbes et al., 1969) (Table 5-4).

*Hodgkins Bank Pinnacles.* The Gulf of Alaska near the Canadian continental shelf is also the location of the Hodgkins Bank Pinnacles (Fig. 2-20). A fragment of scoriaceous basalt was dredged at a depth of 85 m. This is a more or less glassy rock (52.6% glass) with a few plagioclase phenocrysts. Its alkaline affinity is shown by its high $K_2O$ (2.36%) and high $Na_2O$ (4.30%) contents (Table 5-4). The total amount of vesicles comprises 28% of the bulk rock. The surface structure indicates an origin of pillow lava flows (C. G. Engel and Engel, 1963).

*Baja California seamounts.* In this region there are about one hundred large submarine volcanoes and many more smaller ones (Menard, 1964). Most of the basalts dredged from this area are pillow lavas. In addition, large amounts of ash material containing occasional rounded fragments of peridotite and pyroxenite were dredged (A. E. J. Engel and Engel, 1964) (see Fig. 1-1). These ultramafics represent volcanic ejecta brought to the surface by the explosive eruption of volcanic ash (A. E. J. Engel and Engel, 1964). The basalts found in the area are almost all alkaline rocks with a total alkali content higher than 3%. Bomb-shaped fragments (PV 71) and vesicular basalts enriched in plagioclase phenocrysts were found at depths varying between 670 and 1000 m. The $Al_2O_3$ content of these rocks varies between 16 and 19% (A. E. J. Engel and Engel, 1964). The molecular percent of anorthite in the plagioclase phenocryst is about $An_{68}$—$An_{76}$ (samples PV 77, PV 71, and D 6; C. G. Engel and Engel, 1963; A. E. J. Engel and Engel,

1964). Other alkali basalts dredged at shallower depths (402 m) were found on the Henderson Seamount near 26°N in the area of the Baja California seamount zone (Poldervaart and Green, 1965). One analysed rock from this seamount shows a high $K_2O$ content (2.69%) and a high $Na_2O$ content (3.41%) while the $Al_2O_3$ content is lower (14.74%) than that of the other alkali basalts from the same general area (Table 5-4).

*Mathematicians Seamounts.* This chain of seamounts is located at 12—15°N latitude in the central eastern Pacific Ocean and it runs parallel to the East Pacific Rise (Fig. 2-32). Geophysical investigations in the area of the Mathematicians Seamounts suggested that the ridge crest jumped eastward forming the present-day East Pacific Rise and leaving the Mathematicians seamount chain as an old spreading center. In the model of Sclater et al. (1971), the age of the seafloor on which the Mathematicians Seamounts rest would be about 5 Ma. However, for the same area the data presented by Herron (1972) gave an age of 10—15 Ma for the axial shift of the East Pacific Rise from the line of the Mathematicians Seamounts. DSDP preliminary results of the K-Ar isotopic age analyses performed on a trachytic rock (station 241) from the Mathematicians Seamounts area gave a value of less than 1 Ma (Leg 22). If the model of Sclater et al. (1971) is correct, it is likely that volcanism of a sort different from that encountered on the East Pacific Rise has occurred along the Mathematicians seamount chain.

Vesicular fragments of trachytic composition were recovered from a dredge haul and from a piston core taken at a depth of 805 m near the top of an unnamed seamount (15°23'N, 110°56'W). This high volcanic cone rising about 2000 m above the surrounding floor of the ocean is part of the Mathematicians seamount chain. Bottom photographs taken near the sampling area show shattered rock debris (<20 cm in diameter) with sharp edges, scattered on a silty bottom (Fig. 5-4). The structural features of the rocks and their distribution indicate that they are the products of an explosive eruption produced in relatively shallow water. The piston core recovered from the same area as the dredge contains angular pebbles (4—5 cm in diameter) interbedded with a mixture of sand and clay fraction. The sandy fraction consists of glass, feldspar and manganese granules. The clay material is made up of foraminiferal tests, sponge spicules and phyllosilicates. The rock fragments recovered in the dredge haul consist of specimens varying in size between 5 and 15 cm in diameter. Chemical analyses of trachytes from the Mathematicians Seamounts are given in Chapter 1, Table 1-15.

The relative amount of $SiO_2$ content and the differentiation index (D.I.) of Thorton and Tuttle (1960) are used to distinguish the field of tholeiites from the ridge crest and the field of alkali basalt suites from seamounts of intraplate regions (Fig. 5-5). The Hawaiian tholeiite and alkali basalt series (Kuno et al., 1957; G. A. Macdonald and Katsura, 1964) and the alkali basalt—trachyte association of Clarion Island (Bryan, 1967) are plotted

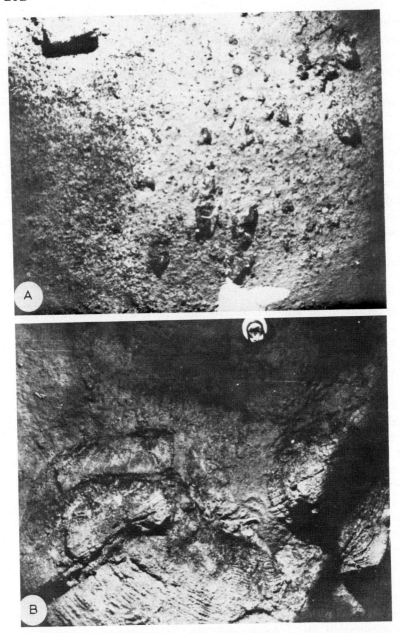

Fig. 5-4. A. Bottom photograph of the flank of a submarine mountain, part of the Mathe-maticians Seamounts (station 241, 5 and 5A), central eastern Pacific Ocean, showing angular fragments of block lavas (courtesy of M. Ewing and C. Fray). This rock assemblage consists of vesicular and scoriaceous material of trachytic composition. B. Bottom photo-graph of the western flank of the East Pacific Rise at station 240A (courtesy of M. Ewing). The lava flow shows a ropey form wrinkled surface (Subbarao and Hékinian, 1978).

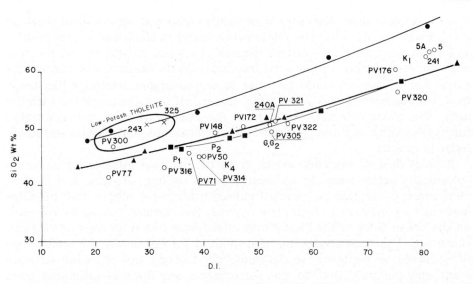

Fig. 5-5. SiO$_2$-D.I. variation diagram of the rocks from the eastern Pacific Ocean. The field of low-potash tholeiites from the East Pacific Rise is indicated; ▲ = average trend of alkali basalt—mugearite—trachyte from Hawaii (Muir and Tilley, 1961); ● = average trend of tholeiite—granophyre—rhyodacite series from Hawaii (Kuno et al., 1957; G. A. Macdonald and Katsura, 1964); ■ = alkali basalt—trachyte trend from Clarion Island (Bryan, 1967). PV 176, etc. = samples from islands and seamounts off the eastern Pacific Ocean (A. E. J. Engel and Engel, 1964; C. G. Engel and Engel, 1966); $K_1$, $K_4$ = trachyte and alkali basalt from the Kodiak Seamount (Forbes et al., 1969); $G_1$, $G_2$ are basaltic rocks from the Giacomini Seamount (Forbes and Hoskin, 1969); 240A is a mugearite from the western flank of the East Pacific Rise; 241, 5A, 5 are trachytic rocks from an unnamed seamount of the Mathematicians seamount chain; 243, 325 are tholeiites from the crest of the East Pacific Rise. (After Subbarao and Hékinian, 1978.)

together with other rock types from seamounts (C. G. Engel et al., 1965; Forbes and Hoskins, 1969; Forbes et al., 1969) and other regions of the eastern Pacific Ocean (C. G. Engel and Engel, 1963; A. E. J. Engel and Engel, 1964; C. G. Engel and Chase, 1965; C. G. Engel et al., 1965). The Hawaiian and Clarion Island rocks show a similar trend of silica enrichment with an increase of the D.I. to the alkali basalt—mugearite—trachyte from the eastern Pacific seamounts (Fig. 5-5). This diagram delineates a field of saturated and oversaturated low-potash tholeiites with SiO$_2$ > 48% and a D.I. between 22 and 30. These tholeiites contain up to 21% olivine and less than 1% quartz in their norm. The oceanic tholeiites and the Hawaiian tholeiite series follow a trend of higher silica enrichment than the alkali basalt trends. The alkali basalt—trachyte trend is divided into three fields consisting of (Fig. 5-5): (1) saturated and undersaturated alkali basalts with normative nepheline >0.3% and normative olivine >9%; (2) saturated and oversaturated mugearite with quartz (0—5%) or olivine <9%; and (3) undersaturated and oversaturated trachytes with olivine <1%, nepheline <4% or quartz (0—10%).

There is a continuous depletion of normative olivine from the alkali basalt to the trachytes (Fig. 5-5). The relationship between alkali basalts, mugearites and trachytes consists of various degrees of crystal separation and reaction with the melt during a process of fractional crystallization. The continuous extraction of olivine could give rise to oversaturated trachytes like those from the Mathematicians Seamounts. Instead, if the olivine reacted with the melt, the residual liquid would have the composition of undersaturated trachytes similar to those from the Kodiak Seamount, and Clarion and Revillagigedo Islands.

A great deal of evidence exists, from islands, that intraplate regions are volcanically active areas. In addition deep-sea drilling, dredging and bottom photographic coverage on elevated volcanic structures suggest that off-ridge magmatism continues in intraplate regions. The elevated volcanoes dispersed on the ocean floor of the Pacific may either represent ridge-crest or off-ridge volcanism. The most straightforward method for estimating the relative age of a submarine feature is to determine the isotopic age of its last eruptive event and compare that to the surrounding sea floor as estimated from marine magnetic lineations. Another way of determining the origin and the age of a volcano is based on the method of oceanic plate flexure (Watts et al., 1980). This method estimates the age at which the lithosphere responds to surface loads. Watts et al. (1980) have calculated the age of the lithosphere at the time of loading by subtracting the age of the load from the age of the underlying sea floor. This was based on an interpretation of gravity anomalies in relation to the response of the lithosphere to surface loads. As the oceanic lithosphere becomes older it cools, thickens and becomes more rigid due to its loading. The deepening of the lithosphere into the asthenosphere may be corroborated by the fact that many isolated volcanic edifices are made up of alkali basalt suites. Indeed the Baja California seamounts, some of the Alaska province seamounts and the Mathematicians Seamounts consist of trachytic rocks as their last eruptive events. Hence it is speculated that in a natural system the degree of magma saturation with respect to silica is a function of load pressure. From laboratory experimental work it has been suggested (D. H. Green and Ringwood, 1967) that alkali basalt melts will be generated under a pressure of 18—20 kbars (60—70 km depth).

Unfortunately details of the various types of lava sequences and the periodicity of volcanic eruption of intraplate submarine features are unknown. Also the extent and compositional variation of intrusive bodies in the sediment and basement rocks of oceanic basins are still to be evaluated. In addition, deep-sea drilling data in various ocean basin provinces is still too sporadic and the presently drilled holes are too far apart to be able to detect trends of compositional variabilities suggestive of local crustal and upper mantle heterogeneities.

## CHAPTER 6

## OCEANIC TRENCHES

It is not the purpose of the present work to give an extensive review of island arc volcanism but in order to understand late tectonic mechanism it is important to mention some basic petrochemical differences encountered between island arcs and mid-ocean ridge volcanism. Island arcs and trenches are important transitional features which link the oceanic lithosphere with the development of continental masses. The trenches, which are associated with sites where the lithosphere is thought to be resorbed, may provide the driving mechanism for the intense seismicity and volcanism of island arcs and continental margins. The occurrence of near-vertical normal faults and deformed sediment in the axial zone of an area of the middle America Trench (Heezen and Rawson, 1977) indicates the displacement of a down-plunging lithosphere. Within the oceanic trenches, two domains of sampling have to be considered: (1) the specimens recovered from the island arcs which are associated with the trenches, and (2) the specimens collected from the trenches themselves. However, this latter domain has been very poorly sampled and up to now only fragmentary data are available from the literature.

Oceanic trenches occur mostly in the Pacific Ocean, around the western continental margins of the Americas and on the border of island arcs. Study of world seismicity shows that most earthquakes are confined to narrow continuous belts that bound large stable areas. In the zones of convergence, seismic activity is at shallow depths but includes intermediate and deep shocks that grossly define the present configuration of the downgoing slabs of lithosphere (Oliver, 1969). The presence of volcanism, the generation of many tsunamis (seismic sea waves), and the frequency of large earthquakes also seem to be related to underthrusting in island arcs. It is suggested that, in the oceanic trench regions, the lithosphere sinks into the asthenosphere under its own weight but encounters resistance to its downward motion below about 300 km (Fig. 6-1; Isacks and Molnar, 1971).

Geophysical and petrological studies of oceanic trenches are mainly concentrated in the Tonga-Kermadec (Pacific Ocean) and in the Caribbean trenches (Atlantic Ocean). Oliver and Isacks (1967) postulated that shallow earthquakes associated with the Tonga-Kermadec arc indicate the existence in the mantle of an anomalous zone whose thickness is on the order of 100 km and whose upper surface is approximately defined by the highly active seismic zone that dips to the west beneath the island arc and extends to depths of about 700 km.

Earthquake foci tend to be confined to a very thin zone, usually less than

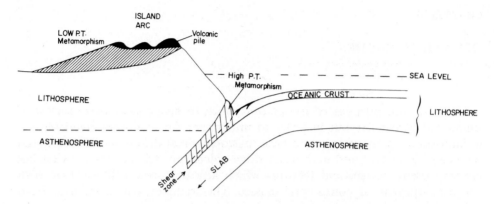

Fig. 6-1. Schematic representation of downgoing lithospheric slab under an island arc formation (after R. N. Anderson et al., 1976). The shear zone with frictional heating is occurring at high pressure and temperature condition of metamorphism. Island arc volcanism and low-grade metamorphism are the results of a dehydration phenomenon of the plunging oceanic slabs and diapirism under the island arc.

50 km and frequently less than 20 km wide on island arcs, associated with oceanic trenches (Sykes, 1967). The study of the focal mechanism could give some idea about the geometry of a zone of convergence between two plates. Shallow focal mechanisms correspond to thrusting either beneath the inner wall of the trench or beneath the outer wall where extension occurs as the result of the plunging of the lithosphere (Oliver, 1969) (Fig. 6-1). The compositions of extrusive rocks from islands associated with circum-Pacific trenches where seismological data are available (Oliver, 1969), suggest that most volcanic activities of islands associated with trenches subject to compressional type of stresses give rise to tholeiitic types of rocks. This is well illustrated by the extrusives found in the Amazi volcano on the Izu-Bonin arc (Kurasawa, 1959), from Talua Island in the Tonga arc (Bauer, 1970; Bryan et al., 1972), in the Asamo volcano on the Northeast Japan arc (Aramaka, 1963), and from the islands south of the Marianas Trench, such as Guam and Saipan.

Other oceanic trenches which are associated with both compressional and extentional mechanisms are made up of calc-alkaline and tholeiitic types of volcanics. Examples in this category are seen in the volcanics of the Kermadec (Brothers and Martin, 1970) and of the north, south and central Kuriles. The amount of calc-alkaline rocks in the central Kuriles suggests an oceanic type of crust, whereas the south and north Kuriles have a thicker and a more continental type of crust (Miyashiro, 1974). However, the eastern Pacific bordered by oceanic trenches shows primarily andesitic types of rock with calc-alkaline affinities. This corroborates the fact that regions of Chile and Central America show focal mechanisms which are mainly extensional above a 400 km depth (Oliver, 1969).

The compressional mechanisms under oceanic trenches are more prominent at depths greater than 400 km. At shallower depths extensional mechanisms are more apparent and are associated with the seaward extension of the trench. Most of the western Pacific oceanic trenches have compressional mechanisms up to 100 km in depth.

Three main volcanic series have been defined as occurring in island arcs and active continental margins (Miyashiro, 1974): (1) calc-alkaline, (2) tholeiitic, and (3) alkalic volcanic series.

The calc-alkaline series is distinguished by its high alkali-lime index ($Na_2O$ + $K_2O$ wt.% and CaO wt.%) relative to the alkalic series. The calc-alkaline series also shows higher rates of $SiO_2$ increase and a lower degree of FeO (total Fe calculated as FeO) enrichment with an increasing FeO/MgO ratio compared to the tholeiitic series. The calc-alkaline and the tholeiitic series follow two differentiation trends. One, the calc-alkaline trend, gives rise to the basalt-andesite-dacite-rhyolite series commonly observed in orogenetic belts (Bowen, 1928); the other, the tholeiitic series, is exemplified by the Skaergaard intrusion in Greenland (Wager and Deer, 1939). The existence of these two trends in island arc volcanism was emphasized by Kuno (1959, 1966). It was shown (Kuno, 1959, 1966) that an increasing alkalinity away from the trench in Japan gave rise to alkali basalt and shoshonitic types of rocks. Joplin (1968) used the term "shoshonite" to indicate the association of K-rich rocks ($K_2O$ + $Na_2O$) found in orogenetic zones. The $K_2O$ content has been used (Dickinson and Hatherton, 1967) to correlate the chemical parameter with depth in the "Benioff zone". In general, a pattern of high $K_2O$ content is related to the increasing depth of earthquakes associated with the "Benioff zone". However, this generalized scheme of spatial ordering due to compositional variations is not valid for all island arc systems. Recently Arculus and Johnson (1978) have suggested that the Lesser Antilles and the Papua—New Guinea arcs do not conform to the above-described scheme. Indeed, temporal sequences of a relative spatial ordering of rocks could differ from one island arc to another.

Miyashiro (1974) has used the $SiO_2$ and FeO/MgO ratios (total Fe calculated as FeO) to distinguish between island arc volcanics in relation to their structures. Fig. 6-2 shows the island arcs and active continental margins, associated with trenches arranged in an approximate order of increasing thickness of crust above the Moho. Apparently, the size of volcanic islands tends to become greater with the increasing thickness of their crust. In general, continental margins tend to have a thicker crust than island arcs (Miyashiro, 1974). The ratio of calc-alkaline series rocks versus tholeiitic series rocks in an island arc tends to increase with the increasing thickness of underlying crust (Miyashiro, 1974). Fig. 6-2 shows that in the area of the Tonga Trench the thickness of the crust is close to that of a mid-oceanic ridge spreading center. The active continental margin area represented by the Kamchatka region, the central Andes, and the Cascades Range where the crustal thickness is about 30—70 km, has mainly rocks of the calc-alkaline series which are principally of andesite, dacite and rhyolite com-

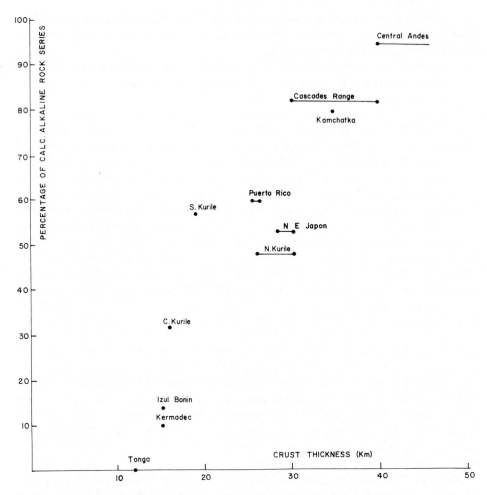

Fig. 6-2. Relationship between crustal thickness and the percentage of the calc-alkaline rock series found in islands. Most of the data are from Miyashiro (1974).

position (Fig. 6-2). The increase in the proportion of calc-alkaline series rocks is accompanied by an increase in the proportion of more silicic volcanic rocks. Miyashiro (1974), compiling data from island arc regions, has shown that the average content of $SiO_2$ increases from about 52—57% for the Kermadec and Tonga Trenches to about 61—65% for the Cascades Range and the central Andes regions. The andesitic rocks from island arc complexes differ from other oceanic andesites found in aseismic ridges or seamounts set on oceanic crust. This difference is illustrated in the trend of their Zr and Ti variation, as shown in Fig. 6-3. However, more data on oceanic andesites (Chapter 1) is needed in order to assess the mineralogical differences between these two provinces.

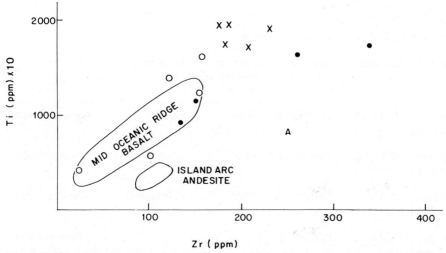

Fig. 6-3. Ti and Zr (ppm) variation diagram of island arc rocks compared to mid-oceanic ridge volcanics from both the Atlantic and the Pacific Oceans. The empty and filled circles and the crosses are volcanics from aseismic ridges. $A$ = average oceanic andesite from the Ninety-East aseismic ridge.

For references see Tables 1-17 and 3-3 (diagram after Pearce and Cann, 1971).

While there is extensive literature on the volcanic rocks from island arcs and active continental margins, very little data on the crustal composition are available from oceanic trenches (Table 6-1). some of the oceanic trenches where samples were recovered are mentioned in the following paragraphs.

JAPAN TRENCH

Particular interest is given to the island arcs associated with the Japan Trench where petrological data are well documented. The Japan Trench is bordered by a double arc system. The crust under the island arc is 28—30 km thick (Kanamori, 1963). The rocks of the inner volcanic zone, tending to be more alkalic types of volcanism, have erupted only in the westernmost part of the Japan Trench, while the tholeiitic and calc-alkaline series are mainly concentrated in the eastern part of the Japan island arc towards the ocean. The calc-alkaline series of the inner volcanic zones (more towards the continent) are usually higher in $H_2O$ and in $Na_2O$ than those of the outer volcanic zone (Miyashiro, 1974). This is also true for the northeast of Japan, as well as the north Kuriles and also for the Cascades Range. Olivine basalts were found on the outer wall (ocean side) of the Japan Trench (Gladkikh and Chernysheva, 1966) (Table 6-1).

TABLE 6-1

Rocks collected from circum-Pacific trenches

| Province | Latitude | Longitude | Depth (in uncorrected meters) | Rock type | Source |
|---|---|---|---|---|---|
| New Britain (island side) | 5°44.6'S | 152°43.4'E | 7600 | basalt | Petelin, 1964 |
| New Britain (island side) | 6°17.3'S | 153°43.8'E | 8000 | argillite | Petelin, 1964 |
| New Britain (island side) | 3°0'S | 151°27.5'E | 3500 | tuff breccia | Petelin, 1964 |
| Solomon (island side) | 12°0'S | 173°8.3'E | ? | | Petelin, 1964 |
| Tonga (island side) | 20°25'S 20°32.2'S 20°16'S 20°39.5'S | 173°16'W 173°25'W 173°12'W 173°27.8'W | 9150—9400 7200—7700 9000—9400 7000 | peridotite, basalt gabbro alkali basalt altered basalt | Fisher and Engel, 1969 |
| Marianas (ocean side) | 11°20.6'N | 142°16.2'E | 10710 | basalt, dolerite | Yagi, 1960; Petelin, 1964 |
| | 13°33.0'N | 146°59.4'E | 5400—5600 | basalt | Dietrich et al., 1978 |
| | 11°16.6'N | 142°9.9'E | 10700 | dolerite | Yagi, 1960; Gladkikh and Chernysheva, 1966 |
| Marianas (island side) | 12°7.8'N | 144°30.6'E | 8100 | basalt, metabasalt gabbro, mylonite, | Dietrich et al., 1978 |
| | 12°16.4'N | 144°21.2'E | 5500 | basalt | |
| Japan (island side) | 38°16.0'N | 143°48.1'E | ? | | Gladkikh and Chernysheva, 1966 |
| Japan (ocean side) | 40°32.1'N | 145°8.0'E | ? | olivine basalt | Gladkikh and Chernysheva, 1966 |
| Peru-Chile | 16°49.6'S | 74°32.7'W | 7356 | basalts | Scheidegger and Stakes, 1977 |

MARIANAS TRENCH

A study sponsored by the International Correlation Program on the project "Ophiolites" has obtained detailed petrographic data on the island arc side of the Marianas Trench. The results, presented by Dietrich et al. (1978) indicate the presence on the island arc slope of an ophiolitic suite with serpentinized and mylonitized peridotites, troctolites, gabbros, rodingitized troctolitic gabbros and basalts. Andesitic rock types, also called boninites after Bonin Island, and having a $SiO_2$ content of 54—57% were also found in association with the other rock types (Table 6-2). Many basaltic rocks have undergone various degrees of alteration, however some fresh specimens were classified as being tholeiitic in composition (olivine-normative tholeiites) (Dietrich et al., 1978). One station (1402; Dietrich et al., 1978) made on the ocean side of the Marianas Trench contains olivine-normative basalts and potassic rocks of the shoshonite type (Table 6-2). These rocks have the appearance of alkali basalts which are particularly high in potash content. Sanidine with labradorite rims is set in a matrix of smectite,

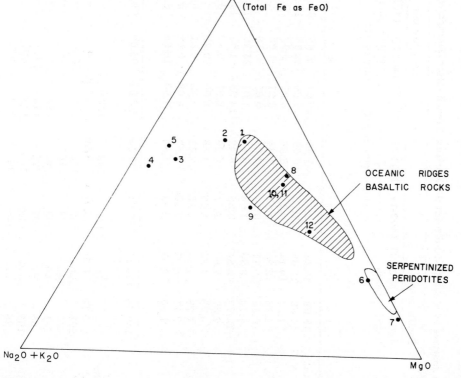

Fig. 6-4. AFM ternary diagram of the Marianas Trench rocks compared to the serpentinized peridotite and to the mid-oceanic ridge basalts. The numbers correspond to the analyses presented in Table 6-2.

TABLE 6-2

Chemical analyses of rocks dredged from the Marianas Trench (after Diedrich et al., 1978)

| | Pacific slope | | | | | Island arc side | | | | | | |
|---|---|---|---|---|---|---|---|---|---|---|---|---|
| | 1402-7 basalt | 1402-19 basalt | 1402-11 | 1402-13 | 1402-Z | 1403-50 troctolite (mylonite) | 1403-26 gabbro | 1403-37 basaltic rock | 1403-48 meta-basalt | 1404-E basalt | 1404-F basalt | 1403 andesitic rock (ave. of 6) |
| | | | shoshonitic rocks | | | | | | | | | |
| | 1 | 2 | 3 | 4 | 5 | 6 | 7 | 8 | 9 | 10 | 11 | 12 |
| $SiO_2$ (wt.%) | 48.09 | 48.58 | 49.41 | 47.45 | 40.57 | 28.88 | 39.12 | 50.28 | 47.59 | 49.89 | 50.21 | 55.73 |
| $Al_2O_3$ | 15.63 | 16.65 | 17.70 | 17.09 | 15.50 | 12.56 | 16.50 | 14.45 | 16.44 | 15.60 | 15.75 | 11.04 |
| $Fe_2O_3$ | 7.24 | 10.14 | 10.41 | 8.86 | 8.87 | 6.93 | 0.68 | 3.45 | 4.36 | 5.25 | 4.84 | 3.44 |
| $FeO$ | 4.96 | 2.85 | 0.87 | 1.44 | 0.97 | 1.33 | 1.70 | 7.51 | 4.31 | 4.45 | 4.85 | 5.15 |
| $MnO$ | 0.15 | 0.17 | 0.14 | 0.08 | 0.05 | 0.16 | 0.05 | 0.22 | 0.18 | 0.15 | 0.17 | 0.16 |
| $MgO$ | 4.70 | 3.90 | 1.86 | 0.69 | 1.01 | 25.29 | 18.17 | 8.27 | 7.04 | 7.80 | 7.72 | 12.43 |
| $CaO$ | 12.70 | 9.84 | 5.58 | 8.42 | 14.08 | 1.55 | 14.25 | 11.67 | 9.86 | 12.38 | 12.30 | 5.97 |
| $Na_2O$ | 2.37 | 2.80 | 2.81 | 3.29 | 2.61 | 0.77 | 0.16 | 2.09 | 4.43 | 2.15 | 2.10 | 1.92 |
| $K_2O$ | 0.73 | 1.32 | 4.06 | 4.56 | 3.00 | 0.28 | — | — | 0.16 | 0.20 | 0.30 | 0.84 |
| $TiO_2$ | 1.78 | 1.94 | 2.86 | 2.71 | 2.46 | 0.51 | 0.08 | 0.96 | 1.10 | 0.88 | 0.87 | 0.21 |
| $P_2O_5$ | 0.17 | 0.35 | 0.96 | 2.87 | 6.08 | 0.12 | 0.02 | 0.07 | 0.10 | 0.09 | 0.07 | 0.04 |
| $CO_2$ | 1.90 | 2.01 | 2.95 | 1.84 | 2.80 | 16.68 | 8.86 | 1.80 | 4.13 | 1.14 | 1.27 | 3.55 |
| $H_2O$ | | | | | | | | | | | | |
| Total | 100.66 | 100.55 | 99.61 | 99.55 | 98.84 | 95.09 | 99.59 | 100.77 | 99.84 | 100.22 | 100.42 | 100.48 |
| Nb (ppm) | 12 | 15 | 138 | 149 | 100 | — | — | — | — | — | — | — |
| Ta | 0.50 | 0.50 | 5.10 | 5.50 | 4.10 | — | — | — | — | — | — | — |
| Cr | 203 | 146 | 162 | 113 | 170 | 259 | — | 145 | 411 | 150 | 153 | 820 |
| Co | 45 | 43 | 38 | 10 | 7 | 30 | — | 46 | 41 | 46 | 44 | 48 |
| Ni | 35 | 45 | 145 | 25 | 20 | — | — | 60 | 122 | 55 | 65 | 280 |
| Cu | 65 | 97 | 97 | 126 | 115 | 141 | — | 152 | 101 | 119 | 108 | 87 |
| Zn | 101 | 210 | 112 | 146 | 115 | 90 | — | 57 | 69 | 80 | 87 | 72 |
| Zr | 86 | 89 | 340 | 360 | 264 | 67 | — | 46 | 100 | 45 | 108 | 33 |

*Norms*

| | | | | | | | | | | | | |
|---|---|---|---|---|---|---|---|---|---|---|---|---|
| Q | 7.07 | 7.86 | 13.56 | 4.91 | 7.03 | 28.20 | 3.67 | 4.91 | — | 5.21 | 5.21 | 16.84 |
| Or | 4.31 | 7.80 | 23.99 | 26.94 | 17.72 | 1.65 | 1.35 | — | 0.94 | 1.18 | 1.77 | 4.96 |
| Ab | 20.05 | 23.69 | 23.77 | 27.83 | 22.08 | 6.51 | 14.54 | 17.68 | 37.48 | 18.19 | 17.76 | 16.24 |
| An | 29.85 | 28.96 | 2.75 | 11.38 | 12.42 | — | — | 30.04 | 22.14 | 32.32 | 32.66 | 6.90 |
| Di | 15.65 | 3.77 | — | — | — | — | — | 12.94 | — | 16.80 | 15.75 | — |
| Hy | 4.87 | 7.96 | 4.63 | 1.71 | 2.51 | 62.98 | 47.77 | 24.10 | 19.88 | 14.17 | 15.55 | 37.52 |
| Ol | — | — | — | — | — | — | — | — | 0.34 | — | — | — |
| Mt | 10.49 | 4.11 | 2.13 | 3.21 | 2.15 | 3.33 | 0.98 | 5.00 | 6.32 | 7.61 | 7.01 | 4.98 |
| Ilm | 3.38 | 3.68 | 2.26 | 6.77 | 14.34 | 0.96 | 0.15 | 1.82 | 2.08 | 1.67 | 1.65 | 0.39 |
| Ap | 0.40 | 0.82 | 1.73 | 1.01 | 1.32 | 0.28 | 0.04 | 0.16 | 0.23 | 0.21 | 0.16 | 0.09 |
| Rt | — | — | — | — | — | — | — | — | — | — | — | — |
| Hm | — | 7.30 | 10.40 | 8.85 | 8.86 | 4.63 | — | — | — | — | — | — |
| C | — | — | 7.67 | 2.56 | 3.40 | 47.11 | 10.90 | — | 0.86 | — | — | 4.44 |
| Cc | 4.32 | 4.57 | 6.70 | 4.18 | 6.36 | 37.93 | 20.15 | 4.09 | 9.39 | 2.59 | 2.88 | 8.07 |
| C.I. | 45.45 | 33.54 | 3.22 | 11.55 | 12.67 | — | 19.10 | 40.85 | 24.16 | 48.59 | 47.42 | 10.03 |

chlorite aggregates and Fe-Ti oxides (Dietrich et al., 1978). The occurrence of alkaline rocks on the oceanic side of the Marianas Trench led Dietrich et al. (1978) to suggest that the rocks are derived from a seamount type of structure. Also the basalt-gabbro-andesite association from the Marianas Trench was plotted on an AFM diagram and shows a continuous trend of compositional variation which is similar to that of other mid-oceanic ridge basalts (Fig. 6-4).

The serpentinite-gabbro-basalt sequence found on the island arc slope of the trench could represent the upper portion of the lithospheric plate exposed. The presence of mylonitized material in association with the serpentinite-gabbro-basalt sequence found on the island arc slope of the trench brought Dietrich et al. (1978) to suggest that the downgoing slab phenomenon had exposed the upper portion of the lithospheric plate. Another alternative explanation could be the fact that the sequence is part of the island arc volcanic process and that it represents differentiated products within a magma chamber. Mylonitization and metasomatic alteration could have developed by simple fracturing along fault planes.

TONGA TRENCH

The first deep dredging in the Tonga Trench was done by the R.V. "Albatross" in 1889 and it recovered red clay sediment from a depth of 5700 m. In 1952 the R.V. "Galathea" cored brown silty tuffaceous products from about 10,000 m depth along the trench axis east of Tonga Tabu Island. Four successful dredges from the Tonga Trench included ultramafics and basaltic specimens (Fisher and Engel, 1969). These samples were taken during the

TABLE 6-3

Chemical analyses of peridotite from the island arc side of the Tonga Trench (Fisher and Engel, 1969)

|  | Dunite |  | Dunite |
| --- | --- | --- | --- |
| $SiO_2$ (wt.%) | 44.62 | Ba (ppm) | 6 |
| $Al_2O_3$ | 0.16 | Co | 130 |
| $Fe_2O_3$ | 0.51 | Cr | 3700 |
| FeO | 7.31 | Cu | 8 |
| MgO | 46.35 | Ni | 3000 |
| CaO | 0.09 |  |  |
| $Na_2O$ | 0.10 |  |  |
| $K_2O$ | <0.02 |  |  |
| $P_2O_5$ | <0.01 |  |  |
| $H_2O$ | 0.10 |  |  |
| Total | 99.14 |  |  |

Nova expedition (R.V. "Argo") in 1967 from the nearshore flank of the trench near its axis at depths of 7000—9400 m (Fisher and Engel, 1969). The deepest successful dredge haul (88-D) was taken at 9150—9400 m deep and it consisted of about 20 kg of peridotites and dunites (Table 6-3). Study was done on one of the largest samples of peridotite, consisting of 86% olivine ($Fo_{93}$) 13% orthopyroxene and less than 1% opaques. Granulated textural features were noticed in some other samples of peridotite.

PUERTO RICO TRENCH

The Puerto Rico Trench is another area where serpentinites and low-potassium tholeiites were found exposed on both the north and south walls (inner walls). This trench is characterized by a strike-slip motion with only small components of underthrusting (Molnar and Sykes, 1969). The trench floor is about 7500 m deep and the walls are made up of scarps. All the material was recovered from the walls of the trench at depths between 5500 and 7300 m. The basalts are tholeiitic in nature and the most crystalline ones show zones of clinopyroxene with reaction rims of another pyroxene (Table 6-4). The core is formed by pigeonite ($2V = 0—20°$) while the rim consists of augite ($2V = 45—47°$). The occurrence of phenocrysts with a pigeonite core in these rocks is the result of the low content of CaO and of the low $CaO/Na_2O$ ratio (Miyashiro, 1974).

TABLE 6-4

Chemical analyses of basalts and serpentinites from the Puerto Rico Trench

| | Basalts | | | | Serpentinites | | |
|---|---|---|---|---|---|---|---|
| | RC8-RD1-1 (ave. of 7) | RC8-RD1-2 (ave. of 5) | CHN34-D3-41 | CHN34-D3-50 | CH19-D10-4 | CH19-D10-6 | CH19-D2-2 |
| $SiO_2$ (wt.%) | 48.9 | 50.90 | 47.25 | 48.49 | 38.66 | 39.34 | 42.59 |
| $Al_2O_3$ | 15.55 | 15.97 | 19.34 | 15.46 | 0.87 | 0.61 | 0.41 |
| $Fe_2O_3$ | 3.91 | 5.78 | 6.24 | 5.07 | 8.22 | 7.25 | 5.44 |
| FeO | 5.37 | 3.66 | 1.42 | 3.75 | 1.09 | 0.94 | 2.14 |
| MnO | 0.19 | 0.15 | 0.06 | 0.12 | — | — | — |
| MgO | 7.64 | 6.76 | 5.47 | 7.25 | 35.43 | 37.28 | 37.09 |
| CaO | 11.80 | 7.48 | 7.44 | 11.41 | 0.14 | 0.15 | 0.29 |
| $Na_2O$ | 2.71 | 3.86 | 2.77 | 2.75 | 0.24 | 0.24 | 0.29 |
| $K_2O$ | 0.05 | 0.68 | 1.27 | 0.36 | 0.03 | <0.1 | <0.1 |
| $TiO_2$ | 1.34 | 2.05 | 1.14 | 1.37 | 0.10 | <0.1 | <0.1 |
| $P_2O_5$ | 0.10 | 0.23 | 0.15 | 0.26 | 0.02 | 0.05 | 0.25 |
| NiO | — | — | — | — | 0.27 | 0.43 | 0.21 |
| LoI | 1.29 | 1.66 | 7.39 | 3.35 | 13.74 | 13.41 | 11.85 |
| Total | 98.85 | 99.18 | 99.92 | 99.69 | 99.23 | 99.70 | 100.85 |

RC8-RD1 samples are from Miyashiro et al. (1974) and located at 20°06.8′N, 65°06.5′W at a depth of 6591 m; CHN samples are from Chase and Hersey (1968); serpentinites are from Bowin et al. (1966).

Detailed bathymetric and petrographic studies were done on the northern wall of the Puerto Rico Trench around 20°16'N and 65°45'W (Bowin et al., 1966; Chase and Hersey, 1968). The main scarp of the northern wall has an average slope of 15° and a local relief of 800—820 km; the steepest slope, of about 24—30°, shows ultramafic outcrops (Bowin et al., 1966). A bottom photograph from the main scarp shows brecciated material and some dredge hauls containing serpentine breccia were collected from the northern wall. From the combination of seismic velocities and the type of rock dredged, Bowin et al. (1966) suggested a possible stratigraphy for the north wall of the Puerto Rico Trench. The main scarp of the north wall changes from Lower Tertiary sedimentary rocks (mudstone, shale, cherts, limestone) to serpentinized peridotites with basalts (Bowin et al., 1966; Chase and Hersey, 1968). The basaltic rocks recovered by the R.V. "Chain" consist primarily of a plagioclase-rich basalt (HPPB) containing plagioclase phenocrysts (up to $An_{85}$) and clinopyroxene with an ophitic texture and an $Al_2O_3$ content of less than 16% (Table 6-4). The dredge hauls taken by the R.V. "Conrad" and described by Miyashiro (1974) also consist of doleritic basalts and low-K tholeiites. The basalts associated with Cenomanian sedimentary zones are similar in composition to those found in the Mid-Atlantic Ridge.

The serpentinized peridotite found on the north wall of the trench consists of antigorite, chrysotyle and relics of orthopyroxene. Talc is also a common constituent, with Fe-oxide minerals concentrated along the borders of former olivine grains.

Recently, results obtained from extensive dredging operations, concentrated mainly on the south wall (island side) of the Puerto Rico Trench, showed the presence of a heterogeneous type of material (Perfit et al., 1980) such as semi-indurate sediment and metamorphosed rocks. The metamorphics were altered to the greenschist-facies assemblages and consist of albite, epidote, amphibole, chlorite and spessartine garnets. In addition, blue-green hornblendes were detected in some amphibolites (Perfit et al., 1980).

It is noteworthy that up to now ultramafic rocks have not been reported from the seaward region away from the trenches of the Pacific Ocean (Fisher and Engel, 1969). Ultramafics associated with gabbroic rocks are found on various island arc sides of the trenches in the Pacific and Atlantic Oceans. When fresh, the peridotites consist of hornblende, augite, olivine and chromite. Gabbroic rocks were also found in Eua (Tonga Islands) and peridotites and norites were reported from the nearshore flank of the Tonga Trench (Fisher and Engel, 1969).

The presence of metamorphic and serpentinized rocks from the oceanic trenches suggests that both the inner walls (island side) and the outer walls (ocean side) are under regional tectonic stresses (Fig. 6-1). It is likely that medium- and high-pressure/low temperature metamorphism will affect the inner walls of the trench during the plunging of the lithospheric plates

(Fig. 6-1), whereas serpentinization of crustal and upper mantle material will be more prominent on the outer walls. Whether the inner walls of the trenches represent reworked and fractured remains of oceanic lithosphere incorporated into the island arc formations (or other similar subaerial regions) is not clear. Further investigations are needed in order to assess structural and petrological relationships between the inner walls, the trench floor and the outer walls; these will give more insight into the formation of island arcs and active margin systems.

## CHAPTER 7

## OCEANIC FRACTURE ZONES

Fracture zones as discussed in this chapter are defined as structural lineations offsetting accreting plate boundary regions. The evolution of fracture zones is intimately tied to that of oceanic ridge systems with which they form the major morphological and tectonic features of the ocean floor. The fracture zones and their transform area are a locus of extensive faulting and are areas where tectonic activity prevails over volcanism. Focal mechanisms for a number of earthquakes have revealed the prevalence of horizontal compressional stresses in the oceanic crust near fracture zones (Sykes and Sbar, 1974). Both compressional and extensional tectonism are prominent, as suggested by the recovery of mylonitized rocks and the presence of a horst-graben morphology which is commonly observed in fracture zones. The tectonic evolution of a fracture zone is probably related to the relative motion of two plates during spreading. Thus a change in the direction of spreading of a plate on either or both sides of a transform zone will alter the geometry of the various segments involved.

The fact that fracture zones and their transform areas vary considerably in size, depth, reliefs, and extension is probably related to the degree of spreading and other complex tectonic mechanisms. However, both small and large fracture zones are able to offset ridge segments and they have considerable similarities as far as their crustal composition is concerned. The North Atlantic fracture zones are, up to now, the most extensively studied and surveys made south of the Azores have shown that transform faults with offsets of more than 20 km occur on the order of one transform every 55 km (Fox et al., 1969; G. L. Johnson and Vogt, 1973; Vogt and Avery, 1974). Fracture zones and their transform areas in both the Indian and the Pacific Oceans are less well known than those from the Atlantic. Only a few Indian Ocean fracture zones show detailed petrological and morphological data. Among those fracture zones which have been recently most extensively sampled is the Alula-Fertak Fracture Zone which delineates a boundary between the Gulf of Aden and the northwestern Indian Ocean (Bonatti, 1978). The Owen Fracture Zone on the Carlsberg Ridge, near 5°N, and other major transform faults offsetting the Central Mid-Indian Oceanic Ridge were also studied (C. G. Engel and Fisher, 1975). The structures and the composition of the exposed outcrops in the Indian Ocean fracture zones are comparable to those encountered in the Atlantic Ocean. Series of basalts, gabbros, serpentinized peridotites, and metamorphics are found.

Since fracture zones represent areas with extensive faulting, it is likely that slices of oceanic crust and perhaps of upper mantle are exposed and hence are more accessible for the study of layer 2 and layer 3. However, there is still some doubt about whether deep crustal and upper mantle material could be exposed in fracture zones by the simple process of faulting. Indeed, submersible studies in a small fracture zone such as Transform Fault "A", near 37°N on the Mid-Atlantic Ridge, has revealed the absence of major vertical motions or normal faulting which could suggest the exposure of deep-seated oceanic crust (Arcyana, 1975; Choukroune et al., 1978). However, Bonatti (1978) has suggested, from the interpretation of seismic profiles across the Owen and the Romanche Fracture Zones, the presence of transverse ridges which represent uplifted upper mantle and crustal blocks. A transverse ridge on the western side of the Owen Fracture Zone has exposed a 2 km thickness of ultramafic rocks (Bonatti, 1978). However, a survey of dredge hauls from North Atlantic fracture zones indicates a sporadic distribution of the various rock types with depth which could suggest a limited extent of crustal exposure, without any obvious evidence for stratigraphic sequences. Fracture zones are made up of a variety of rock types which are often compared with ophiolitic complexes observed in subaerial environments. The term "ophiolite" is used here in the sense defined at the Geological Society Penrose Conference of 1972 and described by Coleman (1977). The ophiolites or their equivalent found in fracture zones consist of ultramafics and mafic rock associations. The ultramafics are made up of harzburgite and lherzolite while the mafics consist of fresh and metamorphosed basalts, and gabbroic rocks.

In a few instances, evidence of recent volcanism has been identified in the transform zones of the St. Paul's Rocks, the Romanche Fracture Zone and another fracture zone near 43°N in the Atlantic Ocean (Melson et al., 1967; Honnorez and Bonatti, 1970; Shibata et al., 1979b). The indications for the presence of recent volcanism of alkali basalts on a crust older than 1 Ma were based on the relative freshness and the fission track ages of the basaltic glass sampled.

The most conclusive evidence for the presence of recent volcanism between two plate boundary regions is that given by the occurrence of a volcanically active ridge system found in the Cayman Trench (Holcombe et al., 1973; Perfit, 1977; G. Thompson et al., 1980) which represents the plate boundary area dividing the North American from the Caribbean plate, both of which are moving in a left-lateral sense relative to one another. The Cayman Trough's glassy basalts have a compositional range (G. Thompson et al., 1980) which indicates a more fractionated rock type than that found for the Mid-Atlantic Ridge volcanics. These basalts also have alkaline affinities as indicated by their high $Na_2O$ content (3.4—4.2%). In addition to recent volcanism, plate boundary regions represented by fracture zones are also the preferential sites for the outcropping of intrusives (sills and/or dykes) of a doleritic nature. Since direct field observations have been very limited up to now, marine geologists have always been handicapped in the recognition of intrusive rocks from the ocean floor.

PROBLEM OF THE RECOGNITION OF INTRUSIVES

On the one hand, deep-sea drilling and conventional dredging do not provide detailed information on the form and field relationships of the sampled outcrops. On the other hand, relatively coarse grain size, holocrystallinity, ophitic or at least subophitic texture, and the absence of a pillow lava morphology, are not necessarily reliable criteria for the recognition of a small intrusive body. For example, the interiors of some pillows and thick lava flows may have well-crystallized zones with subophitic textures; conversely, a small intrusion, or the margins of a larger one, may lack such properties. Further, samples may be too fragmented to display the morphological features of the parent rock.

Despite the problems of identification, there is a good deal of evidence from petrological studies that outcrops of intrusions do occur on the Mid-Oceanic Ridge and on older portions of the ocean floor. Shand (1949), Muir et al. (1964), Bogdanov and Ploskho (1967), Melson et al. (1968), Van Andel and Bowin (1968), Cann (1971a) and others have used the degree of crystallization, textural features and hand-specimen appearance to assess the mode of emplacement of the rocks they discuss. Intrusive rocks have also been identified in drilled cores from the Alula-Fertak Fracture Zone (Dimitriev, 1974), the Mid-Atlantic Ridge (Melson et al., 1976), the Argo Abyssal Plain (Hékinian, 1974a; Robinson and Whitford, 1974), the South Fiji Basin (A. J. Andrews, 1978) and the southern Madagascar Basin (Simpson et al., 1974; Erlank and Reid, 1974). In this last area, diabase intermixed with basalt was found to have a younger K-Ar age than the overlying sediment. In other cases, the cored units diagnosed as intrusives were associated with thick overlying and underlying sediments which have baked or discolored contact zones (Dimitriev, 1974; Hékinian, 1974a), or they may contain recrystallized sedimentary carbonate veins (Melson et al., 1976) or display textures sharply different from those of contiguous basalt units (Robinson and Whitford, 1974). Because of commonly discontinuous recovery, the range of vertical extent of the intrusions encountered in the drilled cores is not known. Inferences suggest a range of about 2—15 m, although some of the evidence referred to seems to leave open the possibility of larger dimensions. However, direct visual observation by submersible could give information useful for the recognition of such intrusions.


FRACTURE ZONE "A" NEAR 37°N

Fracture Zone "A" is a small transform fault when compared to those of other equatorial regions. The maximum depth encountered is about 3100 m and the total width is less than 15 km. The general tectonic lineations of Transform Fault "A" are east-west, and the structure offsets the adjacent segments of the Rift Valley about 20 km in the dextral sense. Reid and Macdonald (1973) showed that the transform fault is active at present and Detrick et al. (1973) described an approximately north-south structural pattern superimposed on the main east-west trending fabric of the area. The study areas of the 1973—1974 FAMOUS operation were two distinct zones of the transform fault (Arcyana, 1975): (1) the median part of the transform fault which lies half-way between the offset portions of the Rift Valley in an area where the crust should be about 1 Ma. old, assuming a mean of 1 cm/yr for crustal accretion to each plate, and where the crust should have been subjected to about 10 km of lateral shear (Fig. 7-1); and (2) the Rift Valley/transform fault intersection zone which is topographically expressed by a large roughly circular depression in which younger and better-preserved pillow flow was found than in the median part (Fig. 7-1).

Detailed description of the structure of the median part of Fracture Zone "A" was made by Choukroune et al. (1978). The axial zone consists of a 3-km-wide, "V"-shaped valley which is comprised of the main deep flanked by the two walls or shoulders of Transform Fault "A", the deepest portion of which is a little above 2800 m in the median part (Fig. 7-1).

*Geological setting of the median transform fault*

Some 38 samples were collected by "Cyana" in the median part of the transform fault. Of these some 9 are sediments or fragments of rock too small to study or analyse. Nearly all the samples were collected within a 0.6-km-wide, NNW-trending corridor running from the base of the southern, inner slope of the Main Deep, across the narrow floor of the deep and across the Northern Outer Scarp in the median part of transform fault "A" (Fig. 7-2). Two samples were collected about 1 km east of the corridor: one high on the edge of the plateau-like area at the top of the Northern Outer Scarp (sample 10), at a depth of about 2100 m (Fig. 7-1), and the other on a west-facing slope of the Main Deep at about 2700 m depth (sample 9). The sample numbers correspond to the last digit of the sample station number shown in Fig. 7-2.

The Main Deep within the area of the corridor is about 0.5 km wide at the 2675-m level, is about 100 m deep and has a narrow and steeper (45°) north-facing east-west scarp. These slopes represent a crowded succession of fault scarps on narrow intervening steps up to about 20 m wide. At the foot of the scarps a scattered, only partially buried, and generally unsorted rubble of pillow flow fragments occurs and it testifies to slumping under an active tectonic regime.

Fig. 7-1. Bathymetric map of a portion of the FAMOUS area showing location (rectangle) of Fracture Zone "A" where detailed submersible study was done. Bathymetric map is from Renard et al. (1975). Upper inset shows local structural setting of mapped area and lower inset the regional location of the upper inset.

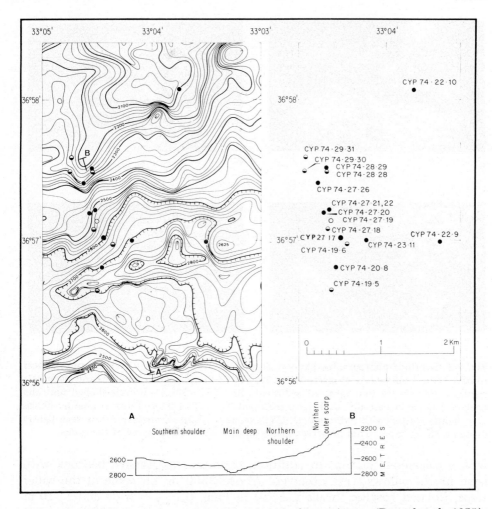

Fig. 7-2. Location of sea-floor stations shown on bathymetric map (Renard et al., 1975). Area covered by map is shown in Fig. 7-1. Filled circles: rock sampling stations with corresponding sea-floor photographs (Figs. 7-3, 7-4, 7-5); half-filled circles: rock sampling stations without accompanying photographs; open circle: photograph station only. CYP74 indicates that station was occupied by the submersible "Cyana" in the year 1974 and the last digit(s) give(s) the station number. The topographic cross-section is drawn without vertical exaggeration and shows the distribution of the physiographic micro-provinces described in detail by Choukroune et al. (1978).

Samples taken at stations 5, 6, 8, 9, 10 and 11 all represent examples of rubble deposits from the Main Deep (Figs. 7-2, 7-3A—D). Sample 6 is from the extreme upper edge, in the transition zone with the Northern Shoulder.

On the Northern Shoulder itself, east-west-trending scarps and scarp complexes 2—3 m high face toward the Main Deep. Rubble of pillow lava accumulates locally at the bases of the scarps. The rubble includes breccias

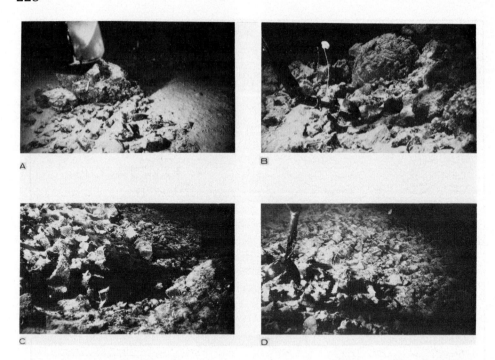

Fig. 7-3. Sea-floor photographs taken at sampling stations (Fig. 7-2) where rocks appear to have typical pillow lava characteristics. A. CYP74-20-8. Angular boulders and fragments partially covered by thin layer of sediment. B. CYP74-23-11. Globule-shaped unit and broken fragments partially buried by sediment. C. CYP74-22-9. Talus of angular debris very lightly dusted with sediment. D. CYP74-22-10. Talus of angular debris very lightly dusted with sediment. Thin preserved glassy surface was seen on some of the rocks.

with a calcareous matrix. In addition, exposures of massive bedrock with large in-situ pillows were observed. Above 2550 m, on some of the wider steps, isolated basaltic "walls", 1—10 m long, 0.1—1 m wide and 1—10 m high have been observed and described by Choukroune et al. (1978). In some cases they are closely crowded and are generally parallel to the regional east-west topographic grain, although direct observations made from the submersibles claim that some trend northeast-southwest to north-south. Some of the basaltic walls are locally expressed as small oval or dome-shaped peaks (Fig. 7-4A—C). Many of these outcrops have a planar anisotropy that affects either the whole unit or only the core. Others are massive and/ or brecciated throughout, or in the outer parts only. These enigmatic features are inferred to be parts of dykes or sills and the designation of "intrusion" or "dyke" is used with these qualifications. The dykes, in most cases, have breached the surface and in others, have nearly done so — and they have been subsequently exposed by the minor erosion that could be expected in a sloping terrain subject to seismic vibrations. Sampling stations 18, 19, 20 and 21 are on the Northern Shoulder at depths between 2500 and 2650 m (Fig. 7-2).

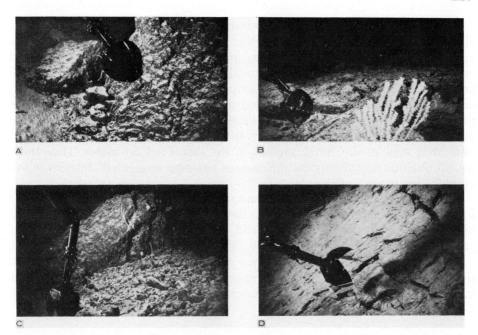

Fig. 7-4. Sea-floor photographs taken at rock sampling stations (Fig. 7-2). A. CYP74-27-22. View is towards north (05°). Sample was taken at the foot of massive outcrop about 2 m high identified as part of dyke (cf. Fig. 7-3). B. CYP74-27-26. Flat surface of massive and jointed bedrock with some broken material with slightly rounded edges and small rock debris. C. CYP74-27-20. Small outcrop identified as part of dyke; within zone of parallel to subparallel dykes, with angular and rounded massive blocks. D. CYP74-28-29. Fault scarp made up of subparallel sheets of intrusives. A massive altered slab 4 cm thick was sampled. The view is towards northwest.

Higher up, above about 2470 m, the base of the Northern Outer Scarp is marked by the appearance of north-south scarps cutting across the dominant trend of the topography. The east-west scarps are steep, often relatively bare of sediment and with considerable accumulations of presumably fault-induced rubble (including brecciated blocks) at the bases of scarps. A few of the scarps are more than 50 m high, evidently resulting from a large component of high-angle normal faulting. The exposed massive is jointed or bedded igneous rock, lying more or less horizontal or dipping steeply towards the south. Based on its field characteristics, the bedded rock is assumed to be intrusive (Fig. 7-4A, D). Samples taken at stations 26, 28, 29, 30 and 31 come from the North Outer Scarp (Fig. 7-4B, D).

One of the more interesting features of the rocks of the Transform Fault "A" is the sheet-like property of sample 29 and, to a lesser extent sample 6, and the cataclastic texture of sample 5 (Fig. 1-8C). At the top of the Northern Outer Scarp, pillow lavas reappear and lie locally exposed on a well-sedimented platform (sampling station 10) (Figs. 7-2, 7-3D).

Very few, if any, of the samples have hand-specimen structures that are diagnostic of pillow lavas. Many of the samples appear to have columnar as opposed to radial jointing and these include samples that were thought to come from pillow lava outcrops on hand-specimen evidence or field evidence or both (e.g. sample 10-C). Other samples are massive and nearly formless, or have only slight indications of jointing or are too small to display their natural structure (e.g. sample 5). Prominent vesicles are rare, except for sample 10C which appears to be part of a broken column, and it is not at all clear that vesicles are never present in the suspected intrusives. Thus there are no obvious morphological differences between suspected intrusives and extrusives as groups of samples.

As noted, breccias are more common among the transform fault samples than in those of the rift; and this reflects the field observations. Examples of breccias are samples 17, 21, 22 and 26 (Fig. 7-5). In all cases individual basaltic fragments in the crystalline breccias are homogeneous, suggesting a local derivation. There may be some association of brecciation with intrusives (injection), but then there is also field evidence of abundant breccias in the Main Deep. The breccia cements are not baked, so cement may be

Fig. 7-5. CYP74-27-17. Fragment of talus breccia from the northern wall of Transform Fault "A" near 37°N on the Mid-Atlantic Ridge (Fig. 7-2). The angular fragments of basaltic glass (dark) are cemented by pelagic sediment intermixed with basaltic debris. *Pl* indicates the zones of palagonitization (light) which surround fresh cores (dark) of glassy basalts.

post-intrusion. The lack of striations and of recrystallization effects suggests that the samples are slope breccia (Fig. 7-5).

*Degrees of rock alteration*

Some of the samples appear to be surprisingly fresh considering that the predicted age of the crust in the sampling area is about 1 Ma, and are indeed fresh compared even with samples from the adjacent inner floor of the Rift Valley. Samples with a notably glassy surface from Transform Fault "A" are 10, 10-B, 10-C, 10-D and 31-C, all of which have $H_2O$ contents of less than 1%, and $Fe_2O_3/(Fe_2O_3 + FeO)$ ratios of less than 0.16. By contrast, other samples (8, 8-1, 9, 10-A, 10-C, 11, 20, 21, 26 and 28) are moderately altered and show the occurrence of secondary products of mineral replacements such as a fair amount of iddingsite replacing olivine and disseminated smectite throughout the matrix.

The remaining samples (5, 6, 29, 30, 31-A, 31-B) are all weathered and altered, with up to nearly 20% of secondary phyllosilicates (14% in sample 6, 19% in sample 29; Table 7-1). X-ray diffraction study indicates that the phyllosilicates are primarily chlorite, smectite (sample 31-A), and mixed layer chlorite-smectite. These minerals occur to a greater or lesser extent in veins and veinlets. In addition, there is replacement of olivine and to a lesser extent clinopyroxene by serpentinite, iddingsite, and in all cases except sample 5 amphibole needles. Broken-up fragments of plagioclase, clino-pyroxene and olivine with undulatory extinction suggest a dynamometa-morphic type of alteration for sample 5. The effect of stress is best indicated by the large crystals which are ruptured, and in which the ruptures have been filled with smectite, sphene, chlorite and, in some cases, serpentinite aggregates (Fig. 1-8C).

*Mineralogy and chemistry of the basaltic rocks*

The bulk-rock modes of the major minerals pyroxene, plagioclase and olivine show that samples 8, 8-1, 10-B, 10-D and 11 (with 10-A as a border-line case) fall within the field of the olivine basalts of the Rift Valley segment (Hékinian et al., 1976), all having less than about 40—41% pyroxene, more than 15% olivine and traces of Cr-spinel (Table 7-1).

The trace element analyses indicate a Cr (500—700 ppm) and a Ni (200—400 ppm) content comparable to the olivine basalts found on the Mt. of Venus located in the inner floor of the Rift Valley (FAMOUS area) (Chapter 2; see also Table 7-2). These are all samples for which there is no field evidence of their having come from intrusives. Samples 30, 31-A, 29, 18-1, 21 (with 28, 20 and 26 as borderline cases) fall outside the bulk olivine field and contain less than 17.5% olivine. Except for the samples 20 and 26, all rocks contain more than 40—41% pyroxene and thus tend towards the

TABLE 7-1

Bulk rock and modal analyses of basaltic rocks collected by "Cyana" in the median part of Transform Fault "A"

| | 5* | 6 | 8 | 8-1 | 9 | 10 | 10-A | 10-B | 10 |
|---|---|---|---|---|---|---|---|---|---|
| $SiO_2$ (wt.%) | 48.73 | 47.53 | 47.88 | 48.60 | 47.50 | 48.06 | 48.19 | 48.11 | 48 |
| $Al_2O_3$ | 15.03 | 15.13 | 14.72 | 15.37 | 15.64 | 15.85 | 15.64 | 15.56 | 15 |
| $Fe_2O_3$ | 5.11 | 4.92 | 1.84 | 1.73 | 1.83 | 1.00 | 2.03 | 1.41 | 1 |
| FeO | 4.80 | 3.91 | 7.28 | 7.30 | 7.83 | 8.04 | 7.59 | 7.74 | 7 |
| MnO | — | 0.17 | — | — | 0.17 | 0.16 | — | — | — |
| MgO | 7.92 | 8.20 | 10.38 | 10.38 | 10.48 | 12.44 | 10.40 | 10.60 | 10 |
| CaO | 10.73 | 11.55 | 12.47 | 12.07 | 11.78 | 11.70 | 11.87 | 11.95 | 11 |
| $Na_2O$ | 2.51 | 2.16 | 2.05 | 1.98 | 2.37 | 1.80 | 2.10 | 2.08 | 1 |
| $K_2O$ | 0.24 | 0.22 | 0.19 | 0.28 | 0.19 | 0.05 | 0.21 | 0.20 | 0 |
| $TiO_2$ | 1.07 | 1.05 | 1.05 | 1.02 | 0.91 | 0.56 | 0.88 | 0.90 | 0 |
| $P_2O_5$ | 0.12 | 0.11 | 0.16 | 0.15 | 0.08 | 0.04 | 0.07 | 0.09 | 0 |
| $H_2O$ | 3.91 | 4.98 | 1.05 | 1.02 | 0.96 | 0.56 | 0.86 | 0.82 | 0 |
| Total | 100.17 | 99.93 | 99.07 | 99.80 | 99.73 | 100.22 | 99.84 | 99.46 | 99 |
| C.I. | 46.68 | 50.89 | 57.42 | 55.91 | 59.59 | 61.82 | 57.25 | 58.05 | 56 |
| FeO*/MgO (mol.%) | 0.662 | 0.561 | 0.483 | 0.479 | 0.504 | 0.403 | 0.508 | 0.477 | 0 |
| FeO*/MgO (wt.%) | 1.19 | 1.02 | 0.86 | 0.85 | 0.90 | 0.72 | 0.91 | 0.85 | 0 |
| *Modal* (vol.%) | | | | | | | | | |
| Plag ⎧ ph | | 3.25 | | | | | | | |
| Plag ⎨ mcrph | | 8.75 | | | | 0.50 | 0.40 | 0.88 | |
| Plag ⎩ mtx | | 12.75 | 37.63 | 35.75 | | 43.15 | 39.88 | 32.39 | |
| Ol ⎧ ph | | | 1.38 | | | 2.40 | 0.58 | 1.31 | |
| Ol ⎨ mcrph | | 1.00 | 4.60 | 1.54 | | 1.05 | 0.97 | 1.97 | |
| Ol ⎩ mtx | | 2.00 | 13.82 | 18.20 | | 15.80 | 13.62 | 19.69 | |
| Px ⎧ ph | | 0.03 | | | | | | | |
| Px ⎨ mcrph | | 5.00 | | | | | | | |
| Px ⎩ mtx | | 22.50 | 11.82 | 37.06 | | 31.11 | 82.29 | 33.48 | |
| Fe-oxides | | 12.50 | 3.99 | 5.26 | | | 6.23 | 7.87 | |
| Spinel | | | tr. | tr. | | | 0.40 | 0.88 | |
| Phyllosilicates | | | | | | | 1.36 | | |
| Mesostasis | | 20.25 | 26.73 | 2.19 | | 0.37 | 4.28 | 1.59 | |
| Ol/(Ol + Plag + Px) | | 0.055 | 0.286 | 0.213 | | 0.205 | 0.173 | 0.256 | |

(Column 5* modal: "altered"; Column 9 modal: "≥ 60% cryptocrystalline")

*Sample numbers represent the last digits of the cruise and site numbers (Fig. 7-2).

FeO* = total Fe calculated as FeO.
Phyllosilicates include serpentine, iddingsite, chlorite and smectite. Mesostasis comprises glass and cryptocrystalline material.

| D | 11 | 18-1 | 20 | 21 | 26 | 28 | 29 | 30 | 31-A | 31-C |
|---|---|---|---|---|---|---|---|---|---|---|
| .79 | 47.26 | 49.54 | 49.30 | 48.10 | 49.18 | 49.40 | 48.14 | 48.96 | 48.40 | 49.96 |
| .68 | 15.49 | 14.40 | 14.80 | 14.63 | 14.77 | 14.42 | 14.40 | 14.16 | 14.17 | 14.40 |
| .35 | 2.24 | 2.03 | 1.90 | 2.41 | 1.88 | 1.79 | 4.41 | 2.96 | 1.89 | 0.89 |
| .68 | 7.77 | 6.62 | 7.39 | 7.12 | 7.69 | 7.21 | 4.62 | 7.67 | 7.76 | 8.87 |
| - | 0.23 | 0.16 | — | 0.15 | 0.17 | 0.16 | — | 0.18 | 0.16 | 0.17 |
| .66 | 10.96 | 9.23 | 10.40 | 10.20 | 9.63 | 10.30 | 9.40 | 7.33 | 9.71 | 9.23 |
| .07 | 11.56 | 11.84 | 12.00 | 11.90 | 11.98 | 12.14 | 11.28 | 11.34 | 10.80 | 11.71 |
| .99 | 2.27 | 2.36 | 2.27 | 1.76 | 2.12 | 1.82 | 2.46 | 2.35 | 2.36 | 2.29 |
| .16 | 0.16 | 0.32 | 0.20 | 0.24 | 0.31 | 0.33 | 0.25 | 0.23 | 0.17 | 0.20 |
| .91 | 0.85 | 1.11 | 1.14 | 1.12 | 1.16 | 1.14 | 1.24 | 1.48 | 1.13 | 1.10 |
| .10 | 0.11 | 0.15 | 0.15 | 0.12 | 0.13 | 0.13 | 0.17 | 0.17 | 0.10 | 0.14 |
| .66 | 1.25 | 1.46 | 1.18 | 1.59 | 1.10 | 0.97 | 3.61 | 3.13 | 3.47 | 0.65 |
| .05 | 100.15 | 99.22 | 100.73 | 99.34 | 100.12 | 99.81 | 99.98 | 99.96 | 100.12 | 99.61 |
| .29 | 59.31 | 49.89 | 54.67 | 52.44 | 52.24 | 52.16 | 48.82 | 43.97 | 48.25 | 49.09 |
| .468 | 0.501 | 0.513 | 0.491 | 0.511 | 0.547 | 0.480 | 0.513 | 0.791 | 0.547 | 0.588 |
| .83 | 0.89 | 0.92 | 0.88 | 0.91 | 0.97 | 0.86 | 0.91 | 1.41 | 0.97 | 1.04 |
| .22 | 0.80 | 3.82 | | 0.50 | 5.50 | | | | 1.13 | |
| .32 | 30.26 | 29.26 | 27.85 | 33.36 | 35.09 | 21.35 | 41.65 | 27.31 | 40.07 | |
| .67 | 1.4 | 2.94 | 4.84 | 3.76 | 1.37 | 2.79 | — | — | — | |
| .57 | 3.81 | 3.19 | 1.45 | 2.57 | 5.55 | 2.36 | — | — | — | |
| .13 | 13.23 | 2.20 | 2.43 | 6.14 | 7.11 | 5.76 | 0.29 | 1.50 | — | |
| | | 2.45 | | | 0.68 | | | | | |
| .65 | 34.27 | 40.69 | 16.22 | 43.17 | 28.89 | 29.60 | 39.49 | 45.16 | | |
| .72 | 3.41 | 7.11 | 9.20 | 8.50 | 5.73 | 6.04 | 5.30 | 9.89 | 6.05 | |
| | 0.60 | — | 0.24 | tr. | | 0.31 | | | | |
| | 3.00 | | | 0.99 | | | 18.50 | 16.13 | 5.86 | |
| .71 | 9.22 | 8.33 | 37.77 | 0.99 | 10.09 | 31.79 | | | | |
| .185 | 0.220 | 0.099 | 0.165 | 0.139 | 0.166 | 0.170 | 0.037 | 0.020 | 0.013 | |

∧ 60% cryptocrystalline

TABLE 7-2

Trace element analyses of basaltic rocks from the median part of Transform Fault "A" in the FAMOUS area

| | Olivine basalt | | | Plagioclase-olivine-pyroxene basalt | | | | Plagioclase-pyroxene basalt | | | |
|---|---|---|---|---|---|---|---|---|---|---|---|
| | 9* | 10 | 11 | 18 | 20 | 21 | 26 | 6 | 30 | 31A | 31C |
| Cr (ppm) | 446 | 658 | 505 | 535 | 541 | 536 | 435 | 212 | 304 | 404 | 399 |
| Ni | 342 | 336 | 258 | 212 | n.d. | 210 | 195 | 87 | 127 | 164 | 157 |
| V | 197 | 149 | 193 | 226 | 237 | 233 | 245 | n.d. | 278 | 242 | 246 |
| La | 3.7 | 1.3 | 4.8 | 6.7 | 6.7 | 6.8 | 7.2 | n.d. | 7.0 | 6.0 | 6.4 |
| Hf | 1.3 | 0.7 | 1.19 | 1.82 | 1.88 | 1.82 | 1.77 | n.d. | 2.35 | 1.85 | 1.7 |
| Ta | 0.36 | 0.08 | n.d. | 0.65 | 0.66 | 0.66 | 0.69 | | 0.65 | 0.48 | 0.6 |
| Th | 0.41 | 0.17 | 0.51 | 0.71 | 0.65 | 0.66 | 0.74 | | 0.73 | 0.49 | 0.6 |

*Sample numbers represent the last digits of the cruise (CYP74) and site numbers (19, 22, 23, 27 and 29) (Fig. 7-2).

Chemical analyses were performed on the bulk rock (X-ray fluorescence) for Cr, Ni and V; La, Hf, Ta and Th analyses by neutron activation (Bougault, 1980).

pyroxene enrichment exhibited by the only intrusive (CYP74-31-38; Arcyana, 1977) seen in the Rift Valley. Samples 30, 31-C and 31-A are considered to be plagioclase-pyroxene-olivine type basalts as defined in Chapter 1 and they have lower Cr (300—400 ppm) and Ni (100—200 ppm) contents than the olivine basalts (Table 7-2). Sample 6 is exceptional in that it also falls in the more pyroxene-enriched group but has no field evidence of being intrusive. This sample (6) is comparable to the more fractionated basalt types (samples 30 and 31-A) with relatively low Cr (212 ppm) and Ni (250 ppm) content (Table 7-2).

Microprobe analyses of two glassy basalt margins (samples 28 and 31-C) confirm the pattern of variability deduced from the bulk-rock compositions (Table 7-1). However, while this is true for some critical oxides such as the $FeO*/MgO$ ratio and $TiO_2$ content others vary with textural change. Indeed, the $K_2O$ content of sample 28 shows a bulk-rock content in potash (0.33%) much higher than its corresponding glassy margin (<0.20%) (Table 7-1).

Pyroxene enrichment representing a more fractionated nature of the basaltic rock is visualized by the distribution of the major mineral constituents of basaltic rock represented by the modal ratio of olivine/(olivine + plagioclase + pyroxene) against the crystallization index (C.I.) of Poldervaart and Parker (1964). The crystallization index that represents the recalculated normative mineral of the major mineral phases entering into basaltic rock (Poldervaart and Parker, 1964), varies between the extrusive flows and the recognized intrusions (samples 26, 28, 29, 30 and 31-A; Fig. 7-6). These intrusive rocks have a C.I. of 43—52 and an olivine/(olivine + plagioclase + pyroxene) ratio of less than 0.2 (Fig. 7-6). Samples 10, 10-A, 10-B and 10-D which were recognized to have been collected from a pillow lava and hence to be from outcrops of an extrusive nature, have a higher C.I. (52—62) and

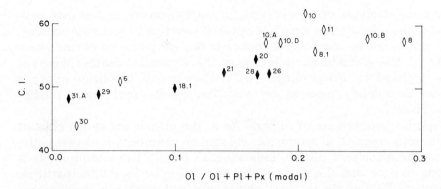

Fig. 7-6. Variation diagram for Transform Fault "A" rocks: C.I. (crystallization index, Poldervaart and Parker, 1964) vs. olivine/(olivine + plagioclase + clinopyroxene) recalculated to 100%. Filled diamonds indicate samples from — or within the immediate vicinity of — outcrops with field characteristics and relationships indicating that they are intrusives.

a higher olivine/(olivine + plagioclase + pyroxene) ratio (0.18—0.3) than the intrusives (Fig. 7-6).

The continuous decrease in olivine content is related to the degree of bulk-rock fractionation (Fig. 7-6). The transition between the extrusive and intrusive rocks is gradual and it is represented by samples 21, 20, 26, 28 and 18-1 which still have chromite and a fair amount of early-formed Mg-rich olivine. The continuous depletion of the bulk olivine content is accompanied by a general increase in $TiO_2$ and by the complete disappearance of chromite (Tables 7-1, 7-2). Hence, most of the intrusive rocks are enriched in titania (>1%) and potash (>0.25%) contents, which is what is expected during differentiation of a basaltic melt.

Alternatively, the concentration of early-formed phases in intrusive flows could be explained by the effect of size sorting due to the increased segregation between the walls and the center of the intrusion, as demonstrated by Komar (1972). However, this may not be applicable here since the massive flows are relatively small (<2 m wide) and the basalts from the presumed intrusive zones are generally aphyric, with less than 13% of phenocrysts (Table 7-1). Also, the haphazard sampling of the intrusive outcrops and their persistently more fractionated nature with respect to the extrusive rocks suggest a compositional diversity reflecting magmatic changes prior to the emplacement of the flows.

*Magmatic history*

The olivine occurs as phenocrysts, microphenocrysts and as euhedral crystals in the groundmass. No zoning or reaction rims were observed. The compositional variation expressed in percent forsterite molecules varies between about $Fo_{73}$ and $Fo_{87}$.

236

Microprobe analyses of phenocrysts, microphenocrysts and matrix constituents of the olivine show a narrow range of variation in forsterite content, and the difference in composition between the early-formed olivine phase and that of the groundmass is less than 2% forsterite. Further analysed samples depleted in modal olivine (<2%) show relicts of olivine minerals which are completely replaced by smectite, calcite and various phyllosilicates (Table 7-1).

The relative proportions of Mg and Fe in the olivine and in the glass of the bulk rock suggest a systematic change in composition between the melt and the associated olivine, indicating an equilibrium condition. It is interesting to note that the value of the slope given by $K_D = 0.30$ (partition coefficient of Fe and Mg between olivine and bulk rock, see Appendix) is close to that determined for the olivine basalts of the rift and their fractionated derivatives (Fig. 7-7).

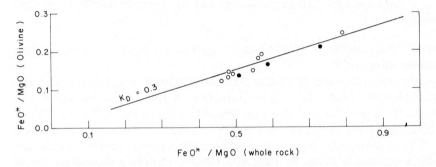

Fig. 7-7. FeO*/MgO (olivine) vs. FeO*/MgO (bulk rock) for sample from Transform Fault "A" (filled circles; this work) and Rift Valley samples 10-14, 13-4 and 7-6 (open circles; Arcyana, 1977; Hékinian et al., 1976). Where there is more than one value for olivines the value representing the highest FeO*/MgO has been taken.

From the above petrographic and chemical observations, it is demonstrated that the compositional variation of the volcanic rocks across the median part of the transform fault does not follow an orderly pattern (Fig. 7-8). The zone of intrusion exposed on several scarps of the south-facing wall between 2300 and 2700 m depth includes both olivine-enriched and pyroxene-enriched basalts. The olivine/(olivine + plagioclase + pyroxene) (modal vol.%) and the $TiO_2$ contents of the intrusives are lower (<0.17) and higher (>1.1%), respectively, than that of the extrusive flows. Fig. 7-7 shows the lack of correlation between the compositional variation of the volcanic rocks and the depth of exposure. This confirms, as previously shown for the major oceanic fracture zones of the Atlantic Ocean (Francheteau et al., 1976), that a simple and continuous layered structure is unlikely to be exposed in fracture zones. It is probable that layer 2 of the ocean crust is made up of a complex series of dykes and sills of different structure and composition intermixed with extrusive flows. Furthermore, the presence of

237

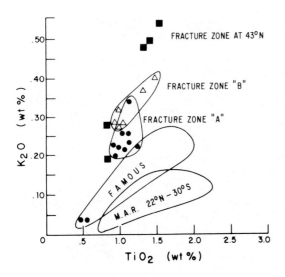

Fig. 7-8. $K_2O$-$TiO_2$ (wt.%) binary diagram of volcanics from North Atlantic fracture zone and from Rift Valley and Rift Mountain provinces of the Mid-Atlantic Ridge (FAMOUS area near 36°50'N and other ridge systems). The data from Fracture Zone "B" and from other segments of the Mid-Atlantic Ridge near 22—23°N and 30°S are from G. Thompson et al. (1980). The data from fracture zone near 43°N are from Shibata et al. (1979b). The least fractionated rocks from Fracture Zone "A" are the highly-phyric plagioclase basalts (samples 31 and 33).

more exposed intrusive zones observed in Fracture Zone "A" than on the inner floor of the Rift Valley near 36—37°N may explain some compositional difference between the volcanic rocks from the fracture zone and those of Rift Valley provinces and other localities on the ridge (Hékinian and Thompson, 1976). In addition, a compositional difference is also observed between Fracture Zone "A", Rift Valley in the FAMOUS area, and Fracture Zone "B" samples. Fracture Zone "B", located south of the FAMOUS area on the inner floor of the Rift Valley at 36°30'N, consists of basalts which are more fractionated than those encountered in Fracture Zone "A", as expressed by the $K_2O$-$TiO_2$ variation of their glassy margins (Fig. 7-8). The glassy basalts from both fracture zones show, in general, a higher $K_2O$ content (0.20—0.35%) compared to the average Mid-Atlantic Ridge basalts (0.16%: Fig. 7-8).

RIFT VALLEY/FRACTURE ZONE "A" INTERSECTION ZONE

The intersection of the Rift Valley with Transform Fault "A" is delineated by a depression with a radius of about 3.5 km at a maximum depth of about 3000 m (Fig. 7-9). This depression is made up of a flat-lying sedimented floor bounded to the north by steep east-west escarpments about 120 m high (Fig. 7-9). Details of the structure and morphology of this zone have been

238

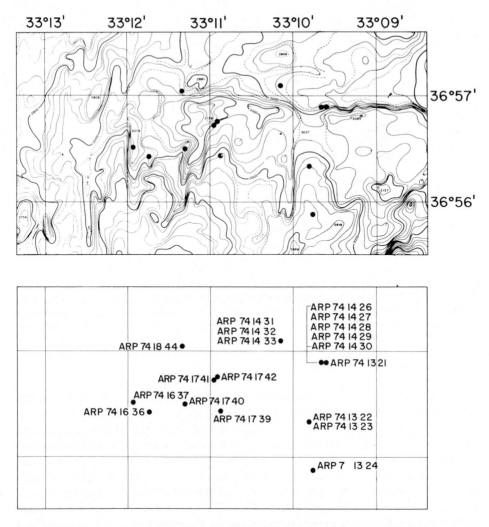

Fig. 7-9. Bathymetric map and sample distribution of the Rift Valley/Transform Fault "A" intersection zone. The sample numbering system is the same as that described for Fig. 7-2.

previously given (Arcyana, 1975; Choukroune et al., 1978). Recent tectonic activity, expressed by the presence of open fissures associated with relatively fresh pillow flows, was detected. Glassy olivine basalts were found to occur within the samples collected from these fissures (samples 31, 32 and 33) (Table 7-3). Ropey texture showing the flow direction appeared in one of the samples (33). These samples collected from the east-west escarpment are also among the most primitive basalts found so far from Transform Fault "A" and they show accumulation of plagioclase and olivine (Fig. 7-9). Samples 31, 32, and 33 are considered to be highly-phyric plagioclase basalts (HPPB) because

TABLE 7-3

Bulk-rock chemical analyses of basalts from the Rift Valley/Transform Fault "A" intersection zone near 37°N in the FAMOUS area

| | Highly-phyric plagioclase basalts | | | Olivine basalts | Plagioclase-pyroxene basalts | | | | | | |
|---|---|---|---|---|---|---|---|---|---|---|---|
| | 31* | 32 | 33 | 36 | 24 | 26 | 27 | 28 | 29 | 40 | 41 |
| $SiO_2$ (wt.%) | 47.30 | 47.36 | 47.22 | 50.10 | 50.45 | 50.47 | 49.73 | 50.37 | 49.60 | 50.17 | 49.72 |
| $Al_2O_3$ | 19.20 | 17.77 | 18.97 | 15.20 | 14.91 | 14.98 | 5.00 | 15.16 | 14.64 | 5.37 | 14.74 |
| $Fe_2O_3$ | 1.21 | 1.06 | 0.98 | 1.30 | 1.51 | 2.43 | 2.68 | 1.66 | 3.34 | 1.81 | 1.50 |
| FeO | 6.07 | 7.07 | 6.64 | 7.47 | 8.19 | 6.77 | 6.61 | 7.04 | 6.55 | 7.20 | 8.20 |
| MnO | 0.13 | 0.14 | 0.14 | 0.15 | 0.17 | 0.16 | 0.17 | 0.17 | 0.17 | 0.16 | 0.17 |
| MgO | 8.60 | 9.56 | 9.21 | 9.30 | 7.90 | 7.37 | 7.67 | 7.63 | 7.74 | 7.27 | 7.62 |
| CaO | 13.85 | 13.37 | 13.61 | 12.60 | 12.71 | 13.49 | 13.25 | 13.45 | 12.95 | 12.81 | 12.30 |
| $Na_2O$ | 1.95 | 1.82 | 2.16 | 2.17 | 2.28 | 2.30 | 2.72 | 2.70 | 2.48 | 2.05 | 2.32 |
| $K_2O$ | 0.13 | 0.11 | 0.17 | 0.22 | 0.27 | 0.25 | 0.26 | 0.16 | 0.33 | 0.39 | 0.28 |
| $TiO_2$ | 0.43 | 0.48 | 0.45 | 1.09 | 1.05 | 1.04 | 1.03 | 1.05 | 1.03 | 1.11 | 1.21 |
| $P_2O_5$ | 0.05 | 0.05 | 0.05 | 0.14 | 0.12 | 0.12 | 0.12 | 0.12 | 0.11 | 0.28 | 0.29 |
| LOI | 0.67 | 0.66 | 0.68 | 0.89 | 0.81 | 1.20 | 1.07 | 1.03 | 1.40 | 1.24 | 1.28 |
| Total | 99.59 | 99.45 | 100.18 | 100.63 | 100.37 | 100.58 | 100.31 | 100.54 | 100.44 | 99.72 | 99.52 |
| Cr (ppm) | 401 | 430 | 412 | 511 | 167 | 123 | 122 | 127 | 119 | 131 | 89 |
| Ni | 176 | 183 | 177 | 160 | 82 | 65 | 68 | 72 | 66 | 77 | 64 |
| V | 142 | n.d. | 143 | 237 | 250 | 242 | 243 | 245 | 240 | 248 | 265 |
| La | 1.8 | 1.7 | 1.6 | n.d. | 5.2 | 6.1 | 6.6 | 6.8 | 6.6 | n.d. | 9.2 |
| Hf | 0.51 | 0.66 | 0.53 | n.d. | 1.5 | 1.6 | 1.6 | 1.6 | 1.5 | n.d. | 2.1 |
| Ta | 0.13 | 0.15 | 0.14 | n.d. | 0.55 | 0.60 | 0.61 | 0.61 | 0.61 | n.d. | 0.89 |
| Th | 0.17 | 0.17 | 0.16 | n.d. | 0.55 | 0.63 | 0.62 | 0.62 | 0.64 | n.d. | 0.94 |

*Sample numbers represent the last digits of the cruise (ARP74) and site numbers (13, 14,
16 and 17).

Major elements and Cr, Ni and V analysed by X-ray fluorescence; La, Hf, Ta and Th
analyses by neutron activation (Bougault, 1980).
LOI indicates that the volatiles (mainly $H_2O$ and $CO_2$) were determined by loss of ignition.

there is more than 25% by volume of plagioclase phenocrysts and macro-
phenocrysts in the glassy chilled margins of the pillow flows. Also, the
relatively high MgO (8.5—9.5%), high Cr (400—600 ppm) and high Ni (100—
200 ppm) contents corroborate the common occurrence of olivine and
spinel in these rocks (Table 7-3). However, the high $Al_2O$ content (17—20%)
of the bulk rocks is in agreement with the abundance of plagioclase content
(Table 7-3). These HPPB differ from those encountered in the adjacent inner
floor of the Rift Valley in their relatively more abundant olivine content.
The majority of the samples collected from or near the east-west escarpment
consist of more fractionated rock types and are classified as plagioclase-
pyroxene basalts comparable to phases encountered in the median part of
Transform Fault "A". They are usually columnar in appearance and have a
subophitic to ophitic texture.

## CHARLIE GIBBS FRACTURE ZONE NEAR 53°N

The Charlie Gibbs Fracture Zone shows a 300-km east-west offset of the Mid-Atlantic Ridge and a north-south offset of about 110 km, separating the Reykjanes Ridge to the north from the ridge system extending to the Azores platform to the south. The median valley of the transform fault shows a discontinuous central ridge system separated by two parallel troughs (Fleming et al., 1970). Detailed bathymetric, structural, and petrological data were presented in previous studies by Olivet et al. (1970, 1974), Hékinian and Aumento (1973) and Campsie et al. (1973b).

The dredged hauls from the Charlie Gibbs transform fault were collected from the central ridge (DR201, DR204) and from the northern wall (DR7, DR203) (Hékinian and Aumento, 1973; Campsie et al., 1973b). One dredge (DR2) was also recovered from the southern wall of the transform fault. Serpentinized peridotites together with basaltic rocks were found to occur on both the northern (DR203) and the southern wall (DR2). A gabbroic rock was also collected using a hydrostatic drill located on the top of the central ridge. All the basalts collected consist of low-K tholeiites and are relatively old. A K-Ar age determination of one sample from a dredge (DR2) gave an age of 47 Ma B.P. which is consistent with the age of the crust extrapolated from magnetic anomalies. A fresh glassy basalt (DR7) recovered from the intersection zone of the transform fault with the Reykjanes Ridge suggests relatively recent volcanic activity in the area.

The basaltic rocks were divided into several types according to their order of crystallinity and the relative abundance of the mineral phases present, as previously described in Chapter 1. Thus four major groups were determined: (1) the plagioclase-pyroxene basalts, (2) the highly-phyric plagioclase basalts (HPPB) ($Al_2O_3$ >17%), (3) the plagioclase-olivine-pyroxene basalts ($Al_2O_3$ = 14—16%; MgO < 9%) and (4) the olivine basalts (MgO > 9%).

Metamorphosed basalts were also found to occur in association with weathered basalts and mylonitized material (DR2 and DR203). The basalts were altered to a low-grade metamorphism with chlorite—prehnite—albite and chlorite—albite—K-feldspar assemblages.

## OCEANOGRAPHER FRACTURE ZONE

The existence of the Oceanographer Fracture Zone was inferred from an offset in the Mid-Atlantic Ridge earthquake epicenter belt in 1967 (Fox et al., 1969). Early surveys have shown that the ridge is offset about 150 km in a right-lateral direction. The depth of the transform area ranges from 1750 to 2400 m with regional slopes as high as 15—25° (Fox et al., 1969). The Oceanographer Fracture Zone is well defined by the 3000-m contour and it is characterized by steep escarpments which rise 1600—2200 m above the floor of the fracture zone. The transform area of the fracture zone is narrow (<1 km) (Shibata and Fox, 1975). The Oceanographer Fracture Zone is morphologically similar to the Atlantis and the Kane Fracture Zones, located at 30°N and 23—25°N respectively, and differs from other equatorial fracture zones such as the Vema, St. Paul and Romanche, which are characterized by a wide transform valley, median ridges and a large ridge-ridge offset (Shibata and Fox, 1975). Both the north and south walls of the transform area revealed the presence of basalts, gabbros, serpentinized peridotite and metamorphosed basalts (Fox et al., 1976).

Seismic refraction studies were conducted on the western limb along the axis of the Oceanographer Fracture Zone and resulted in the identification of three major discontinuities: An upper 2.1-km-thick layer with a velocity of 4.4 km/s, an intermediate layer (about 6 km thick) with a velocity of 6.5 km/s, and a bottom layer below a crustal depth of about 5.5—6 km with a velocity of 7—8.2 km/s (Fox et al., 1976). The compressional wave velocities determined on the dredged rock samples at various pressures down to about 2 kbar are shown in the Appendix.

The relationship between in-situ seismic velocity measurement of the transform fault and the compressional wave velocity measurements on the rocks suggests that a quasi-layered oceanic crust and upper mantle might exist. Indeed, the 4.4-km/s layer could be made up of fresh and altered basalts (2.68—2.85 g/cm$^3$) while the 6.5-km/s layer could consist essentially of fresh and altered gabbroic rocks (2.89 g/cm$^3$). The serpentinized peridotite would represent the material that filled fractures and fissures developed during major tectonic events associated with the fracture zones. This model, suggested by Fox et al. (1976), implies that the serpentinized peridotite had migrated upward from the mantle (7- to 8.2-km/s layer) along fractures of the transform fault system. The dredged basalts from the Oceanographer Fracture Zone are similar in composition to other mid-oceanic ridge basalts erupted on accreting plate boundary regions. They differ from other basaltic rocks found in islands (Hawaii), island arc provinces (Japan) and aseismic ridges in their lower trend of iron enrichment. The basalts recovered in the transform area consist of plagioclase, olivine and/or clinopyroxene. The majority of the basalts contain phenocrysts of plagioclase and olivine set in a microcrystalline matrix (Shibata and Fox, 1975). The groundmass consists of clinopyroxene, plagioclase and iron-bearing minerals. Some specimens have clinopyroxene and plagioclase coexisting together.

High-temperature experimental work was carried out on two types of basalts, the plagioclase-olivine-rich basalts and the clinopyroxene-plagioclase-rich basalts from the Oceanographer Fracture Zone, at temperatures between 1050°C and 1250°C (D. Walker et al., 1978). From the fact that in the first type plagioclase was the first phase to crystallise with olivine and that in the second type clinopyroxene crystallized first, it was concluded that the two types of rocks were not derived from each other by a simple process of crystal fractionation in a magma chamber. The crystallization sequence is spinel, plagioclase at 1150—1250°C, olivine at 1150—1200°C and clino-pyroxene at 1150—1170°C for the plagioclase-olivine type of basalt. The plagioclase-pyroxene basalt shows plagioclase and clinopyroxene crystallizing at 1050—1200°C while olivine is solidifying at 1050—1150°C (D. Walker et al., 1978).

On the olivine-diopside-silica ternary diagram, D. Walker et al. (1978) have used the composition curvature of the liquidus phase boundary diagram to explain the mixing model proposed. Mixing of the two types of residual melts could give rise to a third type of rocks. This mechanism is visualized by the concavity of the three phases curving towards the diopside corner which allows mixtures to be produced within the clinopyroxene field (D. Walker et al., 1978). The crystallization of pyroxene from mixing of two melts could also explain the origin of some plagioclase-pyroxene basalts found in the rift valley in the FAMOUS area (Chapter 2).

## ROMANCHE FRACTURE ZONE

The Romanche Fracture Zone is the longest, largest and deepest fracture zone encountered in the Atlantic Ocean. It was detected over a distance of more than 8000 km from coast to coast between the Gulf of Guinea and Brazil. The transform area of the Romanche Fracture Zone offsets the axis of the Mid-Atlantic Ridge by about 1000 km and is characterized by east-west lineations of the basement morphology (Gorini, 1976). The deepest part of the transform fault, bounded by the northern and the southern walls, consists of a transform valley varying considerably in water depth from about 5000 to 7800 m. The deepest part of the transform valley, attaining about 7800 m, is located at about 17—18°W and is called the Vema Deep.

Because of its size, numerous studies dealing with the structure and petrology have been carried out covering a relatively large geographic surface of the Romanche transform fault. A more detailed survey of a portion of it was made during the 1977 cruise of the R.V. "Jean Charcot". The objective was to map and to sample a small portion of the northern wall of the Romanche transform fault near 18°26'W and 00°01'S. A multi-channel echo-sounder

---

Fig. 7-10. Bathymetric chart of the northern wall of the Romanche Fracture Zone between 18°23' and 18°33'W. A multichannel narrow beam echo-sounder was used (courtesy of L. Berthois, J. C. Duplessy and L. Labeyrie). In addition to the Sea-Beam, the vessel was also equipped with an acoustic transponder navigation system (ATNAV II-AMF) which permits positioning accuracy on the order of 75 m for both the surface ship and that of any towed instrument deployed in the transponder network moored on the ocean floor.

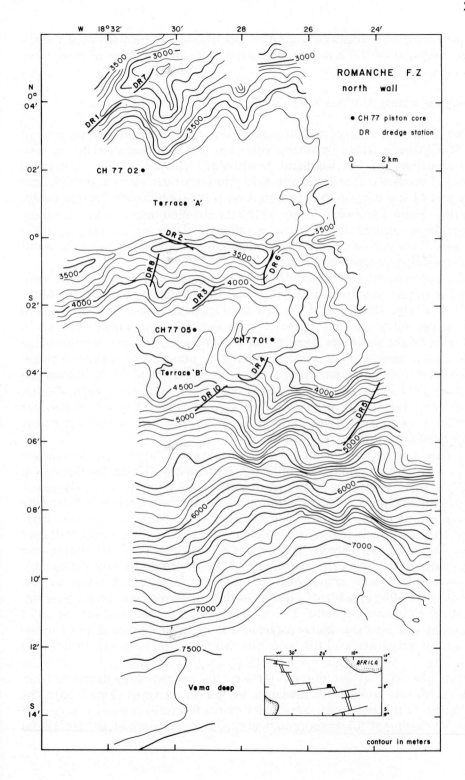

243

ROMANCHE F.Z
north wall

● CH 77 piston core
DR dredge station

0    2 km

Terrace 'A'

CH 77 02 ●

DR 7

DR 1

Terrace 'A'

DR 2

DR 8

DR 6

DR 3

CH 77 05 ●

CH 77 01 ●

Terrace 'B'

DR 4

DR 10

DR 13

Vema deep

AFRICA

contour in meters

coupled with an acoustic transponder navigation system permitted the detailed mapping of the area in combination with the dredging of recognized outcrops.

*Geological setting of the northern wall*

The transform area of the Romanche Fracture Zone offsets the axis of the Mid-Atlantic Ridge by about 1000 km and it is characterized by east-west lineations of the basement morphology (Gorini, 1976). A detailed survey of the Romanche transform fault was performed along a corridor 7 km wide and 35 km long on the northern wall between 2800 and 7660 m depth, near the Vema Deep (Fig. 7-10). From the detailed map of bottom topography (20-m contour lines) it was possible to delineate the main tectonic features of the surveyed area. The average slope of the northern wall is about 15°, but steeper scarps occur locally.

The upper portion of the northern wall consists of two distinct east-west-trending ridges: one located at 2900—3500 m (Ridge I), and one at 3500—4500 m (Ridge II) (Fig. 7-11A). These ridges are bounded by deep scarps with slopes up to 35—40° terminating in isolated peaks which reach up to 600 m in height above the surrounding topography. These east-west-trending ridges are relatively small when compared to other major transverse ridges which were previously recognized along the main strike of the Romanche Fracture Zone (Heezen et al., 1964; Bonatti and Honnorez, 1976; Gorini, 1976). Two major transverse valleys or terraces about 3—5 km wide occur at 3750 m (Terrace A) and 4300—4500 m (Terrace B) at the feet of the east-west-trending ridges (Fig. 7-11A).

Terrace A is slightly inclined towards the west (=1°) and may be enclosed in the east by a junction between Ridge I and Ridge II near the Equator at 0°02'S. Terrace B shows a more rugged topography than Terrace A. Sediment cores with less than 10 m penetration taken on these terraces indicate the presence of nannofossil oozes. The sediment from Terrace A is undisturbed when compared to that of Terrace B and contains Late Pleistocene sediment (core CH77-02). On Terrace B the lower 4 m of core CH77-01 contains an assemblage of mid-Pliocene age (zone NN 16) overlain by mixed Pliocene-Holocene micro- and nannofossils. In contrast, another core taken on Terrace B (CH77-05) consists of Pliocene sediment at the base overlain by undisturbed Holocene sediment. The presence of an undisturbed succession of Oligocene and Miocene fauna suggests that at least Terrace A was formed during the early history of the transform fault near the ridge axis (22—25 Ma ago).

From the edge of the deepest terrace (B), a steep slope dropping from 4500 m down to 7600 m descends to the Vema Deep (Fig. 7-10). The formation of the latter cannot be older than a few million years as suggested from the sediment thickness (about 300 m thick; Heezen et al., 1964) and

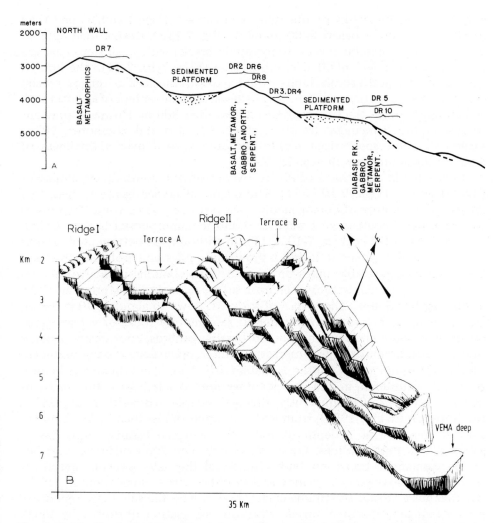

Fig. 7-11. A. North-south profile of the northern wall of the Romanche Fracture Zone near 18°26′W showing the distribution of the major rock types dredged. The projected distances of the various dredges are less than 4 km away from the chosen profile. B. Sketched block diagram showing east-west and north-south faulted structures.

from the exceptionally high sedimentation rate (50 cm/10³ years) in core CH71-08.

A sketched structural diagram (Fig. 7-11B) was made and interpreted from a block diagram constructed by an automated plotter (Berthois and Froidefon, 1978), using the bathymetric contour lines of the charts shown in Fig. 7-10. From the block diagram it is shown that east-west and north-south faultings are prominent structural features found on the south-facing steep scarps of the transform wall. The block-faulted structures are well developed on the flanks of the two main ridges (Ridge I and II) and they become less distinct below 5200 m depth (Fig. 7-11B). Deeper than 5200 m the steep slopes exhibit several topographic breaks with small steps near the 5300-m, 5900-m and 6000-m contour lines (Figs. 7-10, 7-11). The limited extent of the north-south faults could explain the lack of major crustal exposure due to vertical motion. It is believed that the extent of a continuous crustal exposure in the surveyed area is less than 900 m. However, previous studies in the Romanche transform fault (Bonatti and Honnorez, 1976) suggested that serpentinized peridotites are exposed over a thickness of 4 km on the southern wall near 18°40'W.

Dredge hauls were successfully recovered on the south-facing slopes of Ridge I and II (Figs. 7-10, 7-11). The major difference between these two ridges is the presence of homogeneous rock types recovered from the western part of Ridge I consisting of weathered and metamorphosed basalts (CH80-DR1 and CH80-DR7) (Figs. 7-10, 7-11). In addition, the bottom photographs taken along the slope of Ridge I show talus piles and isolated structural features which resemble dikes or massive flows standing up above the surrounding floor similar to those seen from submersibles in Transform Fault "A" in the FAMOUS area. The slopes of Ridge II, down to about 5100 m depth, are made up of heterogeneous material: sedimentary rocks of volcanogenic origin, weathered and metamorphosed basalts, gabbros, anorthosite, pegmatitic gabbros, prehnitized gabbros, serpentinites, ophicalcite and mylonites are commonly found (see Chapter 1, Tables 1-5, 1-6). The lithological heterogeneity encountered in the dredges does not correlate with the depth of exposure (Fig. 7-10). Previously dredged samples (Bonatti et al., 1971; Honnorez and Kirst, 1975), from within a radius of less than 20 km from the studied area and at a depth of 5100 m, recovered basalts, serpentinites, gabbros and metamorphics. Other rocks collected from different portions of the Romanche transform fault also reveal the heterogeneous nature of the outcrops exposed (Bogdanov and Ploshko, 1968; Bonatti et al., 1971). Whether such crustal layering is cyclic, that is with the successive repetition of a basalt-gabbro-serpentinized peridotite association disrupted by block faulting, or whether it represents a continuous lithological sequence as suggested by Bonatti and Honnorez (1976) has not been ascertained.

The heterogeneity of the rock types encountered even within a single dredge suggests an intensive layering of the exposed outcrops on the flank of Ridge II and on the deeper, steeper slope down to at least 5100 m (limit of sampling) (Fig. 7-10). On the basis of dredged material across the Romanche transform fault, Bonatti et al. (1970) suggested the presence of a stratigraphic sequence made up of a 1- to 2-km-thick basalt pile grading downwards into gabbroic rocks and metagabbros, with the base of the section consisting of peridotites. No recent volcanism seems to have occurred in the area surveyed (Ridges I and II). The exposed outcrops were formed during the early stages of magmatic and plutonic activities related to the creation of the crust on the ridge crest. Both low-grade metamorphism and hydration of the oceanic crust are likely to have been enhanced by geothermal processes which took place on the accreting plate boundary region of the Mid-Atlantic Ridge. A survey of the published data indicates that metamorphosed basalts occur at all levels within the transform area of the Romanche Fracture Zone.

*Bulk-rock petrology*

Altered basalts are the most common type of rocks encountered in the studied area, and particular importance is given to them. All the samples studied here consist of crystalline rock which had the original texture and composition of fresh basaltic rock. When considering the order of crystallization and the relative amount of the major mineral phases, the altered rocks from the Romanche transform fault are classified as being one of three major types: (1) the highly-phyric plagioclase altered basalts (HPPB) which contain feldspar megacrysts (samples 3-2, 7, 7-15, 7-1); (2) the moderately-phyric plagioclase altered basalts (MPPB) formed by a moderate amount ($<20\%$) of early-formed feldspar (samples 6-1, 7-5, 7-14); and (3) the plagioclase-pyroxene altered basalts which are aphyric and consist of rare plagioclase phenocrysts ($<2\%$) (samples 8-5, 3-6).

This classification of ocean floor rock is comparable to that used for fresh basalts from the FAMOUS area near 36°N on the Mid-Atlantic Ridge (Hékinian et al., 1976; Arcyana, 1977). Both the highly-phyric and the moderately-phyric rocks have relics of early-formed feldspar, varying in composition from $Ab_{35}An_{65}$ to $Ab_{10}An_{90}$. The clinopyroxene in the altered basalts is fresh and consists of a calcic augite ($Wo_{36-48}En_{35-53}Fs_{10-18}$) comparable to that found in recent accreting plate boundary regions along the Mid-Atlantic Ridge rift valley, as is also suggested by their low Ti/Al ratio. The original texture of the rocks is preserved and consists of subophitic, interstitial, and variolitic types.

However, bulk-rock analyses of ocean floor altered basalts have to be evaluated with caution. The heterogeneous aspect of the rocks is often seen on large fragments which have a preferential concentration of secondary mineral phases filling veins, veinlets, patches and vesicles and which enhance the chemical variability of the bulk samples. Hence, a detailed mineralogical and textural study of the altered rocks should be a prerequisite before adopting any kind of meaningful classification. Here the altered basalts from the Romanche transform fault are divided into two major groups according to the occurrence and abundance of secondary mineral assemblages and to the relative ratio of sodic and potassic feldspar present (Table 7-4): (1) weathered basalts: the plagioclase-clay mineral assemblage; (2) metamorphosed basalts: the albite-zeolite-chlorite and the K-feldspar-albite-chlorite assemblages.

TABLE 7-4

Bulk-rock analyses of samples from the Romanche Fracture Zone near 18°30'W in the Atlantic Ocean

|  | CH80-DR3-3 weathered basalt | CH80-DR2 gabbro | CH80-DR10-2 rodingite | CH80-DR4-1 harzburgite | CH80-DR4-3 harzburgite | CH80-DR10- lherzolite |
|---|---|---|---|---|---|---|
| $SiO_2$ (wt.%) | 48.52 | 49.86 | 27.94 | 39.47 | 37.63 | 38.41 |
| $Al_2O_3$ | 16.28 | 16.72 | 15.57 | 1.11 | 3.31 | 1.38 |
| $Fe_2O_3$ | 5.42 | 2.11 | 3.90 | 6.45 | 10.33 | 7.74 |
| FeO | 3.42 | 3.35 | 15.26 | 0.97 | — | — |
| MnO | 0.15 | 0.12 | 0.37 | 0.08 | 0.11 | 0.09 |
| MgO | 7.23 | 8.69 | 12.61 | 38.19 | 34.92 | 38.85 |
| CaO | 10.95 | 13.92 | 10.01 | tr. | 1.28 | 0.24 |
| $Na_2O$ | 3.21 | 2.21 | 0.03 | 0.06 | 0.33 | 0.48 |
| $K_2O$ | 0.35 | 0.1 | 0.03 | 0.02 | 0.03 | 0.00 |
| $TiO_2$ | 1.53 | 0.33 | 5.38 | 0.05 | 0.30 | 0.20 |
| $P_2O_5$ | 0.21 | tr. | 0.07 | tr. | 0.20 | 0.05 |
| LOI | 2.54 | 1.25 | 6.54 | 13.4 | 12.48 | 13.48 |
| Total | 99.81 | 98.66 | 97.71 | 99.8 | 100.44 | 100.74 |
| Co (ppm) | 50 | 51 | 86 | 72 | 34 | 67 |
| Cr | 322 | 744 | 143 | 2293 | 2365 | 2455 |
| Cu | 111 | 72 | <10 | <10 | — | — |
| Ni | 98 | 142 | 360 | 1728 | 1674 | 2245 |

Samples CH80-DR3-3, -DR2, -D10-2 and -DR4-1 were analysed by UV quantometric method (C.R.P.G. Nancy, France) and samples CH80-DR4-3 and -DR4-4 were analysed by X-ray fluorescence (H. Bougault, R. Hebert, C.O.B.).

More details about these rocks are found in Chapter 9 which deals with metamorphism.

*Ultramafic rocks* from the northern wall of the Romanche transform fault consist mainly of three types of serpentinized peridotite: (1) lherzolite which is found in dredge hauls DR4, DR8 and DR10; (2) harzburgite, found in DR4 and DR10; and (3) werhlite, found in DR10 (Figs. 7-10, 7-11). These ultramafics are abundantly altered and olivine occurs only as ghosts. The orthopyroxene, clinopyroxene and spinel are partially altered and from their compositional variation it was possible to detect differences in some of their chemical constituents. For example, the spinel of the harzburgite is deprived in its $TiO_2$ content (<0.1%) when compared to that of the lherzolite ($TiO_2$ up to 0.8—1.2 wt.%). The chromium content of the spinel from both types of peridotite is about the same, 37—41%. In addition the nickel content of the lherzolite is less important (<2000 ppm) than that of the harzburgite (2300—3000 ppm) (Table 7-4). It is also observed that the orthopyroxene from the Romanche lherzolite falls within the field of other peridotites from subaerial ophiolitic complexes as far as their $Mg/(Mg + Fe^{2+})$ = 90—91 and their $Al_2O_3$ = 2—3% is concerned. Whether these lherzolites from the Romanche transform fault are residues left after partial melting or cumulates from a magmatic melt is not clear. A composition based on the $TiO_2$ and $Al_2O_3$ content of various types of serpentinized peridotites from different tectonic settings is shown in Fig. 12-6 (Chapter 12). The lherzolites from the Romanche area are depleted in both $Al_2O_3$ and $TiO_2$ when compared to other serpentinized peridotites and may represent cumulates having a magmatic origin (Fig. 12-6).

It is interesting to notice that detailed studies reported from both small and large fracture zones such as Transform Fault "A" and the Romanche Fracture Zone do not suggest any evidence for large-scale crustal or mantle exposure due to normal faultings. Indeed the heterogeneous nature of the rocks has, up to now, failed to show any stratigraphic sequence from field evidence. The complex block faulting nature of fracture zones indicates an exposure of outcrops which does not exclude a rhythmic layering of various types of crustal material exposed in the studied areas. However, a diapiric origin for the emplacement of hydrated peridotite bodies might also have contributed to the formation of stratigraphic horizons in fracture zones. The circulation of seawater through fissures and large fractures will enhance the processes of alteration and the uprising of ultramafic and mafic material during faulting. The process of upwelling of relatively large masses must take place during extensive tectonic activities related to the formation of the transform areas. Major faulting and breaking of the oceanic crust is responsible for the outcropping of the various rock types at the same stratigraphic levels in fracture zones.

CHAPTER 8

## OCEAN FLOOR WEATHERING

Alteration of the oceanic crust is related to various phenomena which are involved in the constant change of existing ocean floor material. The purpose in studying altered material from the sea floor is to understand the chemical budget of seawater, the circulation of fluid in the crust, the type of tectonic activity involved and the nature of transport and redeposition of material extracted from the basement rocks. The main processes involved in the alteration of the oceanic crust are related to seawater weathering*, to meta-morphism, to hydrothermal circulations of fluid solutions and to deuteric compositional changes during magmatic solidification. Recognition of the various processes responsible for the alteration of the rocks requires know-ledge of post-consolidation events which have influenced the history of the oceanic crust. This and the following chapters (Chapters 9, 10 and 11) deal with the major mechanisms giving rise to the alteration of the oceanic crust created on accreting plate-boundary regions.

Ocean floor weathering involves a complex of rock formations exposed in a hostile environment where seawater circulation at low temperature favours alteration. Hence the weathering process is time dependent and should be studied in relation to the particular physico-chemical conditions existing at the sampling sites. Obviously, it is important to know if the sample was buried in sediment or exposed to seawater by a fracture of the ocean crust or by the extrusion of lava flows on the surface. Often, organisms encrusting the surface of rock samples could play a certain role in the process of weathering but this is minor when we consider the geological time scale. The circulation of seawater inside fractures and cracks is one of the most important factors facilitating the action of weathering.

It is also evident that the different shapes of pillows (tubiform, massive, bolster, tumulus-like pillows, etc.) could form incoherent piles with empty

---

*The term weathering should not be confused with alteration. Indeed, the process of rock alteration is a general one including the separate processes of metamorphism, diagenesis, and weathering. The problem of ocean floor weathering is controversial and difficult to separate from that of low-grade metamorphism and diagenesis. In addition, the process of deuteric alteration of early mineral phases is easily confused with that of weathering. Early-formed mineral phases are generally altered into a mixture of serpentine and mixed layer clay minerals. Zeolites and mixed layer clay minerals are often found in both low-grade burial metamorphism and during diagenesis. In addition, the overprint of ocean floor hydrothermalism modifies the mineral associations which could have been indicative of weathering processes. Chapters 8—11 are intended to differentiate between the various processes involved, in order to obtain insight into the mechanism causing the alteration of the oceanic crust.

spaces between the flows which are potential channels for water circulation within the upper part of the oceanic crust (Fig. 8-1). However, knowledge of the depths to which weathering processes have reached is still speculative.

In hole 518A of DSDP Leg 51 it was noticed from $^{87}Sr/^{86}Sr$ and $\delta^{18}O$

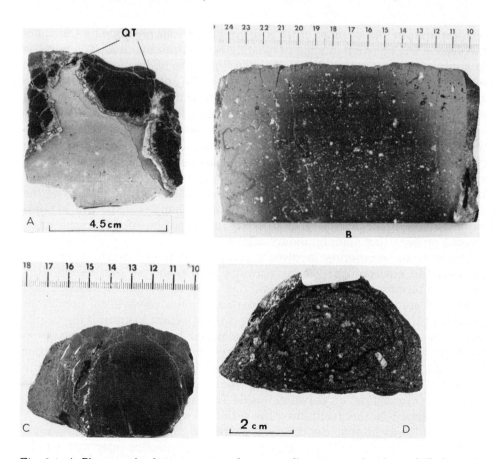

Fig. 8-1. A. Photograph of a contact zone between a limestone and a glassy chilled margin of a basalt from DSDP Leg 22, Site 215 (sample 215-18-1/68—73; Indian Ocean) (Hékinian, 1974a). The glassy margin is surrounded by a palagonite rim. Small patches of quartz (QT) occur between the limestone and the palagonite. This limestone was found trapped between two pillow flows. B. Section of a pillow lava flow (sample ARP73-10-2C) showing a chilled margin and alteration veins (perpendicular to the margin) due to the penetrating of seawater into cracks. This sample was taken by submersible in the FAMOUS area on a crust of about 400,000 years old (Fig. 8-4). C. Cylindrical pillow flow (sample CH4-DR1-58) showing concentric alteration veins around basaltic glass. The white specks consist of pelagic sediment. This sample is from the western wall of the Mid-Atlantic Ridge Rift Valley at 53°00.4'N, 34°59.4'W. D. Section of a plagioclase-pyroxene basalt (sample CH31-DR10-100C) located on the first escarpment of the western wall of the Mid-Atlantic Ridge Rift Valley in the FAMOUS area near 36°50'N. The dark concentric rings are due to the oxidation of Fe-rich compounds by seawater circulation.

studies (S. R. Hart and Staudigel, 1978) that authigenic smectite was formed in equilibrium with seawater throughout the basaltic basement down to 544 m depth (Donnelly et al., 1979). From their strontium isotopic studies on calcite and smectite, S. R. Hart and Staudigel (1978) concluded that the low-temperature alteration processes came to an early halt, maybe less than 20 Ma after the extrusion of the basalt on the ocean floor. Honnorez (1981), studying the weathering of sea floor material, has differentiated between the degree of alteration of samples exposed on the surface which were obtained by dredging or by submersibles and those from the sub-basement obtained by deep-sea drilling operations. Indeed, the samples collected from the surface of the ocean floor undergo a weathering action with a large water/rock ratio which tends to homogenize the degree of rock alteration through time. On the other hand, buried samples with a diffuse type of seawater circulation will alter in a more heterogeneous way.

WEATHERING OF EXTRUSIVE FLOWS

When the volcanics are exposed on the ocean floor for a relatively long period of time (>1 Ma), seawater weathering is more effective on the fragile glassy margins of the pillow flows whereas the more crystalline interior will be relatively less affected.

In hand specimens, the color zonation is often a good indicator of the degree of weathering that ocean floor basalts have undergone. In old volcanic rock (>10 Ma old) a gradual color change is observed from the outer margin of the rock in direct contact with seawater towards the interior of the sample. This color change varies from different shades of light brown to gray as follows:

brownish yellow → yellowish gray → grayish brown → dark-brown gray

In relatively young basaltic flows (<10 Ma old) the color change is sharper and goes from a succession of dark-gray halos to a light-gray fresh core (Honnorez, 1981). The degree of weathering corresponds microscopically to the presence of different types of clay minerals which assume various shades of green and brown. Among these, smectite and mixed layered clays are the most commonly encountered and these fill vesicles, cracks and interstices between minerals.

From chemical observations of weathered rocks, the potash exchange between seawater and the rock is, together with the $H_2O$ content, one of the most effective and noticeable parameters of alteration.

The trace elements in basaltic rocks which most authors have unanimously found to be among the most mobile during seawater weathering are Rb, B and Cs. S. R. Hart (1971), in studying the chemical action of seawater on basalt, showed that the outside margins of pillow flows were enriched in some large ion lithophile (LIL) elements such as K, Cs, and Rb with respect to the inner part of the rocks. However, in other cases, and especially young volcanics (<1 Ma old), it is noticed that these LIL elements, K, Rb, Cs and also $Fe^{3+}$ increase towards the inner part of the pillow flows. In some cases, the $K_2O$ content of the interior of a pillow flow could be as much as three times higher than that of its glassy margin (Hékinian, 1971a; R. B. Scott and Hajash, 1976). The increase of these LIL elements may be explained by the fact that the glassy margins have the most homogeneous surfaces, with only minor fissures or cracks so they act as a less porous medium than the interior of the flow. The more crystalline interior of a pillow flow has discontinuities between crystal boundaries that form pathways for seawater infiltration. Radial jointing in pillow lavas due to thermal contraction during cooling could also represent surface weaknesses through which seawater may infiltrate. This is observed in polished gray bands of hydrations at the margins of the jointing planes (Fig. 8-1).

The concentration of the LIL elements is likely to occur in amorphous material locally filling cracks in mineral interstices and coating vesicle walls. Other authors (S. R. Hart, 1971; Frey et al., 1974) have shown the mobility of the light REE (from La to Sm) in relation to the heavy REE (from Eu to Lu). It was also suggested (Aumento and Milligan, 1971; D. E. Fisher, 1978) that the uranium content increases with the degree of rock weathering. The transitional metals are also considered to be mobile elements when the degree of weathering is pronounced. Analyses of preserved glass set in a palagonitized matrix at the margin and of the most weathered interior of a pillow basalt dredged south of the Austrare Fracture Zone in the Pacific Ocean show a variable Ni content. The Ni content of the fresh glass is about 125 ppm while the brownish yellow interior of the rock has a Ni value of about 69 ppm. Similar observations were also made on DSDP Leg 46 basalts from the Mid-Atlantic Ridge (Ailin-Pyzik and Sommer, 1979) where Cu, Ni and Mn were found to decrease with an increase in the degree of alteration.

CHEMICAL CRITERIA FOR THE RECOGNITION OF WEATHERED ROCKS

The recognition of ocean floor weathering is often a difficult task when using only chemical tests. However, some indications may be obtained when a combined thorough mineralogical and chemical investigation is carried out.

Matthews (1971) used $Fe_2O_3/(FeO + Fe_2O_3)$ ratios plotted against certain oxides such as CaO, MgO, $K_2O$ and $H_2O$ which are particularly sensitive to weathering (Fig. 8-2). This diagram shows that when the oxidation ratios as defined ($Fe_2O_3/(FeO + Fe_2O_3)$)) are greater than 0.55 the basalts cannot be treated as fresh rocks (Matthews, 1971). The weathered basalts showed a loss in CaO and MgO accompanied by a gain in $K_2O$, $H_2O$ and the $Fe_2O_3/(FeO + Fe_2O_3)$ ratio (Fig. 8-2).

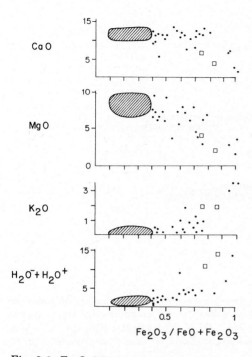

Fig. 8-2. $Fe_2O_3/(FeO + Fe_2O_3)$ vs. $H_2O$, $K_2O$, MgO, and CaO (after Matthews, 1971). The striped areas indicate the unweathered or slightly weathered basalts. The black dots represent the weathered basalts. The most weathered materials are the basalts from Swallow Bank, north east Atlantic Ocean (Matthews, 1971). Open squares indicate highly weathered basalts from the New England seamount chain (P. T. Taylor and Hékinian, 1971).

As mentioned at the beginning of this chapter, it is often difficult to decide whether a rock has been altered by metamorphism or hydrothermalism, or by weathering. This becomes critical when there is an absence of some key mineral phases characterizing the lowest facies of metamorphism. Miyashiro et al. (1971) have shown that low-grade metamorphism of oceanic basalts is primarily characterized by a more rapid increase of the $H_2O^+$ content rather than the $Fe_2O_3$ content. It is observed that low-grade metamorphics can vary considerably in $H_2O^+$ content, from about 1% to 9%, while their $Fe_2O_3$ content does not exceed 3%. This is explained by the oxidation of iron oxide minerals in the matrix due to introduction of seawater into the rock.

An alternative approach for distinguishing metamorphic phases from weathering is to consider the relative variation of potash and soda contents in the rock. Fig. 8-3 shows the plot of $K_2O/(K_2O + Na_2O)$ molecular percent versus total $H_2O$ weight percent variation for fresh and weathered basalts from near 36°N in the Atlantic Ocean (FAMOUS area) and some metamorphics of the greenschist and zeolite facies found in the Atlantic and Indian Oceans.

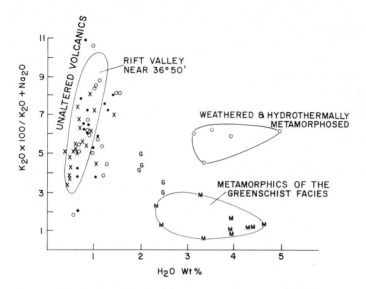

Fig. 8-3. $H_2O$ vs. $K_2O \times 100/(K_2O + Na_2O)$ diagram showing the distribution of unaltered oceanic basalts from the Mid-Atlantic Rift Valley region compared to metamorphic and weathered basalts. $\circ$ = Fracture Zone "A", 37°N, Mid-Atlantic Ridge; $\bullet$, $\times$ = Mid-Atlantic Ridge near 36°N; $G$ = Charlie Gibbs Fracture Zone; $M$ = metamorphics from 22°N, Wharton Basin and Carlsberg Ridge.

This variation diagram (Fig. 8-3) shows that an increase in the soda content causes a decrease of the $K_2O/(K_2O + Na_2O)$ ratio (<3) and when the $H_2O$ content is more than 2% this will characterize metamorphic basalts. Indeed, an increase in the $Na_2O$ and $H_2O$ contents explains the transformation of calcic plagioclase into more sodic varieties and the formation of chloritic minerals, respectively. During seawater weathering, introduction of both $K_2O$ and $H_2O$ will increase the values of the $K_2O/(K_2O + Na_2O)$ ratio (Fig. 8-3). The diagram in Fig. 8-3 also shows a series of unweathered basaltic rocks which have a limited $H_2O$ content (<1.2%) and a considerable variability in their $K_2O/(K_2O + Na_2O)$ ratios (2—11). These unweathered basalts, found on the inner floor of the Rift Valley near 36°N in the Atlantic Ocean, show various degrees of crystal-liquid fractionation which reflects their variations in $K_2O$ content due to the extrusion of different types of lava flows (see Chapter 2 for discussion on rocks from the FAMOUS area).

## PROGRESSIVE WEATHERING OF THE OCEANIC CRUST

Christensen and Salisbury (1975) have suggested that seismic velocities decrease away from the ridge with a progressive alteration of basement rocks. A survey of the literature on shipboard refraction data showing a number of profiles across the Mid-Atlantic Ridge near 50°N, from 30 to 40°N (Le Pichon et al., 1965), and from the Carlsberg Ridge near 5°5′N (Francis and Shor, 1966), indicates that the seismic velocities range from

6—7 km/s $(V_p)$ for a recent spreading zone to about 5—4 km/s $(V_p)$ for an 85-Ma-old crust. A number of factors, such as the rock vesicularity, grain size, fracturing, interbedded sediment and the degree of weathering, could

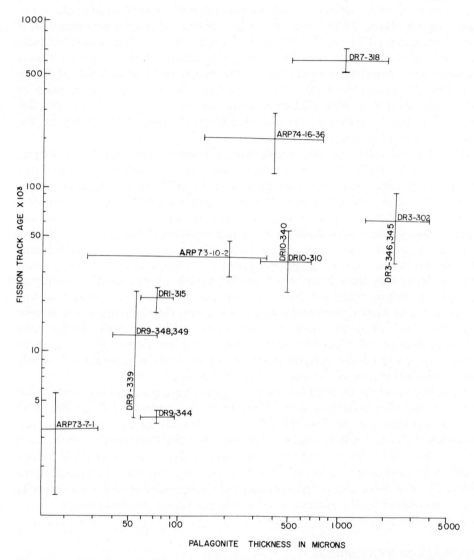

PALAGONITE  THICKNESS  IN MICRONS

Fig. 8-4. Thickness of palagonite coating the surface of pillow flows versus fission track ages (in 1000 years) of basaltic glassy margins. Some of the samples are from dredge hauls (DR) and others were taken by submersible (ARP) on the Rift Valley of the Mid-Atlantic Ridge, FAMOUS area, near 36°50′N. The length of the lines indicates the range of errors due to the method used. In most cases the fission track age and thickness of palagonite were determined on the same samples; otherwise different sample numbers are indicated along each line. The fission track measurements were made by Störzer and Selo (1974, 1976).

affect the measured refraction velocities of layer 2 (R. Hart, 1973). However, rock weathering and probably fracturing are the most important factors of the observed physical variation of the oceanic crust. Using seismic refraction velocities of oceanic layers and on various degrees of weathered rock, it was suggested (R. Hart, 1973) that the major ocean ridge systems consist of a low-velocity cap (3.27—4.85 km/s) on oceanic layer 2 which increases in thickness with distance from the ridge axis. This interpretation suggests that a seawater-rock reaction is operating in the ocean crust at a depth of about 3—5 km. An example of progressive alteration is given by the increase in palagonite thickness with distance from the ridge axis as shown for the samples of basalts collected on the Mid-Atlantic Ridge Rift Valley in the FAMOUS area (Fig. 8-4).

The regional study of weathered basalts from the eastern Pacific Ocean between 15°N and 35°S showed a broad pattern of variation between extrusive flows confined near the axial zone and the adjacent ocean basins of the East Pacific Rise. The rocks from the basin provinces were found on isolated submarine hills with relief of less than 600 m; the samples were collected from depths greater than 3500 m. Seventeen different locations and thirty-seven rock samples were compared during this particular study (Hékinian, 1971a). From the existing magnetic pattern the range in age of the oceanic crust is from less than 2 to about 80 Ma in the general area of sampling. All the specimens collected from the abyssal hill provinces consist of unweathered and highly weathered rocks, while the ridge basalts are relatively less weathered or fresh. However, sometimes within a single dredge haul different degrees of rock weathering exist; such variability, however, stays within the limit of the general trend of an increase in alteration towards the abyssal hill region of the eastern Pacific basins (Fig. 8-5).

$K_2O$, MgO and $Fe_2O_3$-FeO are the oxides which are the most affected by the regional variation reflecting the degree of rock weathering. Near the axial zone of the East Pacific Rise the values of $K_2O$ content, and $Fe_2O_3$/FeO are lowest, <0.5% and <0.3, respectively, and the MgO content is the highest (>7%) (Fig. 8-5). In the abyssal hill region, $K_2O$ and $Fe_2O_3$/FeO values increase to as much as 0.9% and 1.5 respectively, while MgO diminished to about 5% (Fig. 8-5). These various chemical changes observed in basaltic rocks are accompanied by the formation of low-temperature mineral phases.

## MINERALS OF WEATHERING

The process of weathering of ocean floor rocks varies according to the type of material that is exposed to the alteration processes. A glassy margin of a basaltic flow is likely to alter continuously while the crystalline interior will follow a stepwise process (Honnorez, 1981). Within the crystalline interior the matrix, composed of cryptocrystalline and glassy material, alters first

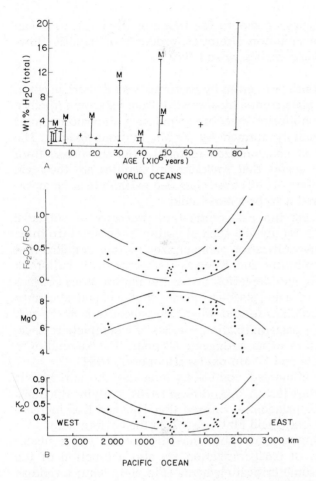

Fig. 8-5. Progressive alteration of the oceanic crust. A. $H_2O$ content versus age of the crust (after R. Hart, 1973). The tie-lines indicate the interior and margin $(M)$ of pillow flow. B. $K_2O$, $MgO$, $Fe_2O_3/FeO$ variation diagram versus distance from the axis of the East Pacific Rise (data from C. G. Engel et al., 1965; Paster, 1968 and Hékinian, 1971a).

together with the filled vesicles. Titamomagnetite is transformed into mag-hemite during the early stage of mineral alteration. Olivine is the next stage to be transformed, into hydrated Fe-oxides and smectite material. The plagioclase and the pyroxene are the most resistant to weathering.

The physical conditions on the ocean floor are different from those of continental weathering. The temperature on the floor of the ocean varies in general between 0 and 5°C. Studying the range of $\delta^{18}O$ calcite veins and clay minerals in sediment cores, T. F. Anderson et al. (1976) found that the temperature of the environment was less than 20°C. The pressure is also very limited and usually does not exceed more than 600 bars. Under such physical conditions, the formation of authigenic minerals is mainly limited to clay and zeolites. Only a few types of zeolite are found as the result of ocean floor weathering. The most common alteration products of basaltic flows encountered on the ocean floor are described below:

*Palagonite.* The term palagonite was given by Sartorius Von Walterhangen in 1846 to an altered basaltic glass from a hyaloclastite from Palagonia (Sicily). Palagonite is a mixture of phillipsite, smectite (saponite and/or montronite) and amorphous material mainly formed by Fe-Mn hydrous oxides. The mineral assemblages found in palagonitized material from the ocean floor are not well known, but it seems that smectite and zeolites are the main constituents. Rodgers and Kerr (1942) have classified palagonite as an amorphous material and considered it to be a mineraloid.

Honnorez (1981), studying the palagonitization processes of subaerial or temporarily submerged flows divided such alteration processes into three stages: an initial stage of palagonitization characterized by the crystallization of intergrowths Na > K-phillipsite and K-Mg-rich smectite with bulk-rock enrichment in K, Na and Mg and depletion in Ca; the mature stage consists of phillipsite and smectite replacing glass accompanied by a chemical change from Na > K- to K > Na-phillipsite and where the smectite is an Fe-rich nontronite; the final stage of palagonitization consists of a complete replacement of the glass by a mixture of an authigenic Ca-poor, K > Na-phillipsite with various types of smectite and Fe-Mn oxides (Honnorez, 1981).

Palagonitization during submerged conditions indicates the enrichment of K, whereas Ca, Na and Mn are lost. A. J. Andrews (1978), studying the transformation of fresh glass into palagonite, showed that $H_2O$ and $K_2O$ increase and $SiO_2$, $Al_2O_3$, MgO, CaO, $Na_2O$ and MnO are lost in the environment.

Scanning X-ray microphotographs show that the various color shades associated with the process of palagonitization are also a function of the chemical redistribution of some critical elements (Fig. 8-6). Thus a reddish light-brown zone enriched in Al and Ti is often surrounded by a lighter yellowish green rim enriched in Mg and Fe (Fig. 8-6). This Mg-Fe-rich outer margin represents an early stage of alteration and occurs at contact with the basaltic glass (Fig. 8-6).

Chemical analyses of palagonite show a considerable amount of $H_2O$ and $Fe_2O_3$ (>20% and >20%, respectively). The silica content varies between about 20 and 40% (Table 8-1). Under polarizing light palagonite occurs in varying colors which were attributed by Honnorez (1969) and Furnes (1975) to the change of temperature. The change in color generally varies from yellow to green to a darker green. Such changes could also be due to the partial dehydration of the rock. Experimental work on the formation of palagonite rims around basaltic rocks has shown that the darker colors are formed at higher temperatures (about 90°C) than the lighter-colored varieties which are formed at lower temperatures (Furnes, 1975). Palagonite is often accompanied by microcrystalline chalcedony, in radial aggregates showing

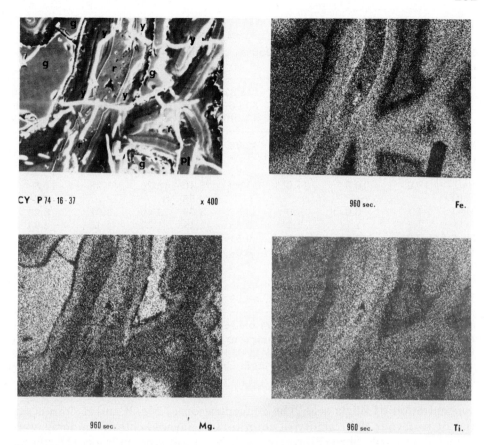

Fig. 8-6. X-ray scanning microphotograph of basaltic glass (*g*) partially altered into palagonite (sample CYP74-16-37). Veins and veinlets of palagonite form reddish brown (*r*) and yellowish green (*y*) zones. The glassy fragments are surrounded by a K-rich zone. *Pl* indicates plagioclase.

polarization crosses (Matthews, 1971). The way in which palagonite is transformed into clay aggregates is not well known. Matthews (1971) suggested that palagonite is associated with some pale-gray clay minerals which appear to be the final stages in the alteration of palagonite. The origin of palagonite is probably related to the progressive weathering of the basaltic glassy margins of extrusive flows. This is substantiated by the correlation between the fission track age and the thickness of palagonite coating the glassy margin of pillow flows (Fig. 8-4).

*Chlorite.* This is a controversial mineral as far as its being the product of weathering is concerned since its formation requires a certain level of temperature. However, chlorite was found surrounding grains of gibbsite in ocean floor sediment (Swindale and Pow-Foong Fan, 1967). The formation of chlorite in the ocean could be due to the instability of gibbsite and to the

TABLE 8-1

Microprobe analyses of palagonite found as alteration products of the glassy chilled margins of basalt

| | Leg 37 332A-40-3 106—109 | Leg 37 332B-21-1 55—59 | Leg 37 335-6-6 22—24 | Leg 37 335-6-1 47—49 | 3745 | ARP74- 14-32 | V18-330 in vein (ave. of 2) | V18-330 in vesicles (ave. of 2) |
|---|---|---|---|---|---|---|---|---|
| $SiO_2$ (wt.%) | 44.4 | 35.9 | 42.6 | 42.36 | 41.96 | 26.36 | 46.07 | 48.07 |
| $Al_2O_3$ | 9.4 | 12.5 | 12.1 | 12.62 | 16.06 | 13.65 | 12.74 | 12.95 |
| FeO | 19.0 | 16.5 | 17.0 | 17.90 | 8.45 | 13.76 | 11.25 | 11.5 |
| MnO | — | — | — | — | 0.08 | tr. | — | 0.38 |
| MgO | 4.1 | 3.0 | 4.9 | 3.87 | 3.69 | 3.49 | 3.91 | 3.23 |
| CaO | 1.1 | 1.1 | 1.1 | 0.55 | 1.49 | 1.33 | 4.45 | 2.94 |
| $Na_2O$ | 0.2 | 0.5 | 1.9 | 0.51 | 1.82 | 1.30 | 1.27 | 0.84 |
| $K_2O$ | 3.3 | 1.9 | 3.0 | 3.64 | 3.48 | 2.32 | 1.28 | 2.01 |
| $TiO_2$ | 2.9 | 1.4 | 2.2 | 1.64 | 1.63 | 0.66 | 0.81 | 0.65 |
| $H_2O$ | n.d. | n.d. | n.d. | n.d. | 20.33 | n.d. | n.d. | n.d. |
| Total | 84.4 | — | — | 83.09 | 98.99 | 63.21 | 81.78 | 82.57 |

Samples 332A, 332B and 335 are from the Mid-Atlantic Ridge (Rift Mountain region) near 37°N (Scarfe and Smith, 1977; Baragar et al., 1977); 3745 is from Swallow Bank in the Atlantic Ocean (Matthews, 1971); V18-330 is from the East Pacific Rise (11°37.5'N, 101°20'W; 3197 m); ARP74-14-32 is from Transform Fault "A" (36°57'N, 33°10'W; Fig. 2-6).

concentration of silicic acid. The concentration of Mg, K and $H_2$ ions is sufficient to favour the stability of chlorite. Interlayered chlorite minerals were detected in some fibro-chlorophaeite found in weathered rocks (Matthews, 1971).

*Chlorophaeite.* This is a clay mineral considered to be the product of alteration of olivine tholeiites in the Thingmuli volcanic series of Iceland (Carmichaël, 1964). Chlorophaeite is also found in deep-sea sediment cores and its presence was attributed to the degradation of volcanic material (Biscaye, 1965). Peacock (1926) describes the occurrence of chlorophaeite from the Icelandic lavas as being spherulites up to 8 mm in diameter, or filling or lining cavities, or as irregular patches in the groundmass, or in veinlets, or as pseudomorphs, replacing olivine, pyroxene, glass, or the mesostasis of basaltic flows. This mineral is isotropic and often birefringent due to the presence of chlorite fibres. Chlorophaeite found in an altered basalt from Swallow Bank (Matthews, 1971) occurs in the matrix as almost colorless or pale-gray translucent material.

Chlorophaeite from ocean floor basaltic rocks drilled during DSDP Leg 37 of the "Glomar Challenger" (Site 332B) was found as interstitial material of the groundmass constituents and lining the walls of vesicles. Microprobe data (Baragar et al., 1977) showed that at least two types of chlorophaeite

TABLE 8-2

Chemical analyses of celadonite from the Indian Ocean (sample 250A, 33°28'S, 39°22'E; Kempe, 1974) and from the Peru Trench (sample 11-28, 16°49.6'S, 74°32.7'W; Scheidegger and Stakes, 1977)

|  | Leg 26 250-25-1 135 | 11-28 |
|---|---|---|
| SiO$_2$ (wt.%) | 55.80 | 54.77 |
| Al$_2$O$_3$ | 5.76 | 3.30 |
| Fe$_2$O$_3$ | 12.25 | 15.03 |
| FeO | 3.42 | 4.41 |
| MgO | 6.78 | 0.36 |
| CaO | 0.36 | 0.36 |
| Na$_2$O | 0.81 | 0.27 |
| K$_2$O | 9.28 | 9.64 |
| TiO$_2$ | 0.29 | 0.19 |
| H$_2$O | 8.87 | 5.40 |
| Total | 103.62 | 100.55 |

TABLE 8-3

Partial microprobe analyses of clay material found in ocean floor basalts

|  | Chlorophaeite | | Saponite | | | | | | | | |
|---|---|---|---|---|---|---|---|---|---|---|---|
|  | Leg 37 332B 33—36 (ave. of 5) | Leg 37 335-6-1 62—131 (ave. of 5) | Leg 37 6-1 (ave. of 6) | Leg 34 14-3 | Leg 34 14-3 | CH4-DR1-11 | CH4-DR2-5 | 36-11 1 | 36-11 2 | FHV-4 2-17 | CH4-D1-15 (ave. of 2) |
| SiO$_2$ (wt.%) | 51.47 | 50.3 | 45.75 | 45.12 | 45.25 | 42.31 | 39.84 | 51.4 | 48.4 | 43.0 | 35.91 |
| Al$_2$O$_3$ | 3.89 | 4.16 | 3.1 | 5.13 | 5.00 | 6.15 | 15.07 | 15.6 | 15.9 | 7.2 | 14.20 |
| FeO | 9.49 | 8.47 | 4.93 | 14.80 | 14.67 | 10.40 | 13.47 | 12.5 | 13.9 | 9.55 | 13.11 |
| MnO | 0.04 | 0.03 | 0.02 | 0.04 | — | 0.09 | — | 0.01 | — | — | 0.19 |
| MgO | 19.6 | 21.9 | 22.09 | 17.06 | 16.22 | 17.29 | 15.56 | 11.4 | 10.7 | 21.0 | 21.08 |
| CaO | 0.61 | 0.69 | 0.32 | 0.21 | 0.19 | 1.51 | 0.71 | 0.7 | 0.6 | 0.4 | 1.41 |
| Na$_2$O | 0.21 | 0.82 | 0.89 | 2.68 | 2.24 | 0.33 | 0.16 | 0.9 | 1.0 | 3.3 | 0.58 |
| K$_2$O | 1.02 | 0.66 | 0.58 | 0.85 | 1.18 | 1.06 | 0.33 | 2.5 | 2.4 | 0.4 | 0.87 |
| TiO$_2$ | — | — | — | 0.23 | 0.19 | — | — | 0.4 | 0.4 | <0.05 | 0.19 |

Leg 37 samples (MAR) are from Baragar et al. (1977); Leg 34 samples (Nazca plate) are from Seyfried et al. (1976); CH4 samples (MAR) are from Jehl (1975); 36-11 samples (MAR) are from Melson and Thompson (1973); FHV-4-2-17 is from Banks (1972); CH4-D1-15 is from the MAR Rift Valley near 53°00.4'N, 34°59.4'W (1875 m).

occur: the Mg-rich and K-Fe-rich varieties. In thin sections this mineral is pale green, yellow and brown (Matthews, 1971; Baragar et al., 1977), while the K-Fe-rich variety is strongly colored in green and deep red.

There are many possible origins for chlorophaeite. It may be formed during a low-temperature crystallization of the basalt's groundmass or as the weathering product of basaltic glass. Upon glycolation, Mg-chlorophaeite expands its lattice (14 Å) and may be considered as a variety of smectite accompanying palagonite. The K-Fe variety of chlorophaeite apparently has a less expandable lattice and it is likely to be formed during low-temperature crystallization. Chemical analyses of chlorophaeite indicate a higher MgO (19—22%) and lower $Al_2O_3$ content (<7%) than palagonite (Tables 8-1, 8-3).

*Celadonite.* Celadonite is bluish-green mica and has been reported as a devitrification product of basaltic glass. Celadonite occurs filling vesicles and in veinlets of altered basaltic rocks. Chemically, it differs from palagonite by its lower $Al_2O_3$ content (<7%) and higher $K_2O$ content (9—10%), and from chlorophaeite by its lower MgO content (<10%) (Tables 8-1, 8-2, 8-3).

*Zeolites.* Zeolites are aluminium silicates containing water, which can be expelled without destroying the crystal lattices (Winchell and Winchell, 1961).
The most commonly found zeolite in both sedimentary environments and as an alteration product of volcanics is phillipsite and it is found as radiating clusters in vesicles, cavities and veins. Phillipsite was found in hyaloclastite from Palagonia (Sicily) (Honnorez, 1972), in deep-sea sediment cores, and it was considered to be the alteration product of submarine volcanism (Arrhenius, 1963). Hormotone is another zeolite which could be attributed to the alteration of palagonite of highly vesicular basalts (Morgenstein, 1967).

*Montmorillonite.* This type of expanded clay mineral may vary in composition by the replacement in its mineral lattice of Mg, Fe and Al. Montmorillonite is commonly found as the product of alteration of volcanic ashes and rocks from the floor of the ocean. The study of relatively fresh lava from the inner floor of the Rift Valley near 36°N in the Atlantic Ocean showed the presence of abundant montmorillonite associated with palagonite and traces of zeolite (Hékinian and Hoffert, 1975). Saponite may be considered to be a variety of montmorillonite (Table 8-3).

*Iron oxides and hydroxides.* The Fe-bearing minerals, such as titanomagnetite, commonly encountered in ocean floor basalts are often altered into a brownish red hematite. Fe-hydroxides are often found lining the walls of vesicles and veins of weathered basalts. Seawater contributes to the hydration of the Fe-bearing minerals and the remobilization of amorphous hydroxide compounds.
The mineral paragenesis of rock weathering in submerged regions is unknown. The environmental conditions, such as small temperature changes, highly oxygenated bottom water, depth of the ocean floor, and sediment blankets which could affect the degree of ocean floor weathering, have not yet been assessed. Another difficulty is raised when comparing ocean floor weathering and the low-temperature mineral constituents created during hydrothermalism.

## CHAPTER 9

## OCEAN FLOOR METAMORPHISM

Metamorphic processes and the extent of metamorphism in the oceanic environment are still not well known. The existence of metamorphic rocks formed under the ocean has been described in detail only since 1966, when rocks were dredged from rift valley walls in the Carlsberg Ridge (Cann, 1969) and near 22°N in the Atlantic Ocean (Melson and Van Andel, 1966). In general, three types of metamorphic regions have to be considered when dealing with the ocean floor: (1) metamorphism which is associated with orogenetic belts and occurs at the convergence of lithospheric plates; (2) metamorphism which takes place along diverging plate boundaries associated with mid-oceanic ridges and fracture zones; and (3) metamorphism in intraplate regions.

The theoretical approach to ocean floor metamorphic petrology is based on the principle of Eskola's (1920, 1939) metamorphic facies. The concept of metamorphic facies as defined by Eskola (1939) refers to a group of rocks characterized by a set of minerals formed under a particular pressure and temperature condition. The mapping of subaerial terranes (Harker, 1932) permitted the recognition of new groups of metamorphic rocks and has helped to infer the temperature of the metamorphic facies. The discovery of the zeolite and the prehnite-pumpellyite facies in the sedimentary tuffs of New Zealand was attributed by Coombs et al. (1960) and Coombs (1961) to burial metamorphism. Similar associations were found elsewhere in the pillow lava spilites of New South Wales (Smith, 1968), in altered lava and pyroclastic rocks of central Puerto Rico (Jolly, 1970), in the spilites of the Virgin Islands (Hékinian, 1971b), and in the pillow lavas of spilitic nature from the Alps and other localities around the world associated with orogenetic belts. A selection of studies on low-grade metamorphics of subaerial regions is found in Amstutz (1974). The relationship between metamorphic terranes in continental areas and on the ocean floor is not yet well established. It is not excluded that the metamorphic processes encountered in continental areas had their origin on the ocean floor and kept their identity during the various orogenetic processes giving rise to subaerial formations. The occurrence of metamorphism on diverging plate boundary regions suggests that the process of alteration starts very early in the history of the ocean floor. Some of the mineral associations found in metamorphosed basalts from recent spreading ridges are comparable to an equivalent alpine type of metamorphism. However, it is not established whether or not the metamorphic terranes found on mid-oceanic ridges have been added to continental masses without being further altered or destroyed by the plunging of the lithospheric plates at converging plate boundary regions.

METAMORPHISM ASSOCIATED WITH DIVERGING PLATES

The study of the metamorphism of volcanics from the ocean floor associated with diverging plate boundary regions has been mainly focussed on the mid-oceanic provinces and fracture zones of both the Atlantic and the Indian Ocean.

## Metamorphics of the Mid-Atlantic Ridge

Metamorphic rocks of the low-grade zeolite facies were found on the western wall of the Rift Valley near 53°N at 1800–2000 m depth (Hékinian and Aumento, 1973). These metamorphosed basalts are found in association with fresh basalts from the flank of an elevated volcanic edifice called the Minia Seamount (samples CH4-DR1-14 and CH4-DR1-3T). The depth of the dredging operation varied between 2000 and 2800 m up to the top of the seamount. Both the fresh and the altered specimens have preserved the typical radial type of jointing characterizing most pillow flow structures. In general, the glassy margin of the metamorphosed rocks is altered into a mixed layer of phyllosilicate (chlorite-smectite) and chlorite, while the inner part of the rock is criss-crossed by quartz veins and veinlets. Most of the plagioclase is altered into a more sodic type ($An_{20}$) (Table 9-1); however, a calcic type of plagioclase is found to co-exist with the more sodic type.

Other low-grade metamorphics were recovered by dredging on the western Rift Mountain region near 45°N from a flank of a seamount called "Bald

TABLE 9-1

Chemical analyses of volcanics with a zeolite-smectite-quartz assemblage from the ocean floor, Mid-Atlantic Ridge near 53°N

|  | D1-3T bulk rock | D2-F bulk rock | D1-3T matrix plag. | D1-3T phen. plag. |
|---|---|---|---|---|
| $SiO_2$ | 49.30 | 43.80 | 57.68 | 50.67 |
| $Al_2O_3$ | 14.25 | 17.15 | 25.79 | 29.50 |
| $Fe_2O_3$ | 2.29 | 6.00 |  |  |
| FeO | 7.16 | 3.96 | 0.17* | 0.72* |
| MnO | 0.16 | 0.17 | — | — |
| MgO | 8.74 | 6.17 | — | 0.46 |
| CaO | 10.30 | 9.39 | 7.86 | 13.38 |
| $Na_2O$ | 2.34 | 2.33 | 7.34 | 3.48 |
| $K_2O$ | 0.10 | 0.46 | <0.02 | <0.04 |
| $TiO_2$ | 1.13 | 1.21 | — | <0.02 |
| $P_2O_5$ | 0.15 | 0.15 | 0.24 | 0.13 |
| $H_2O$ | 2.42 | 6.80 | — | — |
| Total | 98.33 | 98.12 | 99.12 | 98.40 |

*Total Fe as FeO (microprobe analyses).

Mountain" (Aumento and Loncarevic, 1969). Most of these rocks have retained their original igneous textural features. However, some specimens show a marked schistosity in the minerals with only a few poorly preserved relics of the original crystalline phases (Aumento and Loncarevic, 1969). The minerals which are oriented according to the planes of schistosity are tremolite and chlorite aggregates. These were classified as higher-grade meta-basalts than those deprived of oriented mineral associations. A metabasalt less affected by dynamic shearing and containing green hornblende, actinolite, albite, quartz, sphene, and apatite was classified (Aumento and Loncarevic, 1969) as part of the greenschist facies (quartz-albite-epidote) (Table 9-2).

Metabasalts associated with transverse fracture zones, rift mountains and rift valley walls dredged near 30°N in the Atlantis Fracture Zone and near 23°N were reported by Miyashiro et al. (1971). Most of the specimens are non-schistose and have preserved their original pillow lava or tuff-breccia textural features. The great majority of the metabasalts belong to the zeolite and greenschist facies (Miyashiro et al., 1971). Some of these have small amounts of blue-green hornblende together with other greenschist-facies minerals, and are considered by Miyashiro et al. (1971) to be transitional between the greenschist and the amphibole facies.

Metabasalts, basaltic tuffs and low-grade greenstones were also recovered from the east and west walls of the Rift Valley near 22°N between 1500 and 3500 m depth (Melson and Van Andel, 1966). The greenstones containing chlorite, albite and nontronite belong to the zeolite facies. The schistosity shown by many of the rocks, due to the orientation of actinolite, suggests a higher grade of metamorphism belonging to the greenschist facies (Melson and Van Andel, 1966). The greenstones recovered from near 22°N and analysed are those which were classified as albite-chlorite-epidote-actinolite rocks of the greenschist facies (Melson and Van Andel, 1966). Among the two dredges described from the eastern wall of the Rift Valley, some differences were found to occur in the structure and mineral composition. The dredge at shallower depth contains rocks with less epidote, abundant chlorite and showing no schistosity, whereas the rocks from deeper dredge hauls show incipient schistosity and contain more epidote than the previously mentioned samples. The chemical analyses of five greenstones reported by Melson and Van Andel (1966) show a relatively high $Na_2O$ content (2.9—4.5%) and a low $K_2O$ content (0.05—0.11%) (Table 9-2). The $TiO_2$ content (0.9—1.5%) is comparable to the range of the titania variation found in most unaltered Mid-Atlantic Ridge basalts.

*Mid-Indian Oceanic Ridge: metamorphics of the Carlsberg Ridge*

Low-grade metamorphic rocks were dredged from the rift valley walls and from transform faults of the Carlsberg Ridge in the Indian Ocean (Cann, 1969; Dmitriev and Sharaskin, 1975). The material recovered from both the

TABLE 9-2

Chemical analyses of low-grade metamorphics with plagioclase-chlorite-epidote assemblages from the Atlantic Ocean near 22°N (Melson et al., 1966), from 45°N (Aumento, 1968), and from the Romanche Fracture Zone (this work). The samples from the Carlsberg Ridge are from Cann (1969)

| | 22°N (MAR) | | | 45°N (MAR) | Romanche Fracture Zone | Carlsberg Ridge (Indian Ocean) | | | | | |
|---|---|---|---|---|---|---|---|---|---|---|---|
| | 2-5 | 3-2 | 3-3 | 33-13 | CH80-DR8-5 | C7 | C8 | C9 | C10 | W5 | W6 |
| $SiO_2$ | 49.71 | 50.14 | 48.18 | 51.81 | 52.40 | 52.52 | 48.32 | 45.85 | 39.43 | 49.43 | 52.24 |
| $Al_2O_3$ | 15.32 | 16.30 | 15.17 | 14.82 | 13.75 | 14.59 | 15.17 | 14.60 | 15.46 | 15.04 | 15.02 |
| $Fe_2O_3$ | 2.09 | 3.91 | 2.56 | 5.50 | 1.99 | 1.61 | 2.34 | 2.63 | 2.43 | 2.21 | 2.93 |
| $FeO$ | 7.31 | 3.92 | 7.18 | 5.15 | 6.02 | 7.74 | 7.05 | 13.56 | 16.68 | 7.39 | 6.31 |
| $MnO$ | 0.16 | 0.11 | 0.16 | 0.19 | 0.17 | 0.26 | 0.25 | 0.15 | 0.14 | 0.23 | 0.14 |
| $MgO$ | 8.92 | 6.35 | 9.47 | 5.71 | 7.93 | 7.90 | 9.19 | 10.69 | 12.01 | 8.40 | 6.01 |
| $CaO$ | 7.32 | 12.56 | 7.61 | 5.30 | 8.61 | 6.89 | 7.81 | 1.28 | 1.81 | 6.69 | 8.73 |
| $Na_2O$ | 2.93 | 3.11 | 3.27 | 5.09 | 3.58 | 5.20 | 4.41 | 1.54 | 1.22 | 4.45 | 4.02 |
| $K_2O$ | 0.05 | 0.11 | 0.06 | 0.49 | 0.91 | 0.04 | 0.06 | 0.10 | 0.10 | 0.11 | 0.21 |
| $TiO_2$ | 1.51 | 0.91 | 1.74 | 2.16 | 1.31 | 1.86 | 1.76 | 1.59 | 1.78 | 1.94 | 1.83 |
| $P_2O_5$ | 0.17 | 0.14 | 0.15 | 0.31 | 0.13 | 0.05 | 0.07 | 0.17 | 0.18 | 0.19 | 0.20 |
| LOI | 4.45 | 2.40 | 4.42 | 3.20 | 2.37 | 3.49 | 3.92 | 7.81 | 8.57 | 0.00 | 0.00 |
| Total | 99.91 | 99.96 | 99.97 | 99.70 | 99.13 | 100.15 | 100.35 | 99.97 | 99.81 | 100.10 | 100.39 |
| *Norms* | | | | | | | | | | | |
| Q | 1.07 | 1.16 | 0.00 | 2.41 | 0.00 | 0.00 | 0.00 | 9.03 | 0.00 | 0.00 | 2.28 |
| Ne | 0.00 | 0.00 | 0.00 | 0.00 | 0.00 | 0.99 | 1.27 | 0.00 | 0.00 | 0.00 | 0.00 |
| Or | 0.29 | 0.65 | 0.35 | 0.30 | 5.37 | 0.23 | 0.35 | 0.59 | 0.59 | 0.65 | 1.24 |
| Ab | 24.79 | 26.31 | 27.66 | 44.62 | 30.29 | 42.15 | 34.96 | 13.03 | 10.32 | 37.65 | 34.01 |
| An | 28.50 | 30.19 | 26.53 | 16.73 | 18.76 | 16.35 | 21.42 | 5.23 | 7.80 | 20.73 | 22.31 |
| Di | 5.40 | 24.69 | 8.21 | 0.00 | 18.57 | 14.18 | 13.55 | 0.00 | 0.00 | 9.03 | 15.86 |
| Hy | 29.14 | 6.82 | 18.40 | 13.36 | 17.52 | 0.00 | 0.00 | 47.00 | 51.58 | 8.04 | 13.72 |
| Ol | 0.00 | 0.00 | 7.00 | 8.26 | 0.56 | 16.74 | 17.96 | 0.00 | 3.13 | 12.63 | 0.00 |
| Mt | 2.97 | 5.66 | 3.71 | 4.25 | 2.88 | 2.33 | 3.39 | 3.81 | 3.52 | 3.20 | 4.24 |
| Ilm | 2.86 | 1.72 | 3.30 | 0.00 | 2.48 | 3.53 | 3.34 | 3.01 | 3.38 | 3.68 | 3.47 |
| Ap | 0.40 | 0.33 | 0.35 | 0.00 | 0.30 | 0.11 | 0.16 | 0.40 | 0.42 | 0.44 | 0.47 |
| Hm | 0.00 | 0.00 | 0.00 | 0.74 | 0.00 | 0.00 | 0.00 | 0.00 | 0.00 | 0.00 | 0.00 |
| C.I. | 34.49 | 52.30 | 39.16 | 0.00 | 34.74 | 36.71 | 44.35 | 7.92 | 12.19 | 36.45 | 34.84 |

western and eastern walls near 5°S (greenstones, gabbro-mylonites and gabbro-granulites) was found in the same dredge haul from depths varying from 2000 to 4000 m (Dmitriev and Sharaskin, 1975). Other dredge hauls containing similar types of greenstones were recovered near 5°50′N in a transform fault intersecting the Carlsberg Ridge (Cann and Vine, 1966; Cann, 1969). This later area overlaps the previous ones surveyed by R. V. "Vitiaz" and R. V. "Akademik-Kurchatov" (Dmietriev and Sharaskin, 1975). Types of metamorphosed rocks similar to those mentioned above (Hékinian, 1968) were also found in a piston core taken on the rift mountain near 2°50′N. All these rocks are massive, without any foliation or other apparent metamorphic structures. The majority of samples has retained the overall fabric of the primary basaltic rock. The original texture has been preserved and a variolitic, granular, or doleritic appearance characterizes most of these specimens. Such metamorphics have undergone various stages of alteration.

TABLE 9-3

Chemical analyses of rocks with an albite—K-feldspar—chlorite assemblage from the Atlantic and the Indian Oceans

|  | Romanche Fracture Zone | Wharton Basin (Leg 22) | |
|---|---|---|---|
|  | CH80-DR7-3 | 212-39-3 10—14 | 212—39-2 122—124 |
| $SiO_2$ (wt.%) | 47.73 | 46.40 | 45.66 |
| $Al_2O_3$ | 20.40 | 17.85 | 16.14 |
| $Fe_2O_3$ | 1.19 | 7.05 | 5.59 |
| FeO | 4.27 | 2.24 | 3.11 |
| MnO | 0.12 | — | — |
| MgO | 8.83 | 5.40 | 7.87 |
| CaO | 7.18 | 7.56 | 8.68 |
| $Na_2O$ | 3.44 | 2.22 | 3.70 |
| $K_2O$ | 1.13 | 2.06 | 2.34 |
| $TiO_2$ | 0.68 | 0.62 | 0.57 |
| $P_2O_5$ | 0.02 | 0.10 | 0.11 |
| $H_2O$ | 4.92 | 7.75 | 5.33 |
| Total | 99.91 | 99.34 | 99.09 |

Transformation of the initial rocks occurs either by the alteration of surface glassy material or of the totality of the rock. Vesicles are usually filled with chlorite associated with quartz and albite. Veins of epidote, chlorite and sphene are commonly encountered throughout the specimens. The ground-mass could contain chlorite, sphene and actinolite or it could even be made up entirely of chlorite. The major mineral phases such as the plagioclase and clinopyroxene are often transformed in the most altered specimens into albite and calcite-actinolite, respectively. Olivine minerals are often trans-formed into chlorite and actinolite. Some titanomagnetite is also transformed into sphene and leucoxene. Chemical analyses of the Carlsberg Ridge samples are shown in Tables 9-2 and 9-3.

## METAMORPHISM ASSOCIATED WITH CONVERGING PLATES

The transition between the ocean and the continents is best observed in volcanic island arc regions and in other active continental margins, such as in South America, Central America, and North America. These areas are considered to be the locus of plate convergence where the oceanic lithosphere is underthrusted either under the continental or oceanic lithosphere. It was observed (Miyashiro, 1973) that mature island arcs and continental margins of the Pacific Ocean have undergone various degrees of metamorphism. These areas are delineated by the Mesozoic low-pressure belt of Canada, the low- and high-pressure belts in Washington State, the Franciscan high-pressure and the Sierra Nevada low-pressure belts, the late Paleozoic Chilean high- and low-pressure belts, and the Jurassic/Cretaceous high- and low-pressure belts of New Zealand and Japan.

From a review (Miyashiro, 1972) of late Tertiary terrane in Japan and in the Sierra Nevada (North America), it was found that regional metamorphism, going from the zeolite facies to the amphibolite facies, was present. The younger Quaternary volcanic arcs (e.g. Aleutian, Indonesian, Tonga-Kermadec, and Izu-Bonin arcs) have no traces of regional metamorphism.

The metamorphism along convergent plate boundaries was classified by Miyashiro (1972) into paired and non-paired metamorphic belts with contrasting characteristics. A pair is composed of a high-pressure metamorphic belt (with glaucophane) accompanied by basic and ultramafic (ophiolitic) rocks, and a low-pressure metamorphic belt (with andalusite) accompanied by granitic, andesitic and/or rhyolitic rocks (Miyashiro, 1973). In the circum-Pacific area, the high-pressure belts are on the ocean-facing side, while low-pressure areas occur towards the continental zone (see Fig. 6-1). This observation led Miyashiro (1973) to suggest that the low-pressure metamorphism facies is due primarily to heat transferred from the rise of magma under island arcs. On the other hand, the high-pressure metamorphism of the glaucophane and the higher subfacies of the amphibolite facies is due to the rapid descent and underthrusting of an oceanic plate beneath an island arc or a continental margin (Fig. 6-1).

In order to explain the effect of metamorphism with a down-going slab, two important factors were taken into consideration by R. N. Anderson et al. (1976): (1) the water content of the oceanic crust, and (2) the heat distribution. It is known from the distribution of oceanic rocks and their composition that the crust is hydrated to a certain degree. The hydration of the oceanic crust is compatible with the explosive character of active margin volcanism. Indeed, the island arc volcanoes of the Pacific region are generally more explosive than those which have erupted in most oceanic islands and seamounts of intra-plate and diverging plate regions.

It has been noticed that the heat distribution measured in the Kurile-Japan-Izu-Bonin-Marianas system of island arcs (Uyeda and Horai, 1964) and for the Melanesian and Tonga-Kermadec systems (Sclater et al., 1971) shows that high heat flow values occur in the shallow marginal basins on the landward side of the arcs. Low heat flow values are concentrated on the oceanward side of the arcs between the trench and the island arc. This relatively low heat region ($<2 \times 10^{-6}$ cal/cm$^2$ s) is due to the absorption of heat during the dehydration observed in the arc-trench gap (R. N. Anderson et al., 1976).

The mechanism of metamorphism and volcanism in island arc areas and active continental margin areas was explained by R. N. Anderson et al. (1976) as following a two-stage process: (1) a dehydration of the sinking oceanic crust with the absorption of heat, and (2) as underthrusting proceeds there is an increase in temperature either by conduction from the asthenosphere or by a friction of the down-going slab. Metamorphism occurs during the process of dehydration, and volcanism of the island arc type occurs at a later stage when the temperature has increased.

There are some objections to this type of model. It is unclear where metamorphism stops and volcanism starts. If the volcanics from island arcs are derived from metamorphism of the oceanic lithosphere with the addition of water, it is surprising that metamorphic rocks are not found in young volcanic arcs (e.g. Tonga-Kermadec). It is also not obvious that metamorphic terranes are located seaward of island arcs or active continental margins. With rising temperature, metamorphism grades into the formation of anatectic rocks. Granitic or granodioritic batholiths are likely to be formed during such metamorphic melting. However, there is no sizeable trace of such formations in the young island arcs of the circum-Pacific regions.

METAMORPHISM ASSOCIATED WITH INTRA-PLATE REGIONS

The alteration of oceanic rocks associated with basins and fracture zones extending away from accreting plate boundary regions is not well documented; this is mainly due to the difficulties encountered in sampling heavily sedimented regions with conventional methods. However, deep-sea drilling in various oceanic basins during both the JOIDES and the IPOD programs of the "Glomar Challenger" has given some information about the degree of crustal alteration in regions of the Atlantic, Pacific and Indian Oceans. The maximum depths reached during these drilling operations did not exceed 900 m of penetration (sediment plus basement rock) (Legs 51, 52, 53, Sites 417, 418; Leg 69, Site 504). Hence, the pressure-temperature conditions of burial reached during drilling would not be comparable to those of the

zeolite facies of known regional metamorphism in subaerial environments
(<1.5 kbar and <150°C).

*Western Atlantic Basin*

Several DSDP sites are located in the northwestern Atlantic Basin (Sites
105, 384, 386, 417A and 418). These sites, drilled on a crust up to about
108–109 Ma old, show the basaltic basement altered to a very low metamorphic grade of the zeolite facies and to the low greenschist facies. The
typical mineral association of Sites 105, 384 and 386 is the albite-chlorite-
smectite assemblage. Calcite is prominent in these rocks and occurs in veins
and vesicles. The plagioclase of both the groundmass and of the phenocrystal
phase shows extensive replacement by chloritic patches and veinlets; sericite
is also a common constituent of these rocks. Site 417A was drilled on the
southern end of the Bermuda Rise (5468 m depth) (Donnelly et al., 1979).
Nearly all the basalts recovered from these holes are porphyritic with 5–15%
phenocrysts of mainly plagioclase and olivine. Baked contact between the
various flow units was recognized; fresh glassy margins and moderate degrees
of rock alteration were seen. The study of the degree of rock alteration in
hydroxide, celadonite, zeolite and smectite indicates a very low-grade meta-
environment. The occurrence of the mineral assemblage of oxidized Fe-
hydroxide, celadomite, zeolite and smectite indicates a very low-grade metamorphism. In addition, K-feldspar, analcime and calcite are also prominent
constituents of the altered basaltic flows. The oxygen isotope studies indicate
a temperature of alteration of about 30°C (Donnelly et al., 1979). The
contents of $K_2O$ Rb and Cs in the basement of holes 417A and 418A (Legs
51, 52, 53) are the most mobile compounds, diminishing according to depth
down to about 200 m in the borehole. There is a continuous decrease in the
$K_2O$ content from about 9% (water-free basis) to about less than 1% at the
bottom of the hole (Donnelly et al., 1979). Rb and Cs contents vary from
25 ppm to less than 0.2 ppm with depth (Joron et al., 1980). In addition,
the loss of MgO (<7%), of CaO (<10%), and of $Na_2O$ (<2%) is evident in
the most altered sections of the drilled holes.

*Wharton Basin (eastern Indian Ocean)*

Deep-sea drilling during Leg 22 of the "Glomar Challenger" penetrated
about 516 m into the igneous basement below the sea floor at Site 212 in
the Wharton Basin (Fig. 2-2). The rocks recovered consist of a series of
pillow flow units altered by metamorphism. Seven pillow flows, varying in
thickness between 15 and 150 cm, were encountered, and all appeared to
have been metamorphosed to the same degree. Each of the units shows
altered aphanitic margins, and very few relics of preserved glassy material
were seen (Hékinian, 1974a). Sodic plagioclase and K-feldspar are the main

mineralogical assemblages observed (Table 9-3). Chlorite, smectite and muscovite only occur in trace amounts. In addition, the occurrence of altered olivine phenocrysts and fresh spinel suggests that the original rock was an olivine basalt. No apparent changes in the degree of rock alteration were found as the depth in the borehole increased.

The aging of the oceanic crust away from recent accreting plate boundary regions and the depth of penetration into the intra-plate basement have failed so far to show progressive changes in metamorphic grades. Further studies of mineral alteration of selected drill holes are necessary to provide more insight into the mechanism of fluid circulation in this type of oceanic region.

FACIES AND MINERAL ASSEMBLAGE OF METAMORPHISM

The transition from weathering to diagenesis and to metamorphism is ill-defined in ocean floor environments. The lowest metamorphic grade of the zeolite facies has been commonly encountered in volcanics associated with various tectonic provinces of the ocean floor, as mentioned previously. Both the greenschist and the amphibolite facies were reported to occur in thrust-faulted zones, island arcs and fracture zones in all the world's oceans (Chapters 2, 6 and 7).

Even if somewhat similar to that of subaerial types of metamorphism, the classification used here for the various types of progressively metamorphosed rocks does not have any genetic implications in the sense postulated by Eskola (1920, 1939) and Coombs (1961). Instead, the following paragraphs are intended to describe the degree of ocean floor rock alteration as defined by their mineral associations and their geographic distribution. Tentatively, the ocean floor metamorphics are classified in four major categories: (1) the zeolite facies, (2) the prehnite-pumpellyite facies, (3) the greenschist facies, and (4) the amphibolite facies (Table 9-6). Within the metamorphic facies the term mineral assemblage is used when some of the secondary minerals formed suggest a metamorphic paragenesis. It is as yet premature to relate the metamorphic mineral association found in oceanic rocks to space and time within a particular geological formation. Recently a new facies for the ocean floor metamorphosed basalts was proposed by Cann (1979). This was called the brownstone facies; it comprises the altered basalts which become a yellowish brown color when under oxidizing conditions and a dark blue-gray color in a reducing environment. The most common mineral assemblage formed during oxidizing conditions is the K-rich illite resembling celadonite (Kempe, 1974). On the other hand, under reducing conditions Mg-rich smectite and saponite are produced. In the early phase of rock alteration, olivine is the first major compound found in ocean floor basalts to alter into a yellowish green material consisting essentially of smectite. In considering an analogy with the Icelandic geothermal field mineral association, Cann (1979) suggested that the boundary between the brownstone and the zeolite facies is located at a temperature of 50—100°C. Brownstone types of rocks abound in the ocean floor in various geological settings, and they are usually associated with greenstones and low-grade metamorphosed rocks found in transform faults. Clay minerals, such as celadonite, phillipsite, and saponite, characterizing this very low grade of metamorphism are also common constituents of weathered rocks (Chapter 8). Caution must be exercised in using a secondary mineral association of the type mentioned above in differentiating between the very low-grade metamorphosed facies and a typical weathered rock when the geological setting is unknown.

Various names have been assigned to the igneous rocks which have undergone a low or intermediate grade of metamorphism. Terms such as metabasalt, greenstone and spilite have been given to rocks of basaltic composition whose primary mineral constituents have been replaced by low-temperature mineral assemblages. In these types of rock, the original igneous texture was preserved. Metabasalts or metamorphosed basalts will be the term used here to indicate the altered basaltic rocks discussed in this chapter.

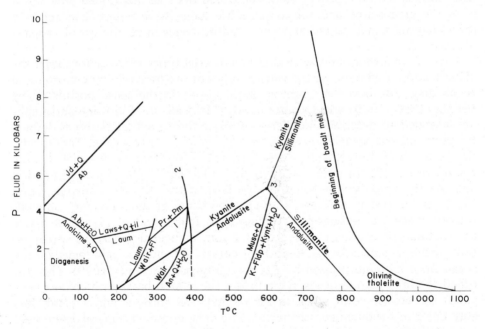

Fig. 9-1. Temperature-pressure variation diagram corresponding to the equilibrium curves for the major metamorphic reactions. $1$ = prehnite + Ca-montmorillonite + quartz = 2 wairakite + fluid. $2$ = 5 prehnite = 2 zoisite + 2 grossularite + 3 quartz + fluid. $3$ = muscovite + quartz = K-feldspar + $Al_2SiO_5$ + $H_2O$. The stability fields were plotted after the experimental works of Yoder and Tilley (1962); Campbell and Fyfe (1965); Evans (1965); Althaus (1967); Richardson et al. (1969); A. B. Thompson (1970) and Liou (1970).

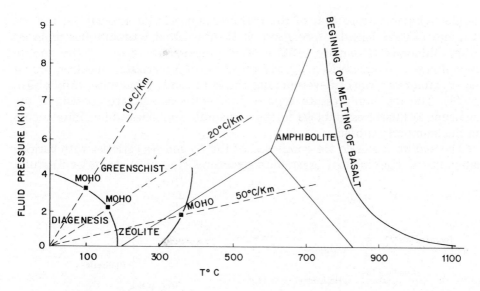

Fig. 9-2. Temperature-pressure variation diagram corresponding to the major metamorphic facies encountered in the ocean floor. The temperature gradient curves were calculated by assuming a thickness of 7 km for the average oceanic crust. The black square indicates the hypothetical temperature of the oceanic Moho.

## Zeolite facies

In a subaerial environment, the zeolite facies is characterized by the mineral assemblage heulandite-analcime-laumonite. From the mineral paragenesis two subfacies are found, the heulandite-analcime and the laumonite subfacies. Smectite was observed to occur in most low-grade metamorphic terranes around the world. As previously mentioned, the zeolite facies of metamorphism was found in altered basalts of island arcs in Iceland, where heulandite, scolectite, smectite, celadonite and phillipsite are common constituents of an altered olivine basalt (G. P. L. Walker, 1960). Metabasalts from the ocean floor characterized by the zeolite-smectite assemblage are the closest representative of the zeolite facies. In addition to smectite and zeolite, the ocean floor metabasalts contain Fe-oxides and hydroxides in the form of piemontite and hematite. Experimental studies (Liou, 1970) have shown that the upper limit of the zeolite facies was tentatively put at about 270° and 2.7 kbar with the following reaction (Figs. 9-1, 9-2):

laumonite = wairakite + fluid

The occurrence of wairakite is rather enigmatic. This mineral, which is the transformation of laumonite, may be a transition between the prehnite-pumpellyite subfacies and the laumonite subfacies of the zeolite facies. Experimental work shows that it should occur at a lower pressure and a similar temperature to that of the prehnite-pumpellyite association. In a St. Thomas (Virgin Islands) formation, it is shown that wairakite has replaced albite minerals (Donnelly, 1966). The low-pressure type of the zeolite assemblage is characterized by the presence of Ca-zeolite, smectite, mixed layer minerals (smectite-vermiculite-chlorite) and celadonite (Miyashiro, 1973). The medium-pressure type of the zeolite assemblage consists of the analcime + quartz assemblage in the heulandite subfacies while albite occurs in the laumonite subfacies.

The zeolite minerals are composed of the Ca and Na varieties with various amounts of $H_2O$ in their crystalline structures (Fig. 9-3). In order to study

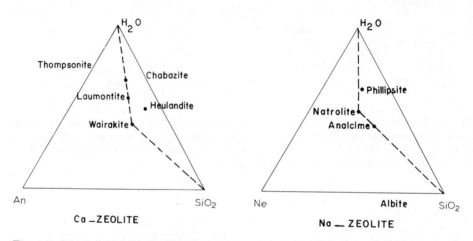

Fig. 9-3. Ternary diagrams showing the composition of various types of zeolite associated with progressively metamorphosed rocks in the zeolite facies (Miyashiro, 1973).

the phenomenon of progressive metamorphism it can be assumed that the rise of temperature and perhaps of pressure tends to form more dehydrated zeolites. The types of zeolite found in progressive metamorphism at various stages indicates the persistence of laumontite towards the higher grade. Analcime and phillipsite are more characteristic of the lower-grade zeolite facies in the slightly zeolitized zone in subaerial terranes. The metabasalts of the zeolite facies in the ocean floor show a marked introduction of $Na_2O$ from the environment (Miyashiro, 1973). In altered basalts from Iceland two zones of metamorphism have been recognized: (1) the analcime-heulandite, and (2) the laumontite zones (G. P. L. Walker, 1960).

*Prehnite-pumpellyite facies*

The recognition of this mineral assemblage in ocean floor metamorphics is not clear. Very few examples of rocks containing pumpelleyite have been reported from mid-oceanic ridges. The occurrence of small amounts of pumpellyite does not always justify the existence of a prehnite-pumpellyite facies in oceanic rock assemblages. However, from regional metamorphism of both continental and converging plate boundary regions it was noticed that first prehnite and then pumpellyite replaced plagioclase phenocrysts. The plagioclase in these rocks is not completely albitized and also contains chlorite and sericite in veins and patches within the plagioclase. Similar associations occur in metamorphic basalts from the Atlantic Ocean (Melson and Van Andel, 1966), where traces of pumpellyite were found in metamorphics from the Mid-Atlantic Ridge near 22°N. Prehnite was also found in schistose amphibolite (sample A150-RD8-AM1; Miyashiro et al., 1971) and was attributed to retrogressive metamorphism. Calcite, smectite and/or mixed layer chlorite-montmorillonite are also a common assemblage in ocean floor metabasalts.

*Greenschist facies*

Most of the ocean floor metamorphosed basalts belong to the zeolite or the greenschist facies. Some of the greenschist facies rocks preserve their original chemical compositions as unaltered basalts except for an increase in their $H_2O$ content, but most rocks in the greenschist facies show a marked decrease of CaO and a variation in the $SiO_2$ content (Miyashiro et al., 1971). The stable mineral assemblages of the greenschist facies at low and medium pressure are: chlorite-actinolite-epidote-albite-muscovite-calcite-biotite (Fig. 9-4). The high-pressure mineral assemblage is characterized by the absence of sphene and hydromica.

It is difficult to assign numerical values to the temperature and pressure necessary for creating the ocean floor greenschist facies. The presence of clay on the one hand and of a chlorite-muscovite assemblage on the other suggests a wide range of stability conditions. A temperature range of above 150°C and below about 550°C and a confining pressure reaching up to about 5 kbar could be considered necessary (Figs. 9-1, 9-2). The metamorphosed igneous suites which fall within the mineral stability field of the greenschist facies can be further divided into: (1) the albite-epidote-chlorite assemblage, (2) the albite-chlorite-actinolite assemblage, and (3) the plagioclase-K-feldspar-chlorite assemblage (Figs. 9-1, 9-2).

*Albite-epidote-chlorite assemblage* rocks are the most common type of ocean floor metamorphics. These are mostly altered basalts which have preserved their original textural features, and they are mainly low-pressure and moderately high-temperature assemblage rocks. Smectite, epidote and mixed layer chlorite-smectite are found as secondary vesicle- and vein-filling material. However, the replacement of calcic plagioclase by albite is the main product of alteration which characterizes this type of rock (Table 9-2, Fig. 9-1). Sometimes, large crystals of plagioclase show traces of K-feldspar scattered throughout. Samples from near 53°N on the Mid-Atlantic Ridge

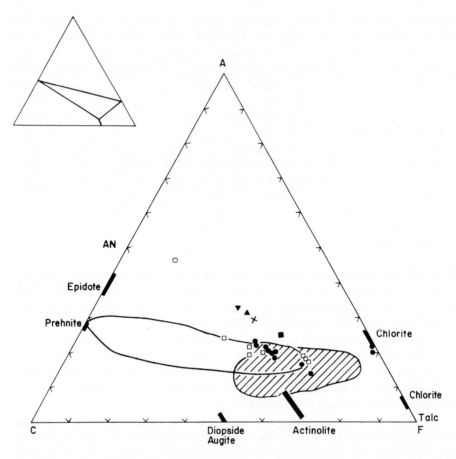

Fig. 9-4. Ternary ACF diagram showing the composition of volcanics metamorphosed to greenschist facies. The altered basalts from the Carlsberg Ridge (•; Cann, 1969) and those from 22°N in the Atlantic Ocean (□; Melson et al., 1966) are compared to the Ordovician metamorphosed basic lava from New South Wales (Australia) (empty field; Smith, 1968) to the spilites from the Caribbean island arc system (Virgin Islands) (striped field; Hékinian, 1971) and to unaltered basaltic rocks from accreting plate boundary region (FAMOUS area). ■ = average of 3 picritic basalts; × = average of 7 plagioclase-pyroxene basalt; ▲ = average of 9 olivine basalt; ▼ = average of 3 moderately-phyric plagioclase basalt; and ○ = average of 2 highly-phyric plagioclase basalt.

(CH4-DR1-56) and from the Romanche Fracture Zone (CH80-DR3-3) show K-feldspar replacing phenocrysts of albite. This type of sample appears to be transitional between the albite-epidote-chlorite assemblage and the albite-chlorite-actinolite assemblage, from which it differs by its smaller amount of K-feldspar replacement and its higher chlorite content. The primary plagioclase and the cryptocrystalline and amorphous groundmass are the most altered material encountered. Some rocks with the albite-epidote-chlorite assemblage also contain relics of calcic plagioclase coexisting with sodic plagioclase ($<An_{10}$). However, it is most often noted that albite is the only plagioclase encountered.

*Albite-chlorite-actinolite assemblage.* The occurrence of abundant actinolite (40–50%) in some oceanic basalts, mostly in those recovered from the Mid-Atlantic Ridge near 22°N (Melson and Van Andel, 1966), suggest the importance of this mineral in the metamorphic alteration of basaltic rocks. Large amounts of actinolite, which seems to have completely replaced clinopyroxene, were found in rocks in a dredge haul from the Romanche Fracture Zone (CH80-DR8-10). Actinolite always occurs as a secondary replacement product of mafic minerals, or filling veins and vesicles, and it is associated with chlorite in some metamorphosed igneous rocks. Occurring as fibrous aggregates, the actinolite is pleochroic to a light-green or light-brownish color. Chemical analyses show high MgO (14–20%), FeO (12–15%), and CaO (10–13%) contents, and a low (2–7%) $Al_2O_3$ content. Actinolite is a common constituent of most greenschist facies rocks; however, because of its high concentration in some specimens it could be used as an indicator of a particular group of rocks made up essentially of albite, epidote, and chlorite. The stability of actinolite may be evaluated by using the following reaction of low-temperature minerals such as chlorite and talc:

chlorite + calcite + quartz = actinolite + $H_2O$ + $CO_2$
talc + calcite + quartz = actinolite + $H_2O$ + $CO_2$

The estimated temperature for the formation of actinolite may be evaluated at about 300–500°C.

*The plagioclase-K-feldspar-chlorite assemblage.* The low-pressure and high-temperature side of the amphibolite and greenschist facies is characterized by the transformation of muscovite to K-feldspar as follows (Fig. 9-1; Evans, 1965):

muscovite + quartz = orthoclase + sillimanite + $H_2O$

The occurrence of both K-feldspar and small amounts of mica in metamorphosed rocks from the Romanche Fracture Zone and from the Wharton Basin (Indian Ocean) suggests a variable temperature range (250–500°C) and low-pressure mineral transformation (Fig. 9-2).

Three types of feldspar co-exist in these rocks (Fig. 9-5): (1) the relic of calcic plagioclase irregularly disseminated within both phenocrystal phases and the matrix; (2) the sodic plagioclase (albite), which is the most abundant component within a single mineral grain; and (3) the K-feldspar, which is also commonly found in the same mineral as a secondary replacement product (Table 9-4). The chemistry of the plagioclase-K-feldspar-chlorite assemblage rocks differs from that of the other greenschist facies rocks by the higher $K_2O$ content ($>1\%$) of the bulk rock and by the variable $NaO_2$ content (2–4%) (Table 9-4).

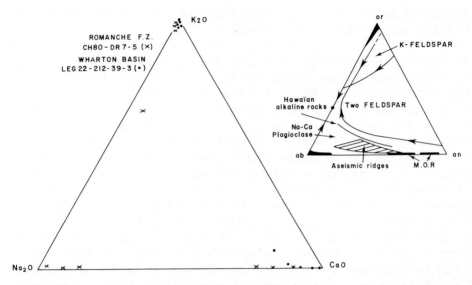

Fig. 9-5. Ternary diagram (Na$_2$O-K$_2$O-CaO, wt.%) and albite-orthoclase-anorthite system at high temperature and low pressure. The diagrams show the field of altered feldspars from metamorphosed basalts located near the K$_2$O and Na$_2$O apex and near the orthoclase and albite apex (inset). The inset diagram shows the solidus boundary lines separating the different types of feldspar fields (after Turner and Verhoogen, 1960, p. 112). The fields of unaltered plagioclase from volcanics associated with various structural settings are also shown. The field of mid-oceanic ridge basalts (*M.O.R.*) also comprises relics of fresh plagioclase associated with altered feldspars.

TABLE 9-4

Microprobe analyses of feldspars from altered basaltic rocks from the Atlantic and Indian Oceans

| | Romanche Fracture Zone CH80-DR7-15 | | | Wharton Basin Leg 22, 212 | | Carlsberg Ridge V14—99C/E | |
|---|---|---|---|---|---|---|---|
| | albite | orthoclase | anorthoclase | plagioclase | orthoclase | albite | plagioclase |
| SiO$_2$ (wt.%) | 66.65 | 63.28 | 65.09 | 51.16 | 64.04 | 67.00 | 52.83 |
| Al$_2$O$_3$ | 20.76 | 19.39 | 19.92 | 29.93 | 18.01 | 20.37 | 28.94 |
| FeO* | — | 0.09 | 0.11 | 0.57 | — | 0.18 | 0.74 |
| MgO | — | — | <0.03 | 0.38 | — | — | 0.17 |
| CaO | 1.08 | 0.02 | 0.81 | 13.78 | — | 0.92 | 12.39 |
| Na$_2$O | 11.02 | 0.60 | 4.97 | 2.59 | 0.06 | 11.36 | 4.35 |
| K$_2$O | <0.08 | 15.07 | 10.45 | 0.90 | 16.73 | <0.02 | — |
| P$_2$O$_5$ | — | — | — | — | — | — | 0.16 |
| Total | 99.60 | 98.45 | 101.38 | 99.31 | 98.84 | 99.84 | 99.58 |

*Total Fe as FeO.

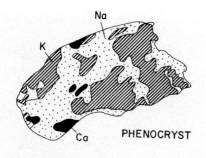

CH 80 - DR 8-5

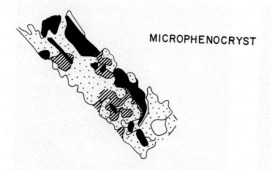

Fig. 9-6. Schematic representation traced after scanning electron microphotograph of plagiclase crystals showing progressive replacement of a calcic plagioclase by a more sodic plagioclase and by K-rich feldspar.

An example of the K-feldspar-albite pair replacing a calcic plagioclase is shown in Figs. 9-6, 9-7 and 9-8. The K-feldspar is always associated with albite and rarely forms a common boundary with the relics of calcic plagioclase. It is inferred that the mineral paragenesis involves a primary replacement of calcic plagioclase by sodic plagioclase (albite), and albite in turn is gradually replaced by K-feldspar:

Ca-plagioclase $(An_{40}-An_{80}) \rightarrow$ Na-plagioclase $(An_{10}) \rightarrow$ K-feldspar.

*Amphibolite facies*

The igneous rocks altered to the hornblende-plagioclase-chlorite assemblage fall within the kyanite-sillimanite zone of the stability field of the amphibolite facies. It was shown experimentally (Althaus, 1967; Richardson et al., 1969) that the triple point kyanite-sillimanite-andalusite occurs at about

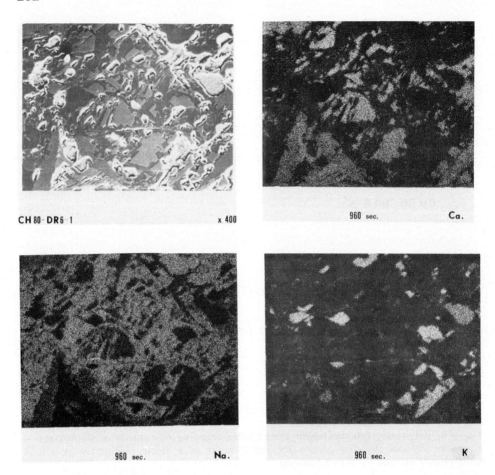

CH 80-DR6-1       x 400

960 sec.      Ca.

960 sec.      Na.

960 sec.      K

Fig. 9-7. Scanning electron microphotograph of feldspar microphonocryst partially replaced by a sodic and potassic feldspar. Notice that sodium is the major phase forming the mineral and that potassium is only forming isolated patches within the feldspar. Sample CH80-DR-6-1 is a metamorphosed basalt taken from the northern wall of the Romanche Fracture Zone.

600°C and at a pressure of about 6 kbar (Fig. 9-2). Hence, the amphibolite facies represents temperatures around 500°C and up to higher than 600°C, and a pressure covering a range above and below 6 kbar. Mineral assemblages characterizing this facies consist of intermediate and/or calcic plagioclase stable with hornblende, tremolite and anthophyllite. The amphibolite found in the ocean floor and probably belonging to this facies of metamorphism has a schistose appearance with alternate layers of plagioclase and hornblende. Amphibolatized basalts and dunite were found on the walls of the Rift Valley in the equatorial Atlantic (Bonatti et al., 1971), and in the Romanche and Vema Fracture Zones (Table 9-5, Fig. 9-9). Sometimes, in

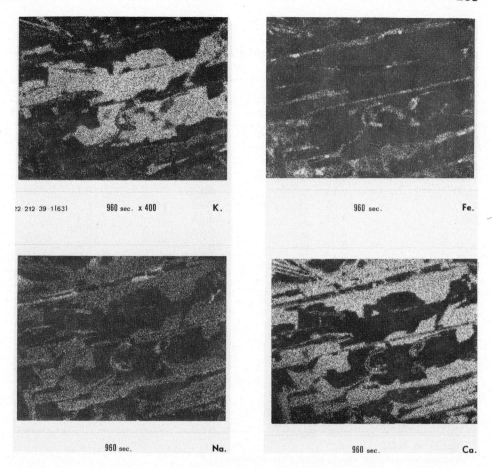

Fig. 9-8. Scanning electron microphotograph of a plagioclase replaced by K-feldspar. Sample 212-39-1/63 was taken during Leg 22 of the "Glomar Challenger" in the Wharton Basin (Indian Ocean).

addition to their foliated nature, amphibole-bearing rocks show a folded type of structure (chevron type). This type of rock found in fracture zone areas (Romanche Fracture Zone, sample CH80-DR10-1) consists of alternating layers of Na-plagioclase, hornblende, and chlorite (Table 9-6, Fig. 9-9). The composition and texture of the rocks suggest that they were originally a gabbro. Their structural appearance (parallel foliation and folded and elongated amphibole grains) suggests a tectonic stress motion which may imply a deformation similar to that encountered in cataclastic types of metamorphism.

TABLE 9-5

Chemical analyses of albite-amphibole-chlorite assemblage rock from the Romanche transform fault (Mid-Atlantic Ridge)

| | CH80-DR10-1 | | |
| --- | --- | --- | --- |
| | bulk rock | amphibole (11)** | plagioclase (9)** |
| $SiO_2$ (wt.%) | 53.57 | 51.55 | 65.44 |
| $Al_2O_3$ | 13.26 | 5.89 | 21.92 |
| $Fe_2O_3$ | 0.82 | | |
| FeO | 5.13 | 9.08* | <0.1* |
| MnO | 0.11 | 0.1 | 0.1 |
| MgO | 10.54 | 17.42 | — |
| CaO | 8.93 | 12.53 | 2.48 |
| $Na_2O$ | 3.40 | 0.93 | 10.04 |
| $K_2O$ | 0.73 | <0.1 | <0.1 |
| $TiO_2$ | 0.47 | 0.40 | — |
| $P_2O_5$ | 0.17 | 0.19 | <0.1 |
| LOI | 2.13 | n.d. | n.d. |
| Total | 99.26 | 98.09 | 99.88 |
| Ba (ppm) | 360 | | |
| Co | 47 | | |
| Cr | 943 | | |
| Cu | 24 | | |
| Ni | 180 | | |
| Sr | 276 | | |
| V | 199 | | |
| Rb | 12 | | |

*Total Fe as FeO.
**Number of samples; values shown are averages.

CATACLASTIC METAMORPHISM

Another type of metamorphism has been noticed mainly in association with the tectonic motion of the oceanic crust. This is called cataclastic metamorphism and it occurs in many rocks collected from fracture zones. One occurrence of contact metamorphism was mentioned by Miyashiro et al. (1971) in a sample from a dredge haul taken at the junction of the Median Valley with the Atlantis Fracture Zone. It is a metagabbro (A150-RD6) showing a gneissic and banded structure and containing brown hornblende. This rock was classified as belonging to the granulite or pyroxene-hornfels facies.

Cataclastic metamorphics consist of mechanically deformed rocks and fragmented rock debris set in a matrix of more or less crushed materials. This type of metamorphism could occur in any tectonic province of the ocean floor which is related to both compressional and distentional stresses.

**TABLE 9-6**

Progressive metamorphism of ocean floor igneous rocks. The data for the Mid-Atlantic Ridge near 45°N, 23 and 30°N and 22°N are respectively from Aumento and Loncaravic (1969), Miyashiro et al. (1971) and Melson et al. (1966). The data for the Carlsberg Ridge are from Cann (1969) and those for the Wharton Basin (Indian Ocean) are from Hékinian (1974a)

| Minerals | "Bald Mountain", Mid-Atlantic Ridge (45°N) | | Mid-Atlantic Ridge near 23° and 30°N | | | Mid-Atlantic Ridge near 22°N | | Indian Ocean | | Romanche Fracture Zone | | |
|---|---|---|---|---|---|---|---|---|---|---|---|---|
| | zeolite | greenschist (low high) | zeolite | greenschist (low high) | amphibolite | zeolite | greenschist (low) | prehnite-pumpelleyite | greenschist (low) | zeolite | prehnite-pumpelleyite | greenschist (low high) |
| Nontronite | | | | | | | | | | | | |
| Natrolite | | | | | | | | | | | | |
| Thompsonite | | | | | | | | | | | | |
| Laumontite | | | | | | | | | | | | |
| Analcime | | | | | | | | | | | | |
| Smectite | | | | | | | | | | | | |
| Mixed layer chlorite-smectite | | | | | | | | | | | | |
| Chlorite | | | | | | | | | | | | |
| Calcite | | | | | | | | | | | | |
| Prehnite | | | | | | | | | | | | |
| Plagioclase | | | | | | | | | | | | |
| K-feldspar | | | | | | | | | | | | |
| Albite | | | | | | | | | | | | |
| Quartz | | | | | | | | | | | | |
| Epidote | | | | | | | | | | | | |
| Sphene | | | | | | | | | | | | |
| Pumpelleyite | | | | | | | | | | | | |
| Tremolite | | | | | | | | | | | | |
| Actinolite | | | | | | | | | | | | |
| Hornblende | | | | | | | | | | | | |
| Talc | | | | | | | | | | | | |
| Muscovite | | | | | | | | | | | | |

Fig. 9-9. Metamorphosed rocks from the ocean floor. A. Sample 212-39-3/0—14, a meta-basalt with albite—K-feldspar—chlorite assemblage from the Wharton Basin (Indian Ocean); *CHL* = chlorite, *BG* = relic of glass, *C* = calcite. The location of this rock is given in Chapter 3 (Table 3-1). B. Sample CH80-DR7, an albite-K-feldspar-chlorite assemblage rock from the Romanche Fracture Zone. C. Sample CH80-DR10-2, a prehnite-plagioclase-chlorite metamorphosed gabbro from the Romanche Fracture Zone. D. Sample CH80-DR10-1, an albite-amphibole-chlorite gneissic-like rock from the Romanche Fracture Zone. The chemical analyses of these specimens are shown in Tables 9-4, 9-5, and 9-6.

Compressional stress fields usually occur in areas of converging plate bound-aries (island arcs), while distentional motions occur in zones of divergence such as oceanic ridges and fracture zones in general. However, the effects of compressional motions have been noticed in some samples from trans-form faults which show a breaking-up of mineral grains.

Based on textural features and on the degree of homogeneity, at least three types of cataclastic rocks are recognized in the ocean floor:

(1) *Weakly deformed cataclastics* consist of tectonic breccias which are mainly made up of rock fragments set in a matrix of finer-grained material.

287

The breccia are usually found in exposed talus piles which are caused by faulting and dislocation of major scarps. Brecciated materials, including various kinds of rocks such as peridotites, basalts, dolerites and gabbros, were found in the Romanche, Vema, and Gibbs Fracture Zones in the Atlantic Ocean (Bonatti et al., 1971; Hékinian and Aumento, 1973). The rocks consist of either peridotite fragments set in a calcic-argonite matrix or basaltic debris cemented by a finer-grained matrix made up of carbonates, minerals, and rock fragments. Zeolites, epidote and occasional prehnite were encountered as alteration products of the metabreccia. The metabreccia are weakly metamorphosed and do not show the strong recrystallization orientation of mineral-rock grain associations. Many samples contain fresh glass which is only partially palagonitized and/or chloritized. These later samples were found in Fracture Zone "A" near 37°N in the Atlantic Ocean at the foot of steep scarps or associated with dikes.

(2) *Mylonitic breccias* are also found in the fracture zones and rift mountain regions of the Atlantic and Indian mid-oceanic ridges. Mylonite breccia are heterogeneous rock fragments composed of different rock types of variable size. Metabasalts, serpentines and gabbros are the most commonly encountered. Mylonitic gabbros from the Romanche Fracture Zone (Bogdanov and Ploshko, 1967) contain hornblende which has granulated in angular crystals. Other mylonitic breccia were encountered in the Kane Fracture Zone and consist of angular fragments of gabbros, basalt and ultramafics. Banding and recrystallization characterize most of the rock fragments, while the matrix is primarily made up of chloritic material. Granulated and banded porphyroclasts are commonly seen. However, the bulk rock does not show any well-developed lamination. Hence, the banding in the individual rock fragments forming the breccia is prior to the phenomena of brecciation.

(3) *Mylonitic coherent rocks* show flow structure (foliation) of cataclastic origin. The structure could be either an alternation of material showing a different mineral composition or various degrees of cataclasis.

Serpentinized mylonitic harzburgites from the Rift Mountain region near 45°N were described as occurring in association with ultramafic intrusions (Aumento and Loubat, 1971). The rocks consist of bent orthopyroxene porphyroclasts, tremolitic amphibole, reddish brown Cr-spinel and lizardite. Plagioclase, harzburgite and lhersolite were also encountered in the Romanche Fracture Zone (Bonatti et al., 1971). These rocks, made up of enstatite, forsterite and accessory plagioclase, were also classified as cataclastics. A mylonitized gabbro from the Ob Fracture Zone in the Indian Ocean was found in association with other deformed rocks, such as pyroxenite. The gabbro is made up of calcic plagioclase, contorted crystals of clinopyroxene, and vermicular intergrowths of spinel and clinopyroxene. The rock also contains banded zones of granulated pyroxene, brown hornblende, tremolite, talc and recrystallized matrix.

Two varieties of mylonitized peridotites were studied from St. Paul's

Rocks. St. Paul's Rocks are a group of islets located in the equatorial Atlantic at about 350 km from the Mid-Atlantic Ridge in the St. Paul Fracture Zone. A 2-pyroxene-peridotite-mylonite and an amphibole-enstatite-peridotite-mylonite were described by Tilley and Long (1967). Wiseman (1966) and Melson et al. (1967) have also described ultramafic rocks which were dynamically metamorphosed. The rocks were originally intrusive in the oceanic crust and their texture was modified by the stresses which caused their upheaval (Washington, 1930). The metamorphosed ultramafics from St. Paul's Rocks consist of olivine ($Fo_{90}$), enstatic clinopyroxene, picotite and a pargasitic amphibole. All these minerals occur as porphyroclasts. When considering the petrogenesis of the samples from St. Paul's Rocks, it is interesting to note that the primary mylonitization was not accompanied by serpentinization. The occurrence of chrisotile and lizardite is associated with later movements which frequently cut across the primary foliation of the rocks. From experimental work (Ringwood et al., 1964) spinel is only stable at temperatures below 1000°C, and hornblende is stable up to 1000°C, provided that the $H_2O$ pressure is equal to that of load pressure (D. H. Green and Ringwood, 1963). Since serpentine would form at 500°C or less (Bowen and Tuttle, 1949), it follows that the mylonitization of the St. Paul's Rocks may have taken place within a temperature range of 500—1000°C.

ORIGIN OF METAMORPHIC ROCKS

With the exception of cataclastic metamorphism, where the samples show obvious signs of stress such as shearing or laminated structural features due to the preferential orientation of some chain and/or sheet silicates, it is often difficult to determine the origin of metamorphic rocks from the ocean floor. The zeolite-smectite assemblage type of metamorphics, grading to those of the hornblende-plagioclase-chlorite assemblage, as described previously, are particularly controversial. The progressive increase of metamorphic grade is suggested by certain characteristic minerals which need precisely limited temperature-pressure conditions for their existence. Laboratory experiments have permitted a determination of the stability relations for certain minerals as follows:

jadeite + quartz = albite
jadeite = albite + nepheline
aragonite = calcite
andalusite = kyanite = sillimanite

These are important relationships which define the various grades of metamorphism. There is, however, some debate about the triple point of the andalusite, kyanite, sillimanite stability field which varies according to different authors between 300 and 595°C and from a pressure of 2.4 to 9.0 kbar (Figs. 9-1, 9-2).

Most of the ocean floor metamorphism occurring on fracture zones and rift valley regions is considered to belong to the low-grade facies. The greenschist rocks are distinguishable from the weathered basalts by their lower $Fe_2O_3$ (<3%) and higher $Na_2O$ contents (>2.9%). Chemically the zeolite facies rocks are more difficult to distinguish from weathered oceanic rocks. Usually they have a higher $Na_2O$ content (>2.9%) than the weathered basalts; also their $K_2O/(Na_2O + K_2O)$ ratio decreases gradually towards the greenschist facies rocks, while in the weathered basalt such a ratio has a tendency to increase with weathering.

The origin of metamorphic terranes in subaerially exposed outcrops such as in continental and converging plate boundary regions (island arcs) could be due to tectonic stresses enhancing various degrees of alteration. Plutonism in this type of unstable environment may also play an important role in the alteration of the surrounding volcanics. Regional and contact metamorphism are important processes for explaining the various degrees of rock metamorphism observed in orogenetic regions. However, in volcanically active diverging plate boundary areas associated mainly with extensive tectonism, it is often difficult to relate or compare the various observed metamorphic groups to those of other subaerially exposed formations. The occurrence of metamorphosed rocks found with serpentinized peridotite in some Atlantic fracture zones has led certain authors (Miyashiro et al., 1971; Bonatti, 1978) to suggest that the metamorphic recrystallization took place at some depth and was later uplifted towards the surface by the intrusion of the sepentinites.

Melson and Van Andel have described the various metamorphic rocks encountered in the Rift Valley walls near 22°N (Mid-Atlantic Ridge) as showing a stratigraphic sequence. Low-grade metamorphics of the zeolite facies overlie rocks that have been altered to the greenschist facies. From such considerations, Melson and Van Andel (1966) suggested that a burial type of metamorphism and subsidence below the 300°C isotherm under more than 2 km of younger rocks was necessary to produce the observed mineral assemblages. On the other hand, the emplacement of the metamorphics in association with their tectonic settings, which have been affected either by uplift, in the case of rift valley wall regions, or by extensional types of fracture such as those encountered in a transform fault area, could suggest a slow shear motion.

If burial metamorphism is the main phenomenon giving rise to the greenschist facies rocks on the ocean floor, the depth of alteration might be calculated from the average oceanic temperature gradient as determined from known heat-flow values. For example, assuming an average thickness of about 7 km for the oceanic crust, it is inferred that with a temperature gradient of about 50°C/km the lower crust could have the composition of the greenschist facies rocks (~350°C) (Fig. 9-2). Similarly, the heat flow values vary greatly for different structural settings. The average heat flow for an accreting plate boundary region of less than 10 Ma is about 2.9 HFU ($10^{-6}$ cal/cm$^2$ s; Le Pichon and Langseth, 1969). Taking a thermal conductivity of $5.5 \times 10^{-3}$ cal/cm s for a diabase (Clark, 1966), the thermal gradient for a region less than 3000 km away from a mid-oceanic ridge axis will be about 52°C/km, while most oceanic basins will have a thermal gradient of less than 29°C/km (Fig. 9-2). However, the effects of the crustal temperature on the phenomena of metamorphism could be intensified by hot fluid circulating in the crust. The degree and rate of increase in the grade of metamorphism in old crust is facilitated by circulating solutions through tectonic fissures and small cracks. The presence of veins, veinlets and interconnecting vesicles filled by secondary minerals such as calcite, epidote, chlorite, mixed layer clays and feldspar are prominent in all low- and medium-grade metamorphosed rocks. An approach to explaining the origin of some of the low-grade facies metamorphics which have preserved their original texture is to take into consideration the effect of hydrothermal fluid circulation in the oceanic crust. The contribution of hot fluid circulating in the oceanic crust could enhance metamorphism in various parts of the ocean floor. The effect of hydrothermalism could be important near active volcanic zones such as accreting plate boundary regions, where relatively high temperature gradients may approach the surface of the ocean floor. Even if concrete evidence has not yet been put forward for the existence of burial metamorphism, this does not exclude its occurrence in certain regions of the ocean floor. Sizeable portions of the oceanic crust need to be exposed and sampled in detail before further speculation can be made on the burial metamorphism phenomena, especially since metamorphosed rocks of both the greenschist and the zeolite facies are frequently found on or near the top of structural highs associated with weathered basalts as described in this chapter.

A phenomenon of autometasomatism might not be excluded as a process for the alteration of oceanic rocks (Chapter 11). However, if autometasomatism occurs, it must be limited to an area where the intrusion of a sizeable magmatic body is prominent. This type of alteration suggests the presence and transfer of relatively abundant volatiles (Chapter 11). Also, the degree of alteration should occur in a certain order. For instance, first-generation phenocrystal phases should alter and then be followed by microphenocrysts and matrix material. This is not what is observed in many metamorphosed basalts from the ocean floor. On the contrary, the degree of albitization is as important in replacing the matrix plagioclase as in the phenocrystal phases. Primary plagioclase is observed to co-exist with a secondary altered plagioclase.

## CHAPTER 10

# HYDROTHERMALISM IN THE OCEAN FLOOR

The circulation of water through fissures, large fractures, and between pillow lava surfaces may reach deep-seated formations to be heated and then to re-emerge on the surface as relatively hot springs carrying various metals in solutions. The study of hydrothermal circulation in the oceanic crust is of primary importance in understanding its effect on the alteration of ocean floor rocks. Another advantage of studying the phenomenon of hydrothermalism is to have a better knowledge of the heat budget of the oceanic crust. D. L. Williams et al. (1974) and Wolery and Sleep (1976) have suggested that 20% of the earth's total heat loss is derived entirely from hydrothermal circulation. The dissipation and/or concentration of fluids in the ocean crust may play an important role in the dynamics of ocean floor volcanism and tectonism. Also, it is not excluded that the concentration of heat through fluid circulation might generate zones of weakness and enhance tensional motions in the crust.

Finally, the economic importance of ocean floor hydrothermalism is not to be overlooked even if at the present time a lack of technology makes the extraction of mineral resources from the sea prohibitive. We are still in the phase of exploration, and in order to ascertain the amount of any metalliferous deposits in the oceans we must have more information on the mechanisms of supply, transport and removal affecting the basement rocks and the sediment-seawater interface.

Detailed surveys of suspected zones of hydrothermalism or exposed hydrothermal sources in relation to their tectonic environment should be the aim of further exploration. It is only recently, less than fifteen years ago, that metalliferous sediment related to hydrothermalism was recognized as occurring in deep ocean floor environments. Only a few well-documented deposits of this kind have been mentioned. However, recent expeditions on the East Pacific Rise indicate that hydrothermalism on accreting plate boundary regions is more prominent than one would have thought. The first active hydrothermal area was found during the 1977 "Alvin" dives in the Rift Valley of the Galapagos Spreading Center near 86°09' west (Fig. 10-1). The difficulty in recognizing hydrothermal products is primarily due to our lack of detailed observations in certain regions of the oceans. I would not be surprised if there were a fair amount of material stored in various oceanographic institutions having the characteristics of hydrothermal material, such as unusually thick manganese crusts, veins of heavy transitional metals, and clays of dubious origins found in association with basaltic rocks. However, since there have not been any general criteria for the recognition of hydrothermal effects on ocean floor material, it is difficult to evaluate the importance of this phenomenon.

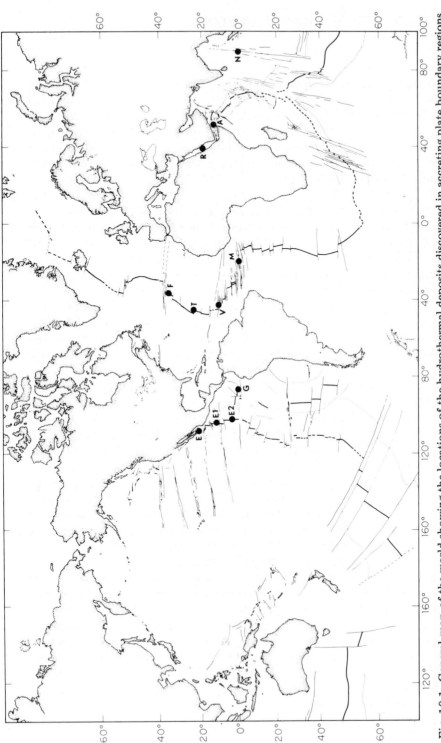

Fig. 10-1. General map of the world showing the locations of the hydrothermal deposits discovered in accreting plate boundary regions. $T$ = Fe-Mn crust of hydrothermal origin located near 26°N on the Mid-Atlantic Ridge (TAG area) (R. B. Scott et al., 1974). $G$ = hydrothermal deposits from the Galapagos area (Scientific Party, DSDP Leg 54, 1977c). $E$ = deposit located on the East Pacific Rise near 21°N (Cyamex, 1978). $A$ = deposits from the Gulf of Aden (Cann et al., 1977). $V$ = presence of disseminated sulfur mineralization in metamorphic rocks from the Vema Fracture Zone (Bonatti et al., 1976); $M$ = disseminated chalcopyrite in metagabbro from the Romanche Fracture Zone. $R$ = Red Sea deposits (Bischoff, 1969); $E1$, $E2$ = sulfides from 13°N and 6°N on the East Pacific Rise; $F$ = FAMOUS area. $N$ = veins of native copper in basalts from the Ninety-East Ridge.

The limited number of sites recognized to be directly or indirectly related to hydrothermalism and the heterogeneous nature of the material recovered make any comparative studies of different metalliferous material difficult. However, in a very broad sense the various types of material recovered could be classified into four main categories:

(1) The hydrothermal deposits sensu stricto, where the structural setting indicates a local origin and the material is hence likely to be the result of direct discharge of deep-seated fluids reaching the floor of the ocean. The structures appear as small ridges or mounds on top of or interlayered with sediment of variable thickness (Corliss and Ballard, 1977; Arcyana, 1978; Scientific Party, DSDP Leg 54, 1977c) or overlying basement rocks (Francheteau et al., 1979).

(2) The metalliferous sediment deposits (e.g. Bauer Deep and Red Sea deposits), where pelagic sediments are intermixed with hydrothermal material sensu stricto and where basement rocks are not prominent in the immediate surroundings.

(3) Hydrogenous deposits (including Mn nodules and Fe-Mn encrustations on rocks), which are due to transport and precipitation of metalliferous material from sea water and may be indirectly related to hydrothermal emanations.

(4) Disseminated hydrothermal material in metamorphic terraces, which are widespread in various settings and are discussed in Chapter 9.

In order to understand the problem of ocean floor basement rock alteration, it is important to have a detailed knowledge of chemical and mineralogical variations of the hydrothermal deposits sensu stricto. These are likely to be the closest to a parent material which has been directly altered by fluid solutions circulating in the basement rocks. It is of primary importance to understand the mode of emplacement and compositional variations of the different types of hydrothermal deposits sensu stricto.

REGIONAL DISTRIBUTION OF HYDROTHERMAL DEPOSITS

Until now the hydrothermal deposits discovered on the ocean floor have been limited to diverging plate boundary zones. Only five deposits of this type have been studied, and they occur in all three major oceans, associated with transform faults and oceanic ridge rift valley systems (Fig. 10-1).

*Galapagos region (eastern Pacific Ocean)*

*Galapagos Spreading Center.* Two fields containing hydrothermal material were found along the Galapagos Spreading Center. One is located on the western branch of the spreading axis west of the Galapagos Islands near 96°W, and the other one occurs east of the islands near 86°W (Fig. 10-2).

294

The easternmost deposit shows evidence of hydrothermalism as defined here, and consists of small mounds around which abundant organisms were found ("Alvin" dives, 1977). Deep-tow survey in the area (Lonsdale, 1977) showed that these mounds are 5—20 m high, 20—50 m wide, conical in shape, and generally elongated in one direction (1—2 km long). These elongated mounds are aligned parallel to the fault block relief of the rise flanks (Lonsdale, 1977). The area where they are found covers a surface of about 75 square miles. During DSDP Leg 54 of the "Glomar Challenger" in the area, a north-south drilling transect across the hydrothermal field was made at about 22 km south of the Galapagos Spreading Center. The estimated magnetic crustal age varies from 0.60 to 0.62 Ma. The mounds are covered by a thick film of pelagic sediment; the surface consists of ripple-like bands of Mn-Fe crust intermixed with yellowish and greenish patches of nontronite. A bottom photograph taken by the "Alvin" (D. L. Williams et al., 1979) shows a variegated colored hill with a scoriaceous aspect which is reminiscent of the hydrothermal deposits found in Fracture Zone "A" in the FAMOUS area (near 37°N in the Atlantic Ocean).

The youngest sediment detectable was 200,000—400,000 years old. Two deep holes (Leg 54) penetrated two different mounds and recovered up to 15—20 m of intermixed Fe-Mn fragments and green clay-rich hydro-thermal mud lying on top of a foraminiferal nanno ooze (Hékinian et al., 1978; Scientific Party, DSDP Leg 54, 1980b) (Fig. 10-3). One of the deep-drilling objectives was to penetrate into the basement in order to locate the root or the vent giving rise to the hydrothermal deposits. This attempt failed since no obvious root zone was noticed. However, it is not unlikely that small fissures and pore structures may channel hydrothermal fluids to

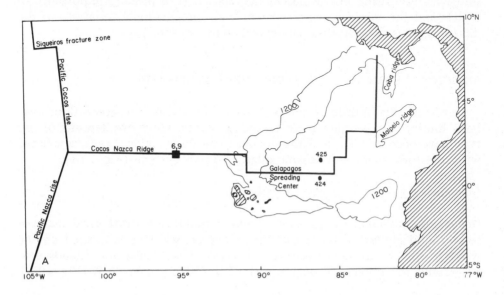

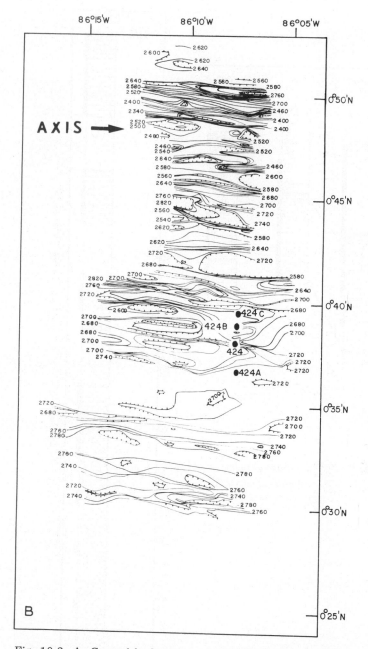

Fig. 10-2. A. General bathymetric map of the eastern Pacific Ocean showing the DSDP
Sites 424 and 425 (Leg 54) near the Galapagos Spreading Center axis. Dredges (D6 and
D9) containing the Fe-Mn crusts and studied by W. S. Moore and Vogt (1976) are reported.
B. Detailed bathymetric map of the Galapagos Spreading Center region based on near-
bottom deep-tow data after Klitgord and Mudie (1974) and Detrick and Klitgord (1972).
The contour intervals are in corrected meters. The holes from Site 424 (Leg 54) are shown.

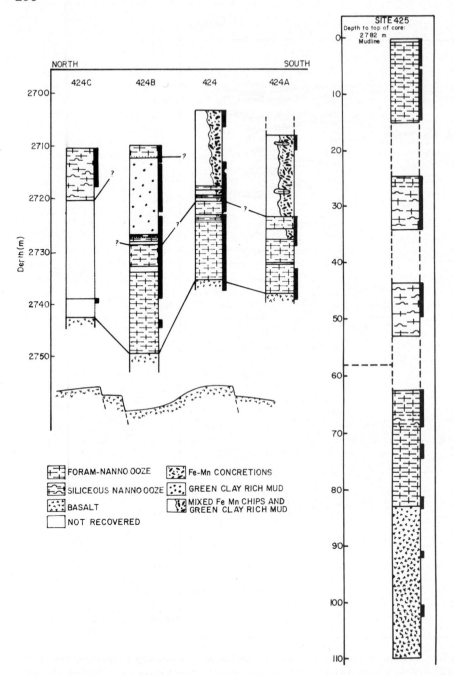

Fig. 10-3. Lithologic columns for the holes drilled in the Galapagos Spreading Center area. Holes 424, 424A, 424B and 424C were drilled approximately 300 m apart from each other at about 22 km south of the spreading axis. Hole 425 was drilled about 65 km north of the spreading axis. Hole 424 and 424A are on or near a topographic high (mound).

the ocean floor surface. Indeed, veins and veinlets of green clay-rich material intermixed with Fe-Mn amorphous material were noticed in a ferrobasalt formation drilled at about 60 km from the Galapagos spreading axis (Site 425 of Leg 54). Other mineralized veinlets were also found in basement rocks overlain by hydrothermal material at Site 424 (Leg 54). The second deep-sea drilling operation undertaken in the mounds area of the Galapagos Ridge (Scientific Party, DSDP Leg 70, 1980d), using a hydraulic piston corer which was able to recover undisturbed sediment sequences, confirmed the inferred lithological sequences obtained during Leg 54. Indeed, the mounds drilled during Leg 70 consist of a pelagic sediment layer followed by a Fe-Mn-rich layer and a Si-Fe-rich layer with intermixed pockets of pelagic sediments overlying a relatively thicker pelagic sediment (Scientific Party, DSDP Leg 70, 1980d). Dredge hauls from the hydrothermal mounds area located south of the ridge axis also brought up various types of Fe-Mn and smectite (nontronite) material (Corliss et al., 1978). The Fe-Mn coatings of basalts dredged from 43 km south of the ridge axis on a 1- to 2-Ma-old crust contain higher Ni, Co, Mn and REE than the deposits from the mounds. Hence it was suggested (Corliss et al., 1978) that the Fe-Mn coatings of these basaltic rocks are of a hydrogenous origin.

*Galapagos Rift Valley.* The Galapagos Rift Valley region consists of active and inactive hydrothermal vents which are set on the basaltic basement. Bottom photographs taken from a submersible show small hills with a scoriaceous structure made up of sulfide material similar to those found near $21°N$ on the East Pacific Rise and discussed later. During the 1977 "Alvin" dives (Crane and Ballard, 1980), it was noticed that these hydrothermal venting areas were associated with a sheet type of lava flow terrane. These sheet flows often terminated with a pillow lava type of formation. The active hydrothermal zone is located within the area of the most recent volcanism (a zone 0.5–1 km wide). The vents are accompanied by abundant marine life which lives in an ambient temperature which does not exceed $20°C$. These vents are believed to be associated with east-west fissures located on the axis of the Rift Valley on the entire 30 km distance covered during the study.

Another site of hydrothermalism was claimed to occur on the western branch of the Galapagos spreading axis at about 30 and 60 km from it (W. S. Moore and Vogt, 1976) (Fig. 10-2). The material recovered consists essentially of Fe-Mn crusts with a banded structure. The sample closest to the spreading axis was considered (W. S. Moore and Vogt, 1976) to be of hydrothermal origin because of its transitional metal content which is lower than that of the hydrogenous deposits. The rapid rate of Fe-Mn accumulation of the crusts (less than 9000 years) is also strong evidence for hydrothermalism, while the site farther away from the Galapagos spreading axis has recorded a later history of hydrogenous precipitation (W. S. Moore and Vogt, 1976).

*East Pacific Rise hydrothermal fields*

Evidence of the presence of hydrothermal events in the Pacific Ocean was gathered recently when sulfide deposits were discovered near the axis of the East Pacific Rise at 21°N (Cyamex, 1978). During a submersible ("Cyana") dive, at least three small hills, averaging about 5 m in diameter and about 10 m in height, were identified to be aligned parallel to the axis of the ridge; these hills were made up of massive sulfides (Francheteau et al.,

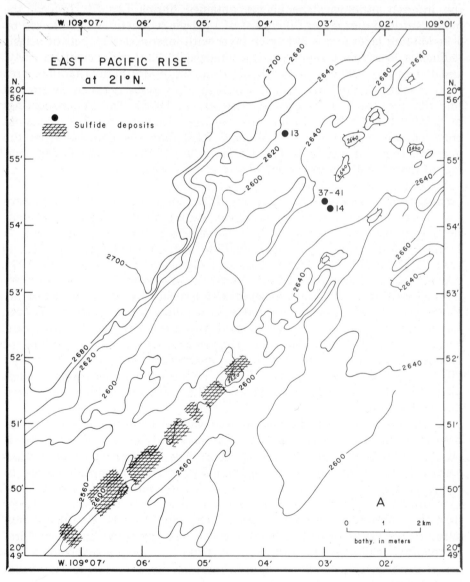

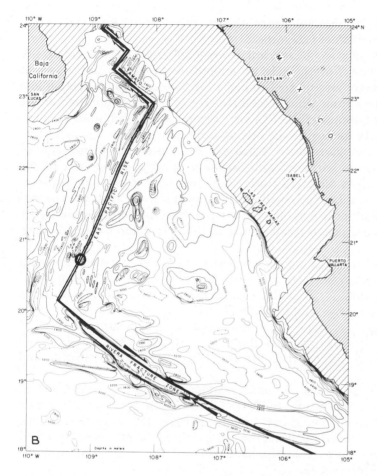

Fig. 10-4. A. Bathymetric map of the East Pacific Rise near 21°N (after Spiess et al., 1980) showing the location of the sulfide ore deposits near the accreting plate boundary regions (contours in meters). The cross-hatched area shows the fields of the hydrothermally active vents visited by "Alvin". The black dots are the zones where non-active vents were seen by "Cyana". General location of the area surveyed near 21°N (Cyamex, 1981).

1979; Hékinian et al., 1980). Further investigation in the area by a bottom-towed camera system (ANGUS: acoustically navigated geophysical under-water survey system) and by the submersible "Alvin" revealed that a zone about 7 km long and 100 m wide consisted of active hydrothermal vents (Spiess et al., 1980). Hydrothermal discharge of fluid was directly observed taking place at depths of 2610—2650 m on the axial zone of the ridge (Fig. 10-4).

From field observation of their structures and from their compositional variabilities, the hydrothermal vents of the East Pacific Rise are further classified into three distinct categories:

(1) *The dark-brown deposits*, which are composed essentially of Zn-bearing sulfides, have formed hills or mounds with gentle slopes (about 10—50 cm in diameter and up to 1 m high) which are crowded together and represent the orifices from which black hydrothermal fluids are discharging at a temperature of about 355°C ("Alvin" dives, 1979). Each individual hill could have as many as up to six orifices or pipes through which hydrothermal fluids are simultaneously discharged. Black precipitates, made up of mainly Zn-sulfides, are scattered as a powdery material at the base of the hills and on the surrounding lava flows. The sheet flows and the pillow lavas in the immediate vicinities (up to about 20 m away) of the vents are covered by a thin film of dark-brown sulfide material. Sphalerite, wurtzite and pyrite are the major mineral associations found in the dark-colored mounds (Fig. 10-5).

(2) *The anhydrite and Cu-bearing deposits* are found in association with tall, columnar chimneys which are more or less cylindrical in shape and which rise up to about 7—10 m above the surrounding floor that is made up of sheet flows. These are found on the flattish floor of a lava lake type of structure. However, this type of sulfide deposit also occurs forming the chimneys of the mounds described above (Fig. 10-6A). The Cu-rich phases consist primarily of chalcopyrite, bornite, high-temperature cubanite and digenite.

(3) *Amorphous silica and Fe-oxide staining* are also the products of deposition of low-temperature fluids (<60°C) which are found partially covering the surface of the glassy basalts between the interstices of the pillow flows (Fig. 10-6B). Marine organisms crowd along these types of vents which are also located in the axial zone of the ridge crest near the high-temperature venting area.

The presence of hydrated and highly oxidized material of hydrothermal origin (gossans), found away from the active vent region (about 2 km from the spreading axis), also suggests a low-temperature alteration and degradation of pre-existing sulfide deposits (Fig. 10-4). Goethite and limonitic material are the major Fe-oxide compounds and they have a powdery and very friable texture. The alteration of the sulfide vents must start very early in the history of these deposits, since a reddish brown goethitic material was found staining the flanks of the active vents.

Fig. 10-5. A. Photograph taken at a depth of 2630 m on the East Pacific Rise near 21°N ("Alvin" dive 981, November 1979) of an hydrothermal mound with active chimneys discharging black columns of fluid (left of photo). The size of the chimneys overlying the mound is about 40—70 cm in diameter. The lighter-colored material forming the chimneys consists mainly of chalcopyrite and anhydrite. B. Fragment of altered sulfide material (gossan) made up essentially of hydrated Fe-oxide products. The sample location is shown in Fig. 10-4 (site 14).

301

Cyp 78.08.14B

2 cm

Other deposits are likely to occur on other segments of the East Pacific Rise. A dredge taken on the flank of a seamount between 2130 and 1700 m at 10°38'S, 109°36'W in the Pacific Ocean recovered abundant fragments of red-yellow friable and powdery material (Bonatti and Joensuu, 1966). The mineralogy and chemistry of this material revealed the presence of abundant Fe- and Mn-oxides. A few pebbles of porphyritic basalt fragments were also recovered, and they show a thin Fe-Mn surface coating (1 mm thick). Bonatti and Joensuu (1966) favored a hydrothermal origin for this material. This conclusion is also suggested by the Fe/Mn (10—15) and Si/Al (3—15) ratios which make the samples similar to other hydrothermal deposits.

Basal sediments in contact with basement rocks from the East Pacific Rise were considered by Boström and Peterson (1966, 1969) to contain fair amounts of Fe, Mn, Cu, Cr, Ni, and Pb as metal oxide precipitates. During DSDP Leg 54 of the "Glomar Challenger", holes near 9°N on the East Pacific Rise also indicated the presence of metalliferous sediment in contact with the basement. The concentration of heavy metals is generally found in crest regions of the East Pacific Rise in zones of high heat flow. It is interesting to note that up to now hydrothermal exhalation seems to occur more readily in accreting plate boundary regions of the East Pacific Rise than in other active spreading ridges in the Atlantic and Indian Oceans.

*Gulf of Aden (Indian Ocean)*

Hydrothermal deposits were found at the northern edge of a small median valley in the Gulf of Aden, in a region where it is broken up into short lengths by successive small transform faults (Cann et al., 1977). Dredge hauls made between 12°31.1'N, 47°39.2'E (2550 m) and 12°35.0'N, 47°39.9'E (2260 m) brought up lumps of brown Fe-Mn concretions, yellowish green crumbly material, and orange powdery Fe-oxide (Fig. 10-7A). In addition, aphyric plagioclase-olivine-bearing basalts were found in the same dredge haul (Cann et al., 1977). Trace element analyses of the basaltic rocks fall within the range of other ocean floor basalts.

Fig. 10-6. A. Fragment of a chimney (sample ALV980-R5) collected from an active vent from the southern portion of the area explored near 21°N (East Pacific Rise). The concave part is coated with chalcopyrite and bornite. The outside of the chimney has a coating of anhydrite. B. Fragment of a pillow lava flow (sample ALV979-R11-1) showing a thin white coating of amorpheous silica and an hydroxide of silica and magnesium. The sample was taken at less than 20 m from an active hydrothermal vent. The rock is made up of concentric layers of fresh basaltic glass.

A

B

*Mid-Atlantic Ridge near 26—30°N (TAG area)*

The area of the Mid-Atlantic Ridge located between 26°N and 30°N includes the Atlantis Fracture Zone and the associated rift valley. Three sampling stations were made in the area: one in the Atlantis Fracture Zone,

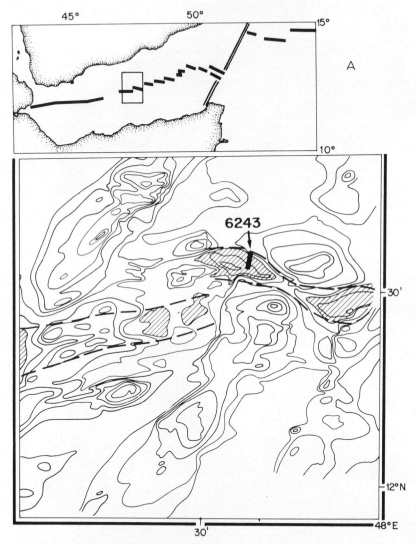

Fig. 10-7. Detailed bathymetric map of regions where hydrothermal deposits were found. A. Map from the Gulf of Aden with the location of the Rift Valley (shaded area; Laughton et al., 1970) and the site of the hydrothermal fields (Cann et al., 1977). B. Map of Transform Fault "A" in the FAMOUS area near 36°N where the two hydrothermal sites are shown (Hoffert et al., 1978).

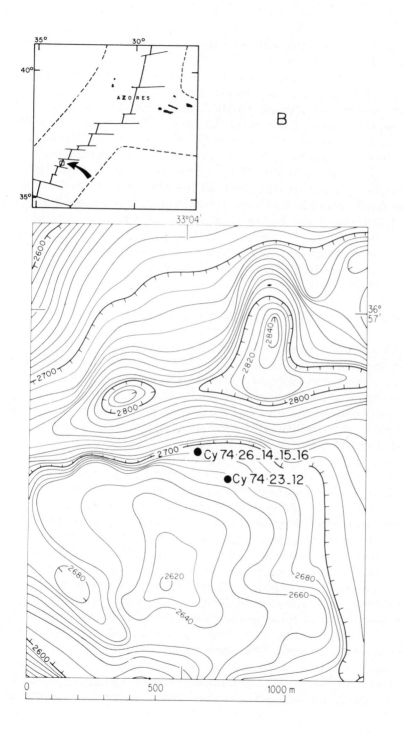

another in the intersection zone with the rift valley, and the third at the base
of the median valley scarp (3400 m).

The material from the eastern side of the Atlantis Fracture Zone consists
of Fe-Mn crusts covering volcanic breccia found at 4200—4500 m depth.
The rift valley samples containing basalts, metabasalts and gabbros were
dredged at a depth of about 3400 m on the eastern flank of the rift valley,
at about 5 km from the inner floor axis (R. B. Scott et al., 1974). The
dredge hauls from the intersection zone contained hydrothermal material
and were recovered at about 3000—3100 m depth (R. B. Scott et al., 1972).
Many samples found in the dredges from all the sites contain Fe-Mn crusts
up to 50 mm thick covering limestone, volcanic breccia and basaltic rocks.
The unusual thickness of the Fe-Mn crust is due to a rapid accumulation
rate measured by radiometric method (130—250 mm/Ma) and a hydrothermal origin was attributed to this type of accumulation (Rona, 1976,
1977). However, since only Fe-Mn encrustations were found, a hydrogenous
origin is not to be excluded.

*Transform Fault "A" near 37°N on the Mid-Atlantic Ridge*

A discovery of two hydrothermal fields in Transform Fault "A" made
during the diving operations undertaken in 1974 permitted the recognition
of small asymmetric ridges elongated in a general east-west direction (Fig.
10-7B). These ridges are associated with faults and are located on the top
of a scarp immediately to the south of the deepest portion of the transform
valley where most of the active left-lateral strike-slip faulting is now occurring (Arcyana, 1975; Coukroune et al., 1978).

The hydrothermal fields are 200 m apart and are at a depth of 2670—
2690 m (Fig. 10-7B). The material delivered through small east-west
elongated fissures covers a surface of about 600 $m^2$ (15 × 40 m). A schematic cross-section of one of the fields shows the ridge covered by a black
stratified scoriaceous-like Fe-Mn material (Fig. 10-8). The structure of the
Fe-Mn cover varies from polygonal near the top of the ridge to scoriaceous
and more uniform on the northern slope. The polygonal, concretion-like
form of the Fe-Mn material appears to be thicker than the more uniformly
distributed scoriaceous-like deposits. These latter also appear associated
with a thick (about 1 m) stratified mud with a variegated aspect (Fig. 10-8).
The two types of material forming the hydrothermal fields are apparently
in stratigraphic continuity with each other and can be differentiated in the
field only by the color changes. The black Fe-Mn concentrations are prominent constituents near the top and on the southern slope, while the
scoriaceous form of the Fe-Mn concentrations and variegated mud-like
material occur mainly on the southern slope where fissure-like vents were
seen.

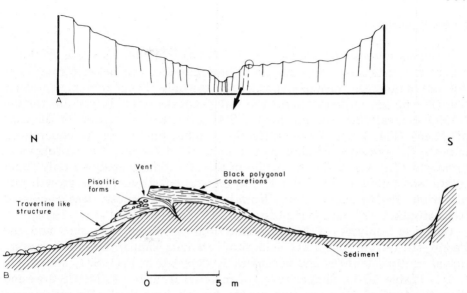

Fig. 10-8. A. Cross-section of Transform Fault "A" near 37°N in the Atlantic Ocean showing the location of inactive hydrothermal vents. These vents were found during submersible dives located on the south wall of the transform fault (Arcyana, 1975). B. Schematic cross section of a hydrothermal deposit visited during dive CYP74-26. The polygonal concretions are made up of Fe-Mn products, the pisolitic concretions of variegated, black, rusty red and yellowish green hydrothermal material (Arcyana, 1978).

CONSTITUENTS OF HYDROTHERMAL DEPOSITS

Hydrothermal deposits from around the world often have material with a poor degree of crystallinity. Amorphous and gel-like associations are commonly intermixed with well- and poorly-crystallized phases of Fe-Mn and clay minerals. The hydrothermal material, except for samples from 21°N on the East Pacific Rise and those from the Red Sea, is made up essentially of Fe-Mn concretions and variegated clay-rich material. The clay material is very friable under the fingernails, and the variegated colors of the product range from yellowish brown through reddish brown, pale olive and grayish green to dark green. The Fe-Mn concretions are essentially black, with occasional inclusions of green clay-rich material. As far as the Fe-Mn crust from the TAG area (R. B. Scott et al., 1974) is concerned, it is fairly homogeneous and no traces of clay-rich material were noticed. Similarly, the Mn-nodules from the Pacific are made up of various dark bands of Fe-Mn compounds and other anisotropic material. For our purposes, the Fe-Mn bands were separated prior to analyses.

*Fe-Mn material*

Electron microscope studies of the Fe-Mn concretions show globular-reticulated structures. The shape of these globules (mamelon forms) does not vary among the various deposits, and the only changes observed are hair size (20—35 $\mu$m in diameter for the Galapagos deposits; 40—60 $\mu$m for the FAMOUS area, 100—400 $\mu$m for the TAG area, and 5—20 $\mu$m for the Gulf of Aden) (Fig. 10-9). X-ray diffraction studies on the Fe-Mn concretions indicate the presence of abundant todorokite (9.7 Å and 4.8 Å peaks) and birnessite (7.2 Å and 3.6 Å), while in the Mn nodules analysed only todorokite was found as the major Fe-Mn mineral constituent. The growth rate of various Fe-Mn encrustations from hydrothermal deposits was calculated to be around 2—20 cm/Ma (Corliss et al., 1978; M. R. Scott et al., 1974).

Chemical analyses of Fe-Mn concretions from the Galapagos and the Fe-Mn crust from the TAG area show striking similarities. Both are enriched in MnO ($\simeq$90%) and considerably depleted in FeO, $SiO_2$ and $Al_2O_3$ (<1%) (Table 10-1). Similar types of material from the FAMOUS area and from the Gulf of Aden are comparable. Indeed, both the Fe-Mn concretions from the FAMOUS area and those from the Gulf of Aden (Cann et al., 1977) have higher FeO (>2%) and $SiO_2$ (>3%) contents than those from the Galapagos and the TAG area (Table 10-1). In this respect, the Gulf of Aden and FAMOUS samples are closer in composition to the Mn nodules of the Pacific ($SiO_2$ = 15—19%; FeO = 6—17%; $Al_2O_3$ $\simeq$ 5%) from which they differ in their lower MnO content (<45%) (Table 10-1). The mineral constituents found in the Fe-Mn concretions are as follows:

*Todorokite* -- (Mn, Ca) $Mn_2O_9 \cdot 2H_2O$. This is the most common mineral phase encountered in all hydrothermal deposits. X-ray diffraction patterns show mainly the 9.6—9.7-, 4.7-, 3.31- and 1.8-Å reflection peaks (Cann et al., 1977; Arcyana, 1975). This material is very dark brown, with a bluish shade on broken surfaces.

*Birnessite* — (Na, Ca) $Mn_7O_{14} \cdot 3H_2O$. It is less commonly encountered than todorokite and can be distinguished from the latter by its X-ray diffraction peaks at around 7.3, 2.4 and 1.4 Å.

*Manganite* -- MnO(OH). This mineral of manganese often occurs as criss-crossed lamellae associated with Fe-Mn concretions. Manganite is recognized on an X-ray diffraction pattern from its major peak intensities of 3.4, 2.6 and 2.3 Å.

*Rancieite* — (Ca, Mn) $Mn_4O_9 . 3H_2O$. This mineral is also a minor constituent of hydrothermal deposits and can be distinguished from todorokite by its strong pleochrism.

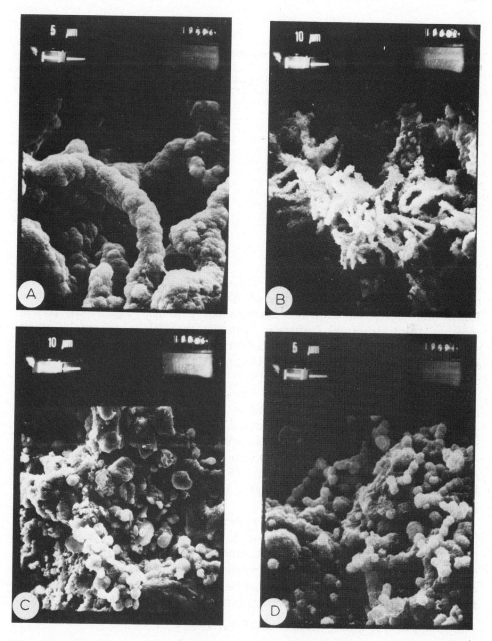

Fig. 10-9. Clay-rich material viewed under a scanning electron microscope (Hékinian and Fevrier, 1979). A. Green clay-rich material (CYP74-26-15-1) from Transform Fault "A" in the FAMOUS area (magnification ×2400). B. Orange clay-rich material (CYP74-26-15) from Transform Fault "A" in the FAMOUS area (magnification ×1300). C. Yellow clay-rich material (6243-12) from the Gulf of Aden (magnification ×650). D. Green clay-rich material (Leg 54, 424-2-cc) from the Galapagos area (magnification ×2500).

TABLE 10-1

X-ray fluorescence analyses of Fe-Mn concretions and Mn nodules

| | Mid-Atlantic Ridge, 26°N (TAG) 73-2A | Galapagos* 424-3-5 35—40 | FAMOUS area CYP74- 23-12 | Gulf of Aden 6243-3 | Northeast Pacific Basin (Mn nodules) | | |
|---|---|---|---|---|---|---|---|
| | | | | | CX 2109 | CP 2171 | A 2249 |
| $SiO_2$ (wt.%) | 0.4 | 0.93 | 16.6 | 6.91 | 15.25 | 16.97 | 19.41 |
| $Al_2O_3$ | 0.91 | 1.01 | 1.7 | 1.51 | 5.51 | 5.79 | 6.46 |
| FeO | 0.15 | 0.36 | 13.3 | 4.21 | 9.32 | 6.06 | 16.64 |
| MnO | 90.24 | 92.25 | 34.57 | 63.04 | 60.84 | 61.24 | 47.73 |
| MgO | 4.22 | 4.15 | 3.7 | 3.12 | 4.16 | 3.94 | 3.59 |
| CaO | 3.05 | 1.78 | 2.3 | 2.78 | 2.84 | 3.14 | 3.26 |
| $K_2O$ | 0.21 | 0.37 | 0.97 | 0.51 | 1.05 | 0.92 | 1.34 |
| $TiO_2$ | 0.01 | 0.02 | 0.08 | 0.10 | 0.7 | 0.48 | 1.64 |
| $P_2O_5$ | 0.04 | 0.12 | — | 0.19 | 0.36 | 0.41 | 0.6 |

*DSDP Leg 54 (Hékinian et al., 1978).

*Clay-rich material.* Viewed under the electron microscope, the clay-rich material shows two distinct types of structure: flaky sheet-like polygonal forms and elongated tube-like agglutinated globules (Fig. 10-9). A less-common third structural feature is found in the Galapagos samples which show tiny fibers with no preferential orientation. The size of the tubular structures varies from 1 $\mu$m (FAMOUS area, Galapagos and Gulf of Aden) to about 50 $\mu$m (FAMOUS area) (Fig. 10-9).

X-ray diffraction studies on the clay-rich material have shown that the major clay-mineral constituent consists of a smectite (12-Å peak) which expands upon glycolation up to about a 17-Å peak. Other minor constituents include goethite (4.5-Å peak) and saponite. In addition, there is a 9-Å peak on most diffraction patterns which was tentatively attributed to a glauconite-celadonite composition (Hoffert et al., 1978) for the FAMOUS area samples. However, it is not excluded that such a peak could correspond to that of a zeolite.

The chemistry of the clay-rich material is the same for all the hydro-thermal localities studied. The clay-rich material is made up essentially of $SiO_2$ (50—54%) and FeO (25—40%), and is depleted in $Al_2O_3$ (<2%) and MnO (<1%) (Table 10-2). The structural formula calculated for all the clay-rich material from the different localities is similar. As an example for the FAMOUS area deposits, the structural formula:

$$(Ca_{0.29}Na_{0.52}K_{0.58})(Fe^{3+}_{3.3}Mg_{0.68})(Si_{6.67}Al_{0.04}Fe^{3+}_{1.29})_8 O_{20}(OH)_4 \cdot H_2O$$

corresponds to an Fe-rich nontronitic clay.

Trace element studies of the clay-rich material published elsewhere (Cann et al., 1977; Hoffert et al., 1978; Hékinian et al., 1978) show that it is depleted in transitional metals content. The Fe-Mn concretions are also depleted in transitional metals content so that the Cu content of the hydro-

TABLE 10-2

X-ray fluorescence analyses of clay-rich material from hydrothermal deposits (sample CYP74-26-15B is shown in Fig. 1-6)

| | Gulf of Aden | | | Galapagos* | | Transform fault "A" | |
|---|---|---|---|---|---|---|---|
| | 6249-24 yellow | 6243-24 green | 6243-12 green | 424-2-4/30—33 green | 424-2-2/11—13 green | CYP74-26-15B green | CYP74-26-15-2 yellow |
| $SiO_2$ (wt.%) | 42.63 | 45.24 | 45.09 | 45.95 | 53.3 | 45.5 | 32.32 |
| $Al_2O_3$ | 0.88 | 0.90 | 0.91 | 0.32 | 0.20 | 0.20 | 2.49 |
| FeO | 29.74 | 28.73 | 28.94 | 25.33 | 27.56 | 34.11 | |
| MnO | — | — | — | 0.06 | 0.17 | 0.10 | |
| MgO | 3.16 | 3.39 | 3.28 | 3.71 | 4.41 | 2.91 | 2.89 |
| CaO | 0.07 | 0.21 | 0.76 | 0.17 | 0.30 | 0.6 | 4.81 |
| $Na_2O$ | 1.88 | 1.96 | 1.85 | 1.43 | 2.08 | 1.71 | n.d. |
| $K_2O$ | 2.43 | 3.12 | 3.28 | 3.65 | 3.39 | 3.23 | 0.67 |
| $TiO_2$ | 0.02 | 0.016 | 0.02 | 0.02 | 0.02 | 0.02 | 0.17 |
| $P_2O_5$ | 0.05 | 0.04 | 0.02 | 0.12 | | | 0.86 |
| $H_2O$ 110° | 12.2 | 9.83 | 9.42 | 6.94 | | 9.88 | 14.30 |
| $H_2O$ 1050° | 6.04 | 6.08 | 6.38 | 8.90 | 6.79 | 6.22 | 10.95 |
| Co (ppm) | — | — | — | — | 3 | 2 | — |
| Cu | — | — | — | — | 30 | 46 | — |
| Zn | — | — | — | — | 53 | 6 | — |
| Ni | — | — | — | — | 10 | 10 | — |

*DSDP Leg 54 (Hékinian et al., 1978).

thermal deposits is less than 200 ppm for the total of the material deposited by hydrothermal activity.

*Fe-oxides and Fe-hydroxides.* The Fe-oxides and hydroxides usually occur as dark and light red and reddish brown powdery, friable, amorphous material. This product could be associated with either sulfide or Fe-Mn concretions and clay-rich material. Limonite is one of the major constituents, together with a more crystalline form of goethite. The chemistry of the Fe-oxide compounds consists essentially of iron oxide (up to 80%) and water. The other constituents are mainly $SiO_2$ (9—20%) and $P_2O_5$ (1—3%).

*Sulfides.* The sulfides found in the ocean floor consist of two types: (1) those disseminated into the various rock types (basalts, serpentinized peridotites and metamorphics; (Table 10-3), and (2) the massive sulfide deposits associated with accreting plate boundary regions such as those from the East Pacific Rise near 21°N (Cyamex, 1978; Spiess et al., 1980; Hékinian et al., 1980) and the Red Sea deposits (Bischoff, 1969).

The massive sulfide deposits consist of friable and porous material varying in color from medium brown to dark gray. The brownish material is intermixed with some oxidized metals, while the darker compounds consist of Zn-, Cu- and Fe-sulfide minerals (Figs. 10-5, 10-6, 10-10). The sulfide deposits from 21°N on the East Pacific Rise display tubiform-shaped material with concentric zonations of various sulfide layers interbedded with amorphous silica. The sulfide phases consist mainly of sphalerite, wurtzite, pyrite, marcasite, chalcopyrite and pyrrhotite (Figs. 10-10, 10-11). The most common form encountered in the sample is an octahedron. Tetrahedra of sphalerite and

TABLE 10-3

Electron microprobe analyses of sulfides disseminated in metamorphosed rocks from the Mid-Atlantic Ridge transform faults

| | Romanche Fracture Zone | | | Vema Fracture zone |
|---|---|---|---|---|
| | CH80-DR10-2 pyrite (ave. of 2) | CH80-DR10-2 chalcopyrite (ave. of 2) | GS7309-93 chalcopyrite (ave. of 3) | P7003-25 chalcopyrite (ave. of 2) |
| S | 50.88 | 32.16 | 34.46 | 34.4 |
| Cu | — | 33.50 | 30.97 | 30.8 |
| Fe | 47.62 | 31.06 | 32.80 | 31 |
| Au | — | 0.26 | n.d. | n.d. |
| Ni | 0.16 | — | n.d. | n.d. |
| Total | 98.66 | 97.01 | 98.23 | 96.7 |

Samples GS7309-93 (0°58.6'S, 24°30.8'W) and P7003-25 (10°51'N, 41°48'W) are meta-basalts (Bonatti et al., 1976); CH80-DR10-2 is a metamorphosed gabbro.

TABLE 10-4

Bulk sample analyses of polymetallic sulfides from the East Pacific Rise near 21°N.

|  | 8-14A2 (U.S.G.S.) | 8-14A2 (C.R.P.G.) | 12-40Aa (U.S.G.S.) | 12-38A12 (U.S.G.S.) | 12-40B (C.R.P.G.) |
|---|---|---|---|---|---|
| S (wt.%) | 30.25 | 30.17 | 35.16 | 52.47 | 45.86 |
| Cu | 0.28 | 0.40 | 1.16 | 0.28 | 1.51 |
| Fe | 8.9 | 1.37 | 12.6 | 45.4 | 1.28 |
| Zn | 49.7 | 22.89 | 49.7 | 0.05 | n.d. |
| SiO$_2$ | 10.7 | n.d. | 2.14 | 0.1 | n.d. |
| Total | 99.83 | — | 98.62 | 98.16 | — |
| Ag (ppm) | 380 | 480 | 290 (121) | 320 (145) | 83 |
| Pb | 640 | n.d. | 330 | 520 | n.d. |
| Cd | 300 | n.d. | 700 | <70 | n.d. |
| Co | <10 | n.d. | 500 | 500 | n.d. |

The analyses were done at the U.S.G.S. and at the C.R.P.G. (Nancy) by atomic absorption spectroscopy for Fe, Zn, Cu, Ag and Pb and by optical emission spectroscopy for Si, Co and Cd. The analyses of sulfur were done by LECO Inc. Silver values shown in parentheses were measured on a separate portion of the sample (analyses, C.R.P.G.) (from Hékinian et al., 1978).

spheroids of chalcopyrite were also seen. Wurtzite often occurs as globular samples (Fig. 10-10). The bulk chemistry of these samples indicates the abundance of Zn (up to 44%), of Fe (up to 49%), and of Cu (up to 1.6%) (Table 10-4). Silver (up to 400 ppm) and traces of gold are often associated with these sulfides. Microprobe analyses of silver grains (about 20$\mu$m in diameter) showed the presence of sulfur, iron, and a small amount of copper (up to 5%). Silver associated with selenium and lead was also detected as isolated grains in the sulfide deposit from 21°N of the East Pacific Rise.

*Sulfates.* Sulfates usually form the minor constituents of most Cu- and Zn-bearing deposits, except for the anhydrite-rich vents. Most sulfates form hydrated compounds of copper, iron and zinc.

Cu-Fe sulfates occur as laths of gray color under reflective light and reddish brown under refractive light, and may represent an alteration product of chalcopyrite. Chemical analyses of several laths show a decrease in the CuO (from 14% to 2%) followed by an increase in the FeO content (from 29% to 71%). It seems that during hydration the Fe was oxidized, which resulted in the subsequent loss of Cu and S from the compound, and ultimately a limonitic product could be formed from such a replacement. Mineral phases such as copiatite and chalcanthite were identified by X-ray diffraction and microscopic work.

Zn-sulfates were identified in association with the other sulfates. White translucent fibers of melanterite and dark-green anisotropic aggregates of goslarite were also detected.

314

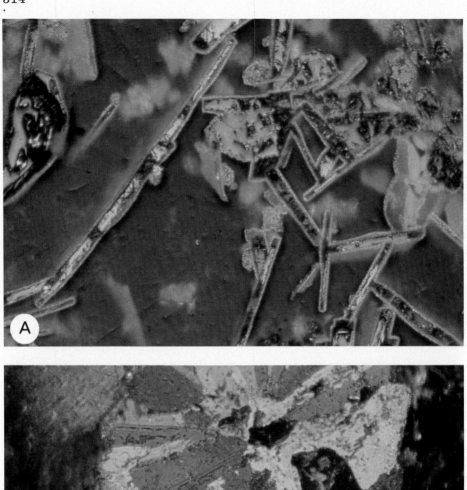

Fig. 10-10. Microphotographs of polymetallic sulfides from the East Pacific Rise near 21°N. A. Sample ALV921-R5 showing lamellae of altered pyrrhotite set in a matrix of sphalerite (dark) (magnification × 120). The alteration product (yellowish red) consists of hydrated Fe-oxide. B. Sample CYP78-12-38A showing spherulite marcasite with a rim of alteration product under polarized light (magnification × 220). C. Sample CYP78-08-14A-2

315

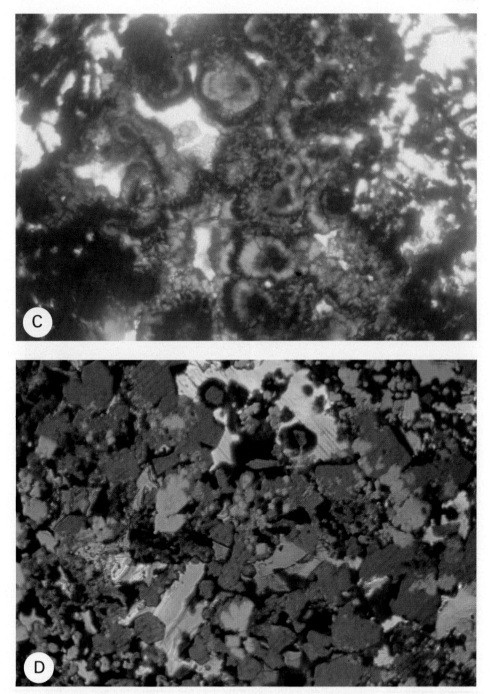

consisting of gel-like (collomorph) material with the composition of sphalerite (magnification ×120). D. Sample ALV914-R8-A consisting of chalcopyrite crystals (dark) intermixed with anhydride (light) (magnification ×120). The samples were taken by the submersibles "Cyana" (CYP) and "Alvin" (ALV).

316

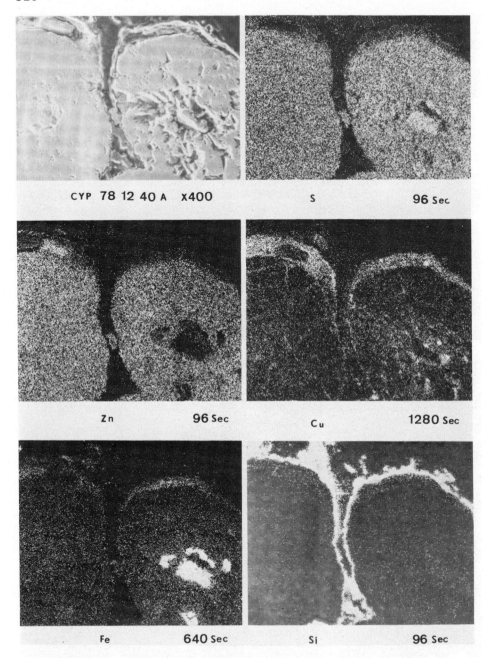

Fig. 10-11. Electron beam scanning microphotographs of globular structure of sphalerite (sample 12-40 Aa, magnification ×400) showing a pyrite forming the nucleus. The outer margin of the globule consists of a Cu-rich phase surrounded by amorphous and hydrated silica and clay.

Barite is a late mineral phase which occurs as radiating laths included in goethite and occasionally in marcasite. Some other sulfate phases identified consist of gypsum, jarosite and atacamite.

In addition to the above types of minerals encountered in hydrothermal deposits, there are others, such as chlorite, zeolite, goethite, hematite, and chrysotile, which also constitute common phases of weathering and metamorphism. Mineralogical differences existing between the various processes of rock alteration in the ocean floor may be evaluated by knowing the type of depositional environment and its bearing on the type of mineral formed.

The mineral associations of weathering and metamorphism affecting the ocean floor rock are mainly of low-temperature types (Table 10-5). Zeolites, smectite and Fe-hydroxide are the major constituents that are found in all types of alterations and in hydrothermal deposits. The hydrothermal deposits are, however, distinguishable from the rocks which have undergone the effects of other types of alteration by the presence of the Fe-Mn minerals (todorokite, manganite, rancieite), gels, Fe-rich smectite, and opaques of Fe-Si clay-like material. Such a difference could be due only to an environmental condition. It is not excluded that the circulation of fluid solution in the oceanic crust, under the confining fluid pressure and temperature, may enhance the alteration of the country rock.

CRITERIA FOR THE RECOGNITION OF OCEAN FLOOR HYDROTHERMAL DEPOSITS

As new hydrothermal fields are discovered it is likely that the criteria for recognizing potential mineral deposits on the ocean floor will be further developed. Rona (1978) has summarized the various geological and geophysical tools used for the prospection of hydrothermal deposits; some of them are noteworthy.

*Petrological criteria.* The recognition of metalliferous sediments of hydrothermal origin is made through the distinctive mineralogical composition of some clay and amorphous material described in this chapter. The high concentration of some metallic elements, lead isotope compositions that resemble oceanic ridge basalts, rare earth patterns which are relatively enriched with respect to normal ridge basalts and seawater, and the presence of abnormal Fe-Mn encrustations on localized areas of the sea floor are also strong indications of hydrothermalism.

*Geochemical criteria* are mainly related to an unusual concentration of some critical elements in the water columns in the vicinities of hydrothermal emanations. An excess of helium (stable isotopes of $^4$He and $^3$He) is a sensi-

TABLE 10-5

Progressive mineral changes during the alteration of ocean floor volcanics. The minerals associated with hydrothermalism are found in Fe-Mn- and Fe-Si-rich deposits associated with volcanic terranes. Massive sulfides deposited on the ocean floor consist mainly of pyrite, sphalerite, chalcopyrite, wurtzite and marcasite in order of abundance

| Minerals | Weathering | Deuterism | Hydrothermalism | Metamorphism |
|---|---|---|---|---|
| K, Na, Ca zeolite | | | | |
| Mg, Fe, Al, smectite | | | | |
| Si-Fe-rich clay | | | | |
| Goethite | | | | |
| Chlorite | | | | |
| Hematite | | | | |
| Chlorophaeite | | | | |
| Mixed layer, smectite | | | | |
| Celadonite | | | | |
| Serpentinite | | | | |
| Sulfides | | | | |
| Hydromica | | | | |

tive indication of mantle-derived volatiles through either magmatic upwelling or hydrothermal fluid circulation. The main source of helium in seawater is from the degassing of mantle material at oceanic spreading centers. Recently it was suggested (Craig and Lupton, 1981) that major differences exist in $^3He/^4He$ ratio variability between atmospheric helium concentration, the oceanic ridge basalts (about 9 times the atmosphere ratio), and the anomalous zone of helium upwelling (about 15--20 times the atmospheric ratio).

Because helium contamination is possible in an atmosphere-ocean interaction, it is important to measure several parameters in order to get an idea about the $^3$He/$^4$He ratios of injected mantle components. Thus, the $^3$He/$^4$He ratios, the absolute $^4$H and the absolute neon concentrations have to be determined (Craig and Lupton, 1981); the He/Ne ratio is used to remove the total atmospheric component. The $^3$He/$^4$He isotope ratio anomalies in natural water are defined as percentage "delta" values relative to the atmospheric ratio (Craig and Lupton, 1981):

$$\delta\,^3\mathrm{He} = \left( \frac{R_{sample}}{R_{atmosphere}} - 1 \right) \times 100$$

where $R = {}^3\mathrm{He}/^4\mathrm{He}$.

The geographic distribution of excess helium is limited to only a few regions of the world's oceans. In the Atlantic a small anomaly ($\delta\,^3$He = 7%) is observed off Bermuda. A $\delta\,^3$He anomaly of 20—22$^\circ/_{\circ\circ}$ was observed on the water columns of the East Pacific Rise near 8°N and 6°S (W. B. Clark et al., 1969, Craig et al., 1975). A helium anomaly was also recorded at 15° south near the crest of the East Pacific Rise in the vicinity of the Garret Fracture Zone where a maximum $\delta\,^3$He value of 50% was found at a depth of 2400 m (Craig and Lupton, 1981). Other areas of relatively high helium concentration associated with hydrothermal emanation occur on the Galapagos Spreading Center on the East Pacific Rise near 21°N and in the Red Sea hot brine area.

Other elements, such as radon, methane, manganese and transitional metals (Zn, Fe, Cu), are also absorbed in deep water masses and could be used as tracers for hydrothermalism.

*Magnetic criteria.* The hydrothermal activity in the oceanic crust may affect the intensity of magnetization. Rona (1978) showed that distinctly low residual magnetic intensities of 200 gammas and of about 650 gammas coincide with the TAG hydrothermal field of the Mid-Atlantic Ridge and with the hot brine deposit of the Atlantis II Deep in the Red Sea.

In addition, metamorphic terranes altered by hydrothermal circulation also have a low intensity of magnetization.

COMPARISON OF HYDROTHERMAL DEPOSITS WITH OTHER METALLIFEROUS PRODUCTS

In addition to their mode of emplacement and their geological settings (for further information see Cronan, 1975; Rona, 1977, 1978), the metalliferous sediments and the hydrogenous deposits, including the Fe-Mn crusts, on ocean floor rocks are chemically different from the hydrothermal deposits sensu stricto. Indeed, the latter are considerably depleted in transitional metal contents (Hoffert et al., 1978; Hékinian et al., 1978) with respect to the other types of deposits. This is also visualized by a binary Cu-Zn distribution diagram shown in Fig. 10-12. With the exception of the sulfides, the Cu and Zn contents of the hydrothermal deposits are generally less than 200 and 400 ppm, respectively (Fig. 10-12), while the metalliferous sediments, represented by those from the Bauer Deep (Bischoff, 1972; Sayles and Bischoff, 1973) and those from the Red Sea (Bischoff, 1969), have Cu and Zn contents higher than 400 ppm (Fig. 10-12). Fig. 10-12 also shows that the Fe-Mn crusts found coating basement rocks have values of Cu (300—600 ppm) intermediate between those of the hydro-thermal deposits sensu stricto and the metalliferous sediments. The average Zn and Cu contents of manganese nodules are about at least 5—10 times higher than those of any hydrothermal deposits sensu stricto. In addition, the growth rate, as determined from the $^{234}U/^{238}U$ radiometric ratio, indi-cates that the accumulation of Fe-Mn crust for the hydrothermal material is on the order of 2—20 cm/Ma (M. R. Scott et al., 1974; Corliss et al., 1978). This growth rate is at least twice as great as the average accumulation rate for Mn nodules.

BASEMENT ROCKS ASSOCIATED WITH THE HYDROTHERMAL DEPOSITS

The volcanics reported to be directly associated with hydrothermal deposits found in the various regions show marked compositional diversities. Major element analyses of these rocks indicate that there is a difference in composition between the basement rocks from the Galapagos area, which are essentially rich in $TiO_2$ ($\approx 2\%$) and total iron oxide content ($\approx 13\%$), compared to both the Gulf of Aden and the FAMOUS area basalts (Hékinian and Fevrier, 1979). The Galapagos rocks are typical ferrobasalts, made up essentially of pyroxene and plagioclase, while those from the Gulf of Aden and the FAMOUS area are aphyric basalts composed of plagioclase, olivine, and clinopyroxene. The degree of alteration of the rocks is variable from place to place. The FAMOUS area samples taken in the hydrothermal field (sample 26-15) and within a radius of less than 2 km, are moderately fresh and only the olivine crystals are completely replaced by clay-rich material (Fig. 1-6D). In the Galapagos area the basement rock underneath the hydrothermal material is surprisingly fresh, with only sporadic veins, veinlets and vesicles filled with green clay-rich material (Scientific Party, DSDP Leg 54, 1977c). However, further north of hole 424, a borehole (425) shows altered basement rock and a bleached zone was noticed (Fig. 10-2). A vein containing a green clay-rich material and a reddish brown amorphous material was analysed (Hékinian et al., 1978) (Table 10-6). The green clay consists of smectite which is rich in FeO ($\cong 49\%$) and in $SiO_2$ (33—50%) and is somewhat similar to the hydrothermal material found in holes 424, 424A and 424B of the other Galapagos sites. The reddish brown material consists

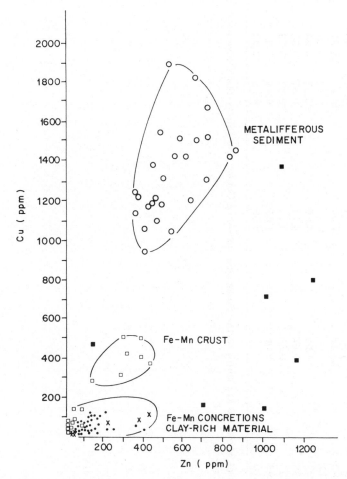

Fig. 10-12. Cu-Zn (ppm) variation diagram showing the distribution of hydrothermal material from the FAMOUS area (□; Hoffert et al., 1978), from the Galapagos Spreading Center (●; Hékinian et al., 1978) and from the Gulf of Aden (×; Cann et al., 1977). The Fe-Mn crusts are those covering the surface of pillow flows. The field of metalliferous sediments are reported after the data from Sayles and Bischoff (1973) (○). ■ = analyses of the Red Sea metalliferous sediments (Bischoff, 1969).

mainly of FeO ($\simeq 7\%$) which has undergone various degrees of oxidation.

Other secondary alteration products of basaltic rocks from the Mid-Atlantic Ridge, near 53°N (Jehl, 1975), near 37°N (DSDP Leg 37, holes 332B, 335; Baragar et al., 1977) and near 1°N (St. Paul's Rock area; Melson and Thompson, 1973), consist mainly of a clay-rich material having a higher MgO content (>15%) and a slightly higher $Al_2O_3$ content (>2%) compared to the hydrothermal deposits. The compositional variation of the clay-rich material of both the ocean floor basement rocks and those of the hydro-thermal deposits is shown in the Mg-Fe-(Na + K) and $SiO_2$-FeO-MgO tertiary

TABLE 10-6

Partial microprobe analyses of secondary clay-rich material found in various basaltic rocks associated with hydrothermal deposits

| | FAMOUS area (Mid-Atlantic Ridge, 37°N) | | Galapagos area (DSDP Leg 54) | | Ninety-East ridge (DSDP Leg 22) | |
| | CYP74-26-15 | CYP74-26-15B1 | 424-5-1/38—40 | 424B-6-1/36—39 | 216-37-4/109—117 | |
| | green | green | green | green | green | green |
|---|---|---|---|---|---|---|
| $SiO_2$ (wt.%) | 47.91 | 44.47 | 48.87 | 49.99 | 48.24 | 51.57 |
| $Al_2O_3$ | 5.12 | 5.95 | 0.22 | 3.34 | 2.57 | 2.63 |
| FeO | 25.43 | 26.62 | 27.48 | 27.95 | 22.48 | 24.51 |
| MnO | 0.19 | 0.11 | — | — | — | — |
| MgO | 5.36 | 6.97 | 4.96 | 4.62 | 3.63 | 4.24 |
| CaO | 0.36 | 0.96 | 0.55 | 0.74 | 0.57 | 0.50 |
| $Na_2O$ | 0.13 | <0.07 | <0.05 | 0.12 | 0.32 | 0.29 |
| $K_2O$ | 6.05 | 3.92 | 5.08 | 4.61 | 6.41 | 7.01 |
| $TiO_2$ | 0.11 | 0.10 | — | <0.09 | <0.09 | <0.02 |

Rocks from the Galapagos area were taken at about 22 km south of the Galapagos Spreading Center; the Ninety-East Ridge rocks were drilled at 1°27.73'N, 90°12.48'E.

diagrams (Figs. 10-13, 10-14). The structural formula of the clay-rich material, found in the basement rocks of the Rift Mountain region near 37°N in the Mid-Atlantic Ridge (DSDP Leg 57, Site 332B):

$$(Ca_{0.19} Na_{0.03} K_{0.1})(Al_{0.07} Mg_{5.01} Fe^{2+}_{0.26})(Si_{6.9} Al_{1.1})O_{20}(OH)_4$$

corresponds to that of a saponite.

The difference in the total FeO content between the clays found in the basement rocks (depleted in FeO) compared to those from the hydrothermal sediment deposited on the sea floor (relatively enriched in FeO) could be due to the degree of oxidation of the material (Fig. 10-14).

It is likely that during the passage of hydrothermal fluids through the basement rocks there must have been some degree of alteration. It is uncertain whether alteration through the deposition of secondary minerals is extensive or limited to interstitial pore fluids concentrated in veinlets,

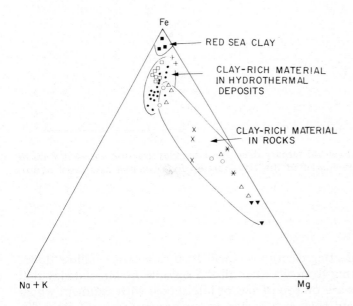

Fig. 10-13. Mg-Fe-(Na + K) ternary diagram of clay material found in various geological environments. The Red Sea clay (■) represents averages calculated from published analyses (Bischoff, 1969). X = clay samples from the St. Paul Fracture Zone (Melson and Thompson, 1973). △, ▼ = secondary clays from basalts drilled near 37°N (Leg 37; Baragar et al., 1977). ＊ = clays from basaltic rocks dredged from the Mid-Atlantic Ridge near 53°N (Jehl, 1975). ○ = clay fractions in basaltic rocks from the Nazca plate (DSDP Leg 34; Scott and Swanson, 1976). □, ● = hydrothermal clays from the FAMOUS area and from the Galapagos area, respectively (Hoffert et al., 1978; Hékinian et al., 1978). + = clay material filling the veins of basement rock of the Galapagos area, Site 425 (Leg 54; Hékinian et al., 1978).

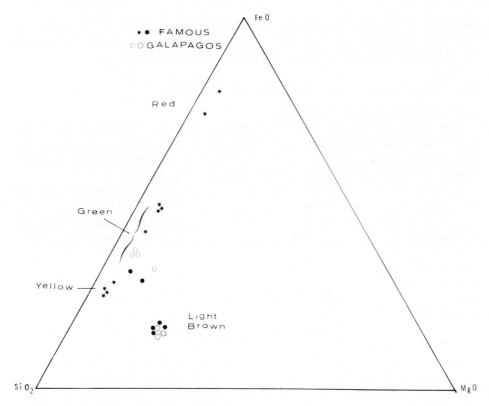

Fig. 10-14. $SiO_2$-FeO-MgO (wt.%) ternary diagram of the clay fraction found in veins of basement basalts (o, •) compared to the clay fraction found in the associated hydro-thermal sediment (*, o ).

vesicles, and interconnecting cavities or small fissures between pillow flows. However, it is interesting to notice that all the sizeable Fe-Mn and clay-rich hydrothermal deposits are on top of and/or intermixed with sediment which might play an important role, as a buffer, in the accumulation of fluid discharge.

All the hydrothermal deposits sensu stricto studied here are similar in composition despite the compositional differences of their associated basement rock and alteration products. This suggests that surface or near-surface environmental factors (such as the amount and type of sediment and/or the relationship of sediment to seawater) are not influential with regard to the type of hydrothermal discharge encountered (i.e. hydrothermal deposits sensu stricto as opposed to metalliferous sediments or hydrogenous material).

ORIGIN OF THE HYDROTHERMAL DEPOSITS

The discovery of hydrothermal deposits associated with diverging plate boundary regions suggests that oceanic ridges may be preferred sites for hydrothermalism. Subaerially exposed ophiolitic complexes around the world, in Cyprus (Fryer and Hutchinson, 1976), in Newfoundland (H. Williams and Stevens, 1974), and in Oman (Glennie et al., 1974), show similar types of hydrothermal deposits to those found in recent oceanic ridge systems. Thus, the so-called "umber" type of product found in ophiolitic complexes corresponds to the Fe-Mn and clay-rich material found in the ocean floor. The sulfide deposits in both cases lie directly on a pillow flow type of basement. The presence of a goethitic zone on the top of the sulfide deposits in ophiolitic complexes resembles the ochre Fe-oxide material found in recent accreting plate boundary regions.

In general, three major types of material are associated with the hydrothermal deposits from the ocean floor: (1) Fe-Mn concretions which are generally intimately associated with an alkali-ferric iron hydrated silicate (nontronitic clay), (2) Fe-oxide and Fe-hydroxide (ochre material), and (3) sulfides. At least two stages of depositional conditions are involved: the first stage, which is probably also the most long lasting and hottest phase ($>300°C$), is related to the precipitation of the sulfides; the latter activity is the vanishing stage of hydrothermalism, with the deposition of Fe-Mn and clay-rich material. The oxidized zone is due to the secondary alteration of the sulfides, which were in a metastable condition in their environment of precipitation. The sequential distribution of recent hydrothermal deposits is comparable to that of subaerial ophiolitic complexes.

Little is known about processes relating magmatism and hydrothermalism on the ocean floor. The presence of shallow-depth temperature gradients near active volcanic ridges may enhance a system of fluid circulation and the resulting deposition of metallic components in accreting plate boundary regions (Fig. 10-15).

The origin of a hydrothermal circulation and precipitation of material into the ocean floor is still controversial and probably each type of deposit represents a particular physico-chemical condition of deposition. The two most attractive theories related to the origin of hydrothermal precipitations are: (1) the leaching phenomenon of pre-existing basement rock, and (2) magmatic segregation related to the cooling of silica melts.

*Leaching and transport of elements from basaltic rocks*

Convection systems involving seawater penetrating into the oceanic crust and transporting leached material from basement rocks have been proposed by several authors (Spooner and Fyfe, 1973; Bischoff and Dickson, 1975; Parmentier and Spooner, 1978) as being responsible for hydrothermal

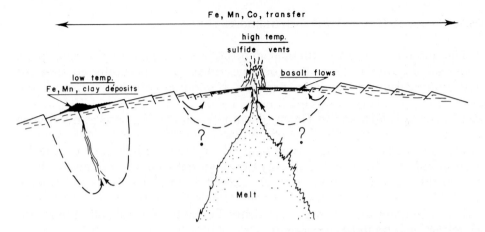

Fig. 10-15. Schematic representation of an accreting plate boundary region (East Pacific Rise type) showing sulfide deposits formed on the axial zone overlying silicate melt. The seawater recharge and discharge zone are represented by dashed lines and arrows. Away from the axial zone of upwelling low-temperature hydrothermal deposits are formed. Fe, Mn, Co and maybe Ni are dissolved in seawater and transported away from the centrally active zone.

circulation. It was also calculated that in order to produce the type of massive sulfide deposits found on the East Pacific Rise near 21°N, a large volume of metal-bearing phases must have gone into solution from pre-existing igneous formations located underneath the ridge axis. For instance, in order to produce a mound of 3 $m^3$ containing 50% Zn, quantitative leaching of about 15,000 $m^3$ of basalt with 100 ppm of Zn is required (Hékinian et al., 1980).

*Magmatic segregates from silicate melts*

The processes involved in segregating sulfide melts is analogous to that known as the "filter pressing phenomenon" related to the crystallization of a silicate magma during which a residual liquid will be expulsed. It is likely that after the solidification of basalt mineral phases from a silicate melt (such as pyroxene, plagioclase and olivine), the solubility of sulfur in the magma will decrease and sulfide melts could then separate. Only a small portion of the sulfide melt is incorporated within the silicate melt and similarly the sulfide melt will trap only a small amount of the low-temperature silicate phases. A sulfide melt segregating from a silicate melt could be injected into cracks or fissures on the ocean floor or could form distinct pockets of accumulation near a magma chamber.

Another alternative to the above hypothesis is to consider the mixing of a hydrothermal fluid with the magmatic segregates. The circulation of heated seawater is involved as a near-surface process (<2 km depth) where it interacts with magmatic residual phases and carries the bulk material onto the ocean floor where metal-rich solutions precipitate.

PRECIPITATION OF THE HYDROTHERMAL SEDIMENTS

The other types of hydrothermal deposits, such as the metalliferous sediment and the low-temperature Fe-Mn-rich and Si-Fe deposits, could be formed either by the degradation of pre-existing sulfide deposits or could represent low-temperature precipitates formed away from accreting plate boundary regions (Fig. 10-15). The temperature enhancing hydrothermal circulation away from a ridge axis (e.g. the Galapagos mounds and the FAMOUS area) could be due either to a regional cooling of the lithosphere or to the local intrusions. Why do hydrothermal deposits occur as isolated features? Do the recharge areas extend over a large region or are these only small-scale convective wells?

Since the hydrothermal sediment mounds are oriented in a preferential direction, it seems likely that they occur on an area of the crust where it is weak and porous. Highly fractured and brecciated zones may be preferential sites for hydrothermal circulation through which low-temperature fluid may circulate.

The uniformity of nontronitic sequences throughout the mounds and their compositional similarities with the clay material found in veinlets of basaltic rocks suggests a local precipitation for these types of deposits. As the hydrothermal fluids rise through a pelagic sediment blanket, it precipitates under particular physico-chemical conditions when at contact with the porous environment (Fig. 10-16). As suggested by Dymond et al. (1980), it is likely that these deposits are formed when an oxidation-reduction gradient

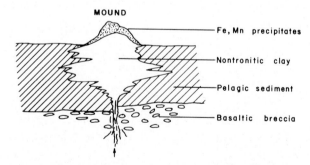

Fig. 10-16. Hypothetical sketch diagram of hydrothermal precipitates away from the axial zone of oceanic ridge systems (e.g. Galapagos Spreading Center and East Pacific Rise near 21°N). The low-temperature deposits consist of Fe-Mn and clay material percolating through porous material.

is formed. The nontronitic layers will represent more reducing conditions of precipitation. Instead the fluid in direct contact with highly oxygenated sea-water will form a surface coating of Fe-Mn material on top of the mounds (Fig. 10-16).

Further detailed studies on the various aspects of these deposits should be carried out on active spreading center regions and away from them, in order to evaluate the extent and relationships between the various types of deposits.

## CHAPTER 11

DEUTERIC ALTERATION

Deuteric alteration is a low-temperature magmatic alteration related to the solidification of a melt. The term deuteric is restricted to reactions involving changes in primary mineral phases during the process of magmatic crystallization. This type of alteration is not to be confused with metasomatism which is the process involving the replacement of a mineral phase by another due to the introduction of material from an external source. The agent of deuterism is the volatile material dissolved in the magma. Often the term autometasomatism may be used instead of deuterism when the rock crystallized is altered by its own late-fluid fraction trapped within the formation. It is often difficult to distinguish between minerals that are formed during late magmatic reactions from those deposited by fluid solutions penetrating the pre-existing rocks. The minerals or the amorphous compounds formed at the end of crystallization are located in the interstices of the matrix minerals, in vesicles or lining the walls of veins and/or cavities. Early-formed minerals could react with the residual melt and form a low-temperature phase. The neoformed material is considered to be part of the original melt which gave rise to the bulk-rock constituents. The composition and amount of the deuteric material depends on the degree of cooling (temperature drop), on the availability of the residual melt and on its composition. The amount of deuteric material encountered in most basaltic lavas does not usually exceed 10—15% of the bulk rock. However, up to about 25% of chlorophaeite and saponite were found in basaltic flows in the ocean at DSDP Site 335 (Leg 37; Baragar et al., 1977).

Because most of the newly formed phases are phyllosilicates and amorphous hydrated compounds, the physical change during deuteric alteration is accompanied by a decrease in density. The original texture of the rock is usually preserved, which suggests equal volume replacement (Wilshire, 1959). The best criteria for recognising deuterically altered minerals is the pseudomorphism of primary minerals. A residual fluid phase present during magmatic solidification may act on material leached from early-formed mineral phases which will then be deposited elsewhere. Modal analyses of the altered material and bulk-density measurements are necessary to determine the degree of rock alteration. Wilshire (1959) has shown the compositional changes that volcanic rock could assume during alteration in terms of milligrams per cubic centimeter. It was shown that compositional changes in basic rocks due to a replacement by carbonate is accompanied by a loss of $SiO_2$, MgO and CaO. The $Fe_2O_3/FeO$ ratio generally increases (Wilshire, 1959). If the replacement of the mineral phases is due to the presence of mixed layer clay, then the bulk rock shows a loss of $SiO_2$ and $MgO$, and a moderate increase in $CaO$, $Al_2O_3$ and the $K_2O/Na_2O$ and $Fe_2O_3/FeO$ ratios.

During the late stage of solidification the residual fluid percolating through the crystalline network is separated into two phases: a liquid and a gaseous phase. The magmatic gases contribute to the process of alteration of the early-formed minerals. Recently, magmatic gases were extracted and analysed from the ocean floor basalts on the Mid-Atlantic Ridge near 36°N (Chaigneau et al., 1980). The main components of these magmatic gases are $HCl$ (100—1000 ppm), $CO_2$ (200—800 ppm), $SO_2$ ($\leqslant$175 ppm), $H_2$ ($\leqslant$50 ppm) and $N_2$ ($\leqslant$200 ppm). The total volatile content of fresh oceanic volcanics occurs as trace amounts (700—2000 ppm) in the bulk rock. J.G. Moore and Schilling (1973) have also shown that the water content ($H_2O^-$ + $H_2O^+$) in most fresh basalt does not exceed 0.5 wt.%.

MINERALS OF DEUTERIC ALTERATION

The minerals and the amorphous compounds encountered in rocks altered by deuterism are all of low-temperature assemblages such as chlorophaeite, celadonite, saponite, K-Na zeolite, goethite, talc, magnesite, calcite, chlorite, tremolite, Fe-Si amorphous material, serpentinite and iddingsite.

The early-formed minerals, microphenocrysts and phenocrysts, of olivine and pyroxene are the most sensitive to deuteric alteration. Olivine crystals are transformed into serpentine, chlorite, talc, Fe-oxide minerals and iddingsite. Saponite and goethite were also found replacing olivine. Clinopyroxene is less sensitive than olivine to deuteric alteration. This could be in part due to the fact that pyroxene minerals usually occur at a later stage of crystallization than olivine in basaltic rocks. Talc, chlorite and calcite are the main alteration products of pyroxene. Hydrous minerals such as chlorophaeite, saponite and celadonite are found filling vesicles in the interstices of matrix minerals, and in veins of basaltic rocks. However, the occurrence of these hydrous minerals, sometimes associated with manganiferous carbonate minerals found in veins, may also suggest a hydrothermal origin where fluid solutions come from an external source and impregnate the basaltic rocks. During deuteric alteration, the mafic minerals are easily altered and replaced by carbonate, while hydrous minerals such as zeolite may often replace plagioclase. Plagioclase is also frequently replaced by sericite, montmorillonite and chlorite. The opaques (titanomagnetite) could be converted to ilmenite, leucoxene and hematite.

The effect of deuteric alteration on rocks is often difficult to distinguish from that of hydrothermalism. In addition to the order of mineral alteration, textural relationships between the low-temperature replacing products and the primary phase should be carefully studied. For instance, it is likely that the presence of veins and veinlets in rocks are signs of extensive fluid circula-

331

tion from the surroundings (Chapters 9 and 10). On the other hand, an early transformation of olivine phenocrysts in a fresh glassy matrix into hydroxide compounds is primarily due to a deuteric type of alteration. Thus, deuterically altered rocks are easily identified in young rock formations before other secondary types of alteration have taken place.

## CHAPTER 12

# INFERRED COMPOSITION OF THE OCEANIC CRUST AND UPPER MANTLE

## INTRODUCTION

The oceanic crust can be divided into two major domains: (1) the accreting plate boundary zone, which is primarily an area where oceanic crust is created at a rate of about 1—17 cm/yr, and (2) the passive crust which, after its creation on the accreting plate boundary region, has moved away and is associated with a region of relatively stable seismicity. The accreting plate boundary zone is related to partial melting and differentiation processes where liquid and solid move with respect to each other. The passive zones of the lithospheric plates, which are more stable than the accreting plate boundaries, could be the locus of solid-solid reactions such as the type encountered in metamorphic terranes. Assuming a thickness of about 75 km for the lithosphere-asthenosphere boundary forming the plate, a pressure of about 20 kbar will occur at the bottom of the plate. If metamorphism is already occurring within the oceanic lithosphere-asthenosphere, it is likely that such a process is more efficient in passive plate regions where load pressure is more effective than temperature. In the accreting plate boundary, where the oceanic crust is from 5 to 10 km thick, the temperature is more effective than the pressure. Accreting plate boundary zones will be the most likely areas for hydrothermal alteration of the oceanic crust.

Knowledge about the deep portions of the oceanic crust and upper mantle is still limited. Direct investigations through sampling deep stratigraphic sequences have, up to now, failed to give any petrological indication of the underlying material of the oceanic crust. The maximum penetration of the crust by deep-sea drilling is less than 1500 m. Also, the origin and the extent of exposed and uplifted areas of upper mantle and crustal material in major fracture zones are controversial.

Geophysical prospection of the ocean floor has established the occurrence of several reflectors which represent physical discontinuities in the oceanic crust. In addition to marine seismic refraction studies, petrological studies combined with laboratory measurements of seismic velocity on rock samples are prerequisites to making any inferences about the composition of crustal and upper mantle material.

Early studies (Hill, 1957; Raitt, 1963) of numerous seismic refraction profiles suggest that the average structure of the oceanic crust is uncomplicated. Raitt (1963) and later Christensen and Salisbury (1975) divided the oceanic crust into layers with different compressional wave refraction velocities. In order to simplify the issue, only the crust and upper mantle beneath oceanic basins were considered, leaving aside aseismic ridges, fracture zones, linear

TABLE 12-1

Generalized and schematic composition of the oceanic crust and upper mantle associated with accreting plate boundary regions older than 0.5 Ma*

| Oceanic layer | 0.5- to 20-Ma crust | | | | >20-Ma crust | | <20-Ma crust (author) | | | Iceland | | |
|---|---|---|---|---|---|---|---|---|---|---|---|---|
| | depth (km) | geophysical discontinuity $V_P$ (km/s) | laboratory-measured velocity $V_P$ (km/s) | rock type | depth (km) | rock type | depth (km) | geophysical discontinuity $V_P$ (km/s) | rock type | depth (km) | $V_P$ (km/s) | rock type |
| 2 | 2 | 4–6 | 3.7–6.5 | basalt | | | | 3.7–5.8 | basalt / metabasalt | | 2.8 | basalt |
| | | | | dikes | | basalt | 2 | | dike / metagabbro | | 5.1 | dikes |
| | | | 5.2–6.3 | metabasalt | 3 | | | | | | | |
| | | | | serpentinite | | dikes metabasalt | | 6.4 | layered complex (gabbro serpentinized peridotite) / metamorphics (medium grade) | 4 | | plagiogranite |
| 3 | 4 | 6.5–6.8 | 3.7–6.6 | amphibolite | | | | | | | | |
| | | | 5.5–7.2 | metagabbro | | gabbro | 5 | 7.1 | dike | | 6.5 | gabbro |
| | 6 | 7–7.7 | | gabbro | 7 | | | 7.5 | ultramafic (lherzolite-dunite) / metamorphics (high grade) | | | |
| Moho | 8 | | | serpentinized peridotite | | ultramafic | | | | 10 | | layered gabbro |

*The rock types are rearranged after the data taken from Fox et al. (1973, 1976) and Christensen and Salisbury (1975). The data from Iceland are extrapolated from Sigurdson (1977). The geophysical discontinuities used by the author are taken from Raitt (1963).

chains of islands and other discontinuous features. Christensen and Salisbury (1975) classified the crust and upper mantle in four layers:

(1) Layer 1 is less than 1 km thick and is made up of sediment ranging in velocity from 1.5 to about 2.5 km/s.

(2) Layer 2 has a thickness of 1—2.5 km and shows a wide range of velocity (from 3.5 to 6.4 km/s) but with the most common values around 5.1 km/s. Further subdivision of this layer into layer 2A, corresponding to a highly magnetized basaltic zone, and layer 2B, unmagnetized intrusives of basalt and gabbro, was suggested by Talwani et al. (1971).

(3) Layer 3 has a variable thickness of 3.4—6.3 km and a velocity range of 6.4—7.7 km/s (most values lying on a peak of 6.7 km/s).

(4) Layer 4 or upper mantle structure has a velocity in the vicinity of 8.1 km/s.

Laboratory velocity measurements on ocean floor rocks show that various types of material could be candidates for the composition of the oceanic crust. Most ocean floor geologists agree that the upper part of the crust (less than 10 Ma old) is made up of relatively fresh basalt and constitutes the seismic layer 2 of about 1 km thick. Seismic discontinuities within layers could be interpreted as being due to the association of basalt, metabasalt and metagabbro complexes (Table 12-1). Seismic layer 3 corresponds to the gabbro-serpentinized peridotite association and has a maximum thickness of about 6.4 km (Fox et al., 1973; Christensen and Salisbury, 1975). However, it is likely that this generalized scheme of crustal and upper mantle compositional variation may vary from place to place depending on structural, tectonic and magmatic processes. Thus, it is not excluded that off-ridge intrusions due to diapirism or to magmatic events could alter the above simplified and extrapolated relationship between seismicity and rock composition. In addition, spreading and creation of new crustal material imply the presence of a melt underneath the oceanic ridges, while lateral spreading and cooling suggest a relatively more rigid crust. Hence, compressional wave velocity propagation in these two regions is likely to differ.

Two major structures are recognized at the accreting plate boundaries: an unrifted and a rifted ridge system. Various interpretations are given for the origin of their observed structural differences.

*Unrifted oceanic ridge systems*

The term "unrifted" refers here to ridge systems which are deprived of any prominent rift valleys. Instead, they could show small-scale uplifted blocks a few tens of meters higher than their corresponding axial zones, such as those described on a segment of the East Pacific Rise near 21°N (Chapter 2). The structure of oceanic ridge systems is complex and varies along the various segments of their accreting plate boundaries. Very few data on surface morphology or from geophysical prospection, or petrological descriptions of the rocks are available from the different segments of the East Pacific Rise up to the present. The ridge segments of the East Pacific Rise which have been studied in the most detail are located near 9°N and 21°E are described in Chapter 2. The intermediate (3—4 cm/yr half spreading rate), fast (5—7 cm/yr) and ultra-rapid (7—9 cm/yr) ridge segments of the East Pacific Rise display predominant axial blocks varying between 10 and 20 km in width. The over-burden necessary to produce a positive crestal feature is related to the presence of a partially molten magma chamber underneath the ridge, and the axial block above the adjacent sea floor is supported by buoyant forces within the magma column (Rea, 1975; Rosendahl, 1976). By computing the elevation change between the flank and the crest in the axial block region of the East Pacific Rise, and taking into consideration the various densities for layer 2 (2.63 g/cm$^3$), for layer 3 (2.90 g/cm$^3$) and for layer 4 (3.15 g/cm$^3$), a maximum elevation of 230 m is computed (Rosendahl, 1976). The mean observed elevation of 260 m agrees with that calculated and suggests that isostatic uplift can produce the observed horst structure in the East Pacific Rise near 8—9°N (Rosendahl, 1976). Sleep (1975) showed a mathematical model in which temperature gradients in fast-spreading ridges (5 cm/yr) are less steep than those found in slow-spreading ridges (Fig. 12-1). Such an interpretation put a limitation on the size of magma chambers and agrees with the theory of a relatively large magma chamber existing beneath the East Pacific Rise (Rosendahl, 1976). Seismic refraction studies on the East Pacific Rise near the Siqueiros Fracture Zone (8—9°N) have shown that a low-velocity (LVZ) could exist in the oceanic crust. The topographic variation from the crestal zone down to the flank of the East Pacific Rise is related to an apparent thinning of the crustal structure within a distance of about 11 km from either side of the ridge crest. The apparent thickness of the crust derived from structural sections, without taking into consideration the presence of an LVZ, appears to be decidedly great (about 11 km) for a normal oceanic crust (Rosendahl et al., 1976). In order to compute thinner oceanic crust, it was therefore necessary to consider an LVZ with a velocity of about 5 km/s (Rosendahl et al., 1976). Some limitations were set in order to obtain the geophysical model of the axial zone. The LVZ must be thicker at the crest and thinner towards the base of the axial block, and its maximum thickness must be less than 4 km (Rosendahl et al., 1976). In order to explain the presence of the LVZ, it was postulated that its composition consisted of the melt and solid fraction of a basalt. Beneath the East Pacific Rise, the LVZ has a wedge-shaped appearance, a horizontal dimension of about 10 km, and a depth of 3.5 km (Rosendahl et al., 1976). However, recent interpretations of multi-channel seismic reflection profiles made across the East Pacific Rise near 8—9°N led Stoffa et al. (1980) to infer the existence of a zone of magma injection only 3 km wide, located near the ridge crest. From the same data it was also shown (Stoffa et al., 1980) that the Moho occurs at a depth of only 4.5—5 km. Another area where an LVZ was detected at a

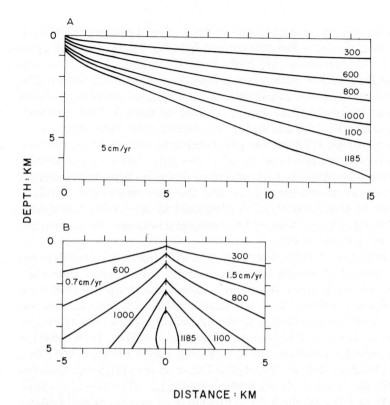

Fig. 12-1. Isotherms (in °C) of accreting plate boundary regions spreading with a half rate of less than 5 cm/yr (A) and 2 cm/yr (B) (after Sleep, 1975).

depth of about 2.5 km is located underneath the East Pacific Ridge crest near 21°N (I. Reid et al., 1977).

*Oceanic ridges with axial valleys (rifts)*

The Mid-Atlantic and Indian Ocean ridges have axial valleys which vary in size along the ridge systems. The inner floor of the rift valley studied in detail near 36°N in the FAMOUS area, located southwest of the Azores, has an average width of about 1—3 km (Arcyana, 1975; Ballard et al., 1975); additional information on the FAMOUS area may be found in Chapter 2. Although it appears that the axial valley of the Mid-Atlantic Ridge has a zone of partial melt underneath it, the width of the magma reservoir may be much less than the mean width of the Rift Valley (cf. Chapter 2). The thermal constraints shown by Sleep (1975) seem to limit the axial intrusive zone of slow-spreading ridges (1.5 cm/yr). From mineralogical studies related to experimental work, it is likely that the crystallization of a basaltic melt occurs in a narrow range of temperature variation (e.g. 1100—1200°C). The computed isotherm shown by Sleep (1975) for a slow-spreading ridge has a steeper gradient than that of a fast-spreading ridge (Fig. 12-1). The isotherms of 1180 and 1100°C increase within a distance of less than 4 km from the axial zone of the Rift Valley, which has a spreading rate of 1.5 cm/yr. The structure of rifted mid-oceanic ridges differs from that encountered in basin regions by the slightly lower (6.37 km/s) velocity of layer 3. The thickness of layer 3 was found to be 4.78 km and that of layer 2 only 1 km (Whitmarsh, 1973). The absence of an LVZ in this particular area may indicate the lack of a molten phase beneath the inner floor of the Rift Valley at the present time. The evidence of earthquake epicenters in this zone is probably related to the relative motion of the inner floor and the wall structures. A micro-earthquake survey in the FAMOUS area (Spindel et al., 1974) has shown that tectonic events are located along the boundary between the inner floor and the adjacent Rift Valley walls.

Several authors (Cann, 1970b; Greenbaum, 1972) have suggested that the oceanic crust is formed in narrow magma chambers. Dike swarms and extrusives underlying the inner floor of the rift valleys make up the thin crust overlying the magma chamber. Reversed seismic refraction lines carried out on the inner floor of the Rift Valley in the FAMOUS area did not show any evidence of an anomalous LVZ in the crust (Whitmarsh, 1973). The formation of a graben-like structure within the Rift Valley could be attributed to a subsidence phenomenon due to the roof collapse of a hypothetical magma chamber beneath the inner floor on mid-oceanic ridge rift valleys. Isostatic equilibrium could be reached during the upwelling of magma in mid-oceanic rift valleys. Osmaston (1971) suggested that the rift valley is kept from attaining isostasy by viscous drag due to the upwelling of magma against the colder margins of newly created crust. As a response to the load reduction during eruption, a portion of a plate will rise between the axis of crustal emplacement and the colder margins. Hence, during the creation of new crust, a new faulted step will be added onto the side of the inner floor of the rift valley. Needham and Francheteau (1974) have suggested that during accretion, the lithospheric plate will thicken, and growth of the inner floor in width and in vertical relief will be limited by the uplift process. Also, the amount of basalt extruded over a given number of square kilometers at the ridge axis may influence the structure of the rift valleys. If there is a small amount of magmatic delivery on a slow-spreading ridge, it is expected that the oceanic crust will be subjected to intermittent freezing and breaking at the axial region of the ridge system. In the Mid-Atlantic Ridge, where spreading is slow, newly emplaced dikes have sufficient time to cool and solidify at depth. Hence, the constant cracking and breaking of slow-spreading ridges would produce a rugged topography.

DEPTH OF MAGMA GENERATION

On the basis of what is presently known of the earth's history, it is likely that most of the oceanic crust is created at the ridge axes through partial melting. The mechanism of partial melting which is responsible for bringing newly formed melt to the surface is not well assessed. Is there a continuous partially melted zone beneath the ocean in the upper mantle, or does partial melting only occur in localized zones underneath oceanic ridges and/or underneath isolated and elevated volcanic edifices? Are the partially melted regions of the earth a residue of a molten stage inherited from a primordial condensation of the earth, or have they a mechanical origin due to stress release during tectonism? Radioactive decay or thermal energy liberated during chemical reactions are additional mechanisms which could explain partial melting.

The depth at which magma is generated is controversial, and two independent approaches are taken into consideration in order to explain the creation of oceanic crust: (1) the geophysical approach, based mainly on seismic wave propagation and gravity anomalies, and (2) the field of stability of the mineral phases forming rocks as viewed under various laboratory pressure and temperature conditions.

*Geophysical approach*

The crust and the mantle transmit both primary and secondary seismic waves at all depths. Changes in seismic wave velocities suggest that some areas may be partially melted. The drop in both shear waves ($V_S$) and compressional velocities ($V_P$) in the upper mantle region (at about 50—100 km depth) beneath the ocean is interpreted as being due to a zone of partial melting. Based on experimental studies, it was shown that an abrupt drop in velocity was observed as melting starts (Murase and McBirney, 1973). Small amounts of melt, approximately 6% (Birch, 1969), could explain the existence of the LVZ. However, the shear velocity is inversely proportional to the density, hence an increase in density could decrease the values of $V_S$.

It has been shown that the LVZ is characterized by the relatively high attenuation or dissipative function of seismic waves (D. L. Anderson and Archambeau, 1964). This is expressed by $Q$ values, defined as $2\pi$ times the ratio of the stored energy to the dissipated energy per cycle. The variation of attenuation with propagation direction from an earthquake in a highly attenuating zone ($Q = 50$—70) extending between 300 and 400 km beneath Japan and Chile-Peru indicated a zone of extensive melting. Kauzel (1972) found that there is a zone of unusually high attenuation beneath the ridge crest of the East Pacific Rise. It was speculated that the low $Q$ extends from about 20 to 60—70 km depth (Kauzel, 1972). The variation of attenuation with propagation direction from an earthquake on the Charlie Gibbs Fracture

Zone and the adjacent ridge crest indicates the presence of a zone with low Q (10 km or less) not deeper than 50—100 km (Solomon, 1973). Solomon (1973) suggests that this zone may be identified as a region of melting at a temperature above the anhydrous solidus of mantle material.

The study of shear velocities in the Pacific (Forsyth, 1977) shows that young sea floor (<0.5 Ma old) is characterized by an LVZ on the order of 4.1 km/s and as having a thickness of about 60—170 km. The cooling of a slab as its age increases results in the LVZ being found at a greater depth. It was found (Forsyth, 1977) in the region of the western Pacific and the Nazca plate area that for a crustal age of 80—160 Ma, the LVZ region begins at a depth of about 100 km. The LVZ has an average shear velocity of 4.25 km/s beneath such old crust. The increase in shear velocity with depth might suggest that cooling of the lithosphere is happening in regions deeper than 80 km. If any partial melting takes place underneath old oceanic crust, this must occur at a depth of not less than 100 km.

Thus the melting depth and the process of melting are likely to vary according to the tectonic environment. Diverging plates in oceanic ridge regions are a locus of tensional stress release. Converging boundary zones, where the oceanic lithosphere plunges underneath island arcs or underneath continental areas, are likely to be sites of frictional stresses. The concept of plate tectonics has revealed the existence of compressional zones between plates of the same density. Overriding or thrust-folding structures are likely to give rise to compressional stress and melting. Intrusives of gabbros, serpentinized peridotites and dolerites are found in thrust-folded regions such as the Gorringe Bank, the Macquarie Ridge and the Palmer Ridge regions.

The frictional melting of a portion of a slab produces magma directly below lines of observed active volcanoes which are parallel to the trenches in island arcs. The increase of alkalinity behind the trench on the island arc side is in agreement with the fact that magma is generated in a deeper portion of the mantle during the plunging of a lithospheric slab under island arcs.

*Mineral phase stability*

Another approach for speculation on the depth of melting in the mantle is to consider the stability field of mineral phases which are tied to pressure-temperature variation for a given parental source. It is well established that the major mineral phases giving rise to the rocks forming the ocean floor consist of olivine, plagioclase and clinopyroxene. The stability field of these minerals varies under certain pressure and temperature conditions. Olivine is less affected by pressure than by temperature changes. However, above a pressure of 30 kbar (>100 km) forsterite tends to disappear, and garnet is the stable phase. Experimental work carried out on the system olivine-anorthite-quartz (Kushiro, 1972) showed that at low pressure (<1 atmosphere) two reaction points exist: one involving forsterite, spinel, anorthite and liquid, and the other one forsterite, anorthite, Ca-poor pyroxene, and liquid. When the pressure is increased up to 10 kbar ($\approx$40 km), the fields of spinel and Ca-poor pyroxene expand in relation to anorthite and forsterite. The transformation of olivine ($Fo_{70}$ wt.%) and anorthite ($An_{73}$ wt.%) was found to take place above 5—8 kbar (Emslie, 1971). The reaction for such a mineral association is as follows:

olivine + anorthite = clinopyroxene + spinel
(low pressure)          (high pressure)

This transformation could be applied to the stability field of troctolitic gabbro which, above 9 kbar (28 km), is unstable (Emslie, 1971).

An investigation on joined plagioclase-diopside-enstatite permitted Emslie (1971) to establish that at a pressure above 15 kbar (about 55 km depth) the field of plagioclase decreases in size and the pyroxene field increases (Fig. 12-2). This implies that if partial melting occurs at a pressure of about 15 kbar, the liquid removed will have the composition of a tholeiitic basalt. Subsequent shallow-depth fractionation, coupled with the speed of magmatic uprise, might give rise to various types of cumulates. If a melt formed at 15 kbar is rapidly brought to 1 atmosphere, it is likely that abundant plagioclase crystals floating in the liquid will solidify near surface conditions and form a cumulate of plagioclase-rich rocks. A slow rise of a melt will permit

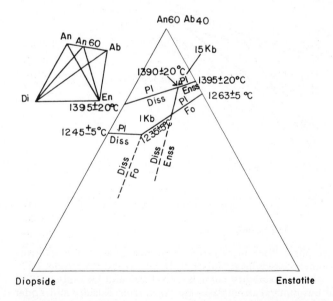

Fig. 12-2. Plagioclase ($An_{60}Ab_{40}$)-diopside-enstatite ternary diagram representing the liquidus relations at 1 atmosphere and 15 kbar dry (after Emslie, 1971). The inset shows the position of the $An_{60}Ab_{40}$-diopside-enstatite join in the system Di-En-An-Ab.

the material to come to equilibrium at shallow depths and form cumulates with various proportions of plagioclase and clinopyroxene, which are directly related to the increase in the stability of the plagioclase field (Fig. 12-2). Yoder and Tilley (1962), studying mainly Hawaiian basalts, have found that they are only stable at shallow depths of less than 5 km (<2 kbar) and at a temperature of about 900—1150°C. Other experimental work, using an oceanic basalt as the starting material, has put some limitations on the depth of origin for ocean ridge basalts.

An olivine basalt and a plagioclase basalt from the Rift Valley near 25°N in the Atlantic Ocean were studied under different pressure-temperature conditions (Kushiro and Thompson, 1972). It was found that the plagioclase and the olivine co-exist in the liquidus up to about 7 kbar and then they are replaced by clinopyroxene. At a higher pressure, olivine is no longer found in the liquidus (Kushiro and Thompson, 1972) (Figs. 12-3, 12-4). It is speculated that both the olivine basalts and the plagioclase basalts are derived from a magma at a depth shallower than 25 km (<8 kbar) under anhydrous con-

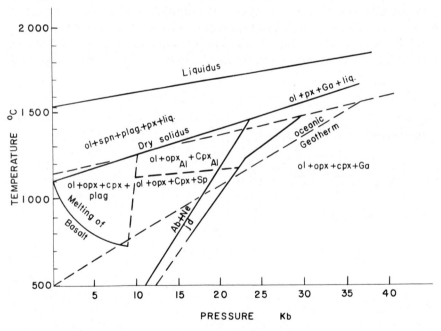

Fig. 12-3. Pressure-temperature phase diagram of garnet peridotite and pyrolite showing the stability field of various mineral assemblages (after D. H. Green and Ringwood, 1970 and Wyllie, 1970). The Ab + Ne = Jd stability curve is from Birch and Lecomte (1960). The stability fields for the various mineral assemblages are taken from Kushiro and Yoder (1966) and D. H. Green and Ringwood (1967). The melting curve of basalt is from Yoder and Tilley (1962). The curve for the liquidus corresponds to that of a peridotitic komatiite (olivine and subcalcic augite) (after Arndt, 1976).

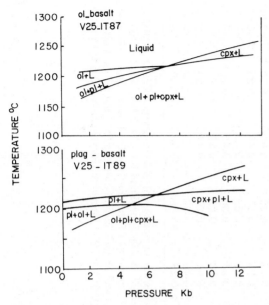

Fig. 12-4. Pressure-temperature experimental diagram of Mid-Atlantic Ridge basalt from the eastern wall of the Rift Valley near 25°N (3155–3274 m depth) (after Kushiro and Thompson, 1972).

ditions. Similar experiments carried out on an olivine basalt glass from the inner floor of the Rift Valley near 36°50′N showed the presence of an ortho-pyroxene-clinopyroxene assemblage at a pressure higher than 10 kbar and a temperature greater than 1270°C (Bender et al., 1978). Bender et al. (1978) concluded that clinopyroxene fractionation at high pressure may give rise to most types of basaltic melts erupted in the FAMOUS area. However, their data do not explain the low and constant values of the $TiO_2$ and the $Na_2O$ content in their model of fractionation.

## SOURCE MATERIAL FOR BASALTIC MAGMA

Determination of a parental material which, on melting, would generally yield a basaltic magma, is a difficult task since no continuous crustal and upper mantle sections have been sampled. It is likely that structural settings and a thorough mineralogical examination of the extruded basalt and the unmelted residue left behind could give important information on the nature of the immediate parental material and its degree of partial melting. Eclogitic nodules in South African kimberlites and peridotite nodules and garnet lherzolite in alkali basalts from oceanic islands are supposed to have originated at great depth in the mantle (Yoder and Tilley, 1962; Boyd and Nixon, 1975). The equilibrium temperature and depth of origin inferred from the

TABLE 12-2

Chemical analyses of rocks which probably represent the composition of the upper mantle

| | Pyrolite | Eclogite | | | | Peridotite | | | | |
|---|---|---|---|---|---|---|---|---|---|---|
| | | glass | 35090 | 62-2 | 6618 | i | 1 | 2 | W | 97-W |
| $SiO_2$ (wt.%) | 45.16 | 46.23 | 50.21 | 49.26 | 48.41 | 42.11 | 43.89 | 43.94 | 44.25 | 44.15 |
| $Al_2O_3$ | 3.54 | 14.52 | 13.78 | 8.63 | 10.41 | 6.59 | 2.89 | 1.90 | 2.65 | 3.59 |
| $Fe_2O_3$ | 0.46 | 0.54 | 1.49 | 3.24 | 2.58 | 3.62 | 1.04 | 1.46 | 1.13 | 1.60 |
| FeO | 8.04 | 11.80 | 11.32 | 5.09 | 5.61 | 5.38 | 6.89 | 6.76 | 7.11 | 5.23 |
| MnO | 0.14 | 0.30 | 0.18 | 0.17 | 0.18 | 0.16 | 0.13 | 0.14 | 0.09 | 0.10 |
| MgO | 37.47 | 12.45 | 6.69 | 19.40 | 17.78 | 30.00 | 41.11 | 43.36 | 41.07 | 34.84 |
| CaO | 3.08 | 13.00 | 11.09 | 12.44 | 12.17 | 3.39 | 2.35 | 1.44 | 1.07 | 5.50 |
| $Na_2O$ | 0.57 | 0.81 | 2.12 | 0.88 | 1.24 | 0.25 | 0.07 | 0.12 | 1.52 | 0.19 |
| $K_2O$ | 0.13 | 0.01 | 0.44 | 0.06 | 0.11 | 0.17 | 0.00 | 0.01 | 0.13 | <0.02 |
| $TiO_2$ | 0.71 | 0.07 | 1.69 | 0.62 | 0.62 | 0.23 | 0.17 | 0.10 | 0.25 | 0.01 |
| $P_2O_5$ | 0.06 | 0.03 | — | 0.02 | 0.04 | 0.05 | 0.08 | 0.02 | — | — |
| $Cr_2O_3$ | 0.43 | 0.24 | — | 0.31 | 0.43 | 0.28 | 0.50 | 0.42 | 0.10 | — |
| NiO | 0.20 | — | — | — | — | 0.12 | 0.21 | 0.26 | — | — |
| $H_2O$ | — | — | 0.37 | 0.40 | 0.43 | 8.94 | 0.55 | 0.44 | 0.83 | 3.52 |
| Total | 99.99 | 100.00 | 99.43 | 100.52 | 100.01 | 100.29 | 99.88 | 100.38 | 100.20 | 98.78 |

Pyrolite: 1 basalt + 3 dunites (D. H. Green and Ringwood, 1967). Eclogites: glass = eclogite material (Ringwood, 1967), 35090 = eclogite from Scotland (Yoder and Tilley, 1962), 62-2 = eclogite from Hawaii (G. A. Macdonald and Katsura, 1964), 6618 = hypersthene eclogite nodules from Hawaii (Yoder and Tilley, 1962). Peridotites: i = peridotite from the Troodos Massif, Cyprus (Moores and Vine, 1971); 1, 2 = peridotite mylonite with enstatite and diopside (1) and amphibole and diopside (2) from St. Paul's Rocks (Tilley, 1966); W = dunitic rock containing jadeiitic pyroxene (Washington, 1930); 97-W = lherzolite from the Indian Ocean fracture zone located near 17°S (C. G. Engel and Fisher, 1969).

mineralogy for the garnet lherzolite, eclogite and peridotite are about 900—1400°C and 70—200 km (Fig. 12-3). The serpentinized peridotites found on the ocean floor are equivalent to continental peridotites. However, because of their pronounced degree of alteration, it is often a difficult task to make any inference as to their relationship with primordial mantle material.

*Peridotitic source*

Peridotite consists mainly of olivine, orthopyroxene, clinopyroxene and garnet. Depending on the various proportions of these minerals, peridotitic rocks may be called different names, such as lherzolite, harzburgite, dunite or websterite. The terms garnet, plagioclase or spinel may be used as a prefix when these three minerals delineate a pressure and temperature field of stability for peridotitic rocks. Experimental studies on peridotitic material were carried out on continental or island-derived samples (Table 12-2). O'Hara and Mercy (1963) suggested that the garnet peridotite found in nodules of kimberlites may have originated in the upper mantle. Similarly, feldspathic peridotite and nodules of lherzolite found in basaltic rocks were presumed to be mantle material (Kushiro and Kuno, 1963). Also, eclogite and various types of peridotites, such as harzburgite, lherzolite, websterite and dunite, have been found in the kimberlite pipes of South Africa (Yoder, 1975). From experimental studies (Ito and Kennedy, 1967) on garnet peridotite with pressures of up to 40 kbar, it was shown that the stability field of the olivine + orthopyroxene + Cr-spinel assemblage occurs at low pressure, while garnet appears at a higher pressure (above 15 kbar) on the solidus (Fig. 12-3). This experiment indicates that the maximum limit of stability at which spinel lherzolite could be found is 23 kbar (80 km depth). At low pressure, no plagioclase was observed during the experimental runs which gave rise to the lherzolite assemblage. This is a dilemma, since plagioclase is the main mineral phase of basaltic rocks. However, plagioclase could have been formed if, during partial melting, the liquid phase with the composition of garnet and olivine were separated from the residue (mainly pyroxene). Another problem with the peridotite source material for basalts is the absence of $K_2O$-bearing phases. Amphibole could be one of the components giving rise to the potash content in the rock. On the other hand, the low $K_2O$ content of oceanic basalts may reflect the absence or the limited amount of amphibole in the parental source. All the peridotitic material found in oceanic environments (in rift mountain regions of oceanic ridges, in fracture zones, in thrust-faulted areas or in trenches) is serpentinized to various degrees. Also, the different kinds of serpentinized peridotite do not occur at mantle depths but, as discussed in previous chapters, occur in the relatively low-pressure regions of less than 3 kbar ($\simeq$10 km depth).

*Eclogitic source*

Experimental work by Yoder and Tilley (1962) has shown that basaltic rocks could be converted to eclogite under relatively high pressure and temperature conditions. The idea that eclogite might be the high-pressure form of gabbro was first put forward by Fermor (1913). Eclogite is formed predominantly of clinopyroxene and garnet, and has a density of 3.3—3.6 $g/cm^3$. Clinopyroxene consists of an omphacite which is a solid solution of mainly diopside, jadeite, acmite ($NaFe^{3+}SiO_6$), and Tschermak's molecule ($CaMgSi_2O_6$). Garnet consists of pyroxene, almandine, and varying proportions of grossularite-andradite. The non-essential minerals include amphibole, orthopyroxene, plagioclase, Fe-oxide minerals and/or rutile. Eclogite is found in both volcanic and metamorphic terranes. Eclogite in volcanic terranes occurs as nodules included in alkali basalt (Kuno et al., 1957). The fact that Hawaiian lavas erupted a considerable volume of tholeiites followed by alkali basalts and nepheline basalts led Yoder and Tilley (1962) to suggest that alkali types of magma are derived from a deeper source than tholeiitic magma. Omphacite (jadeite and Tschermack's molecule) is characteristic of high-temperature and pressure conditions. The stability field of jadeite is found in phase transition studies involving the reaction:

$$NaAlSi_3O_8 \rightleftarrows NaAlSi_2O_6 + SiO_2$$
albite          jadeite          quartz

at about 25 kbar and 1200°C (Birch and LeComte, 1960).

Eclogites are stable over a wide range of pressure-temperature conditions. D. H. Green and Ringwood (1967) suggested the existence of eclogite at temperatures as low as 200°C. The lowest pressure condition at which eclogite is stable is about 6kbar. The transformation of gabbroic rocks to eclogitic material may take place at a pressure greater than 7 kbar (Fig. 10-3). The assemblage of plagioclase-pyroxene and spinel is stable below 7 kbar, while the soldius of garnet at high pressure is due to the breakdown of aluminous pyroxene, according to the reaction (D. H. Green and Ringwood, 1967):

$$4(Mg, Fe)SiO_3 + CaAl_2Si_2O_8 \rightleftarrows (Mg, Fe)_3Al_2Si_3O_{12} + Ca (Mg, Fe)Si_2O_6 + SiO_2$$
orthopyroxene          anorthite          almandine-pyrope          clinopyroxene

This corresponds to the transformation of a quartz-tholeiite type of melt at pressure less than 12—14 kbar and a temperature of 1000—1200°C (D. H. Green and Ringwood, 1967). At low pressure the garnet is almandine and becomes more pyrope-rich at higher pressure. The following reaction between olivine and plagioclase occurs at 7 kbar pressure and at 1100°C:

$$2(Fe, Mg)SiO_4 + CaAl_2Si_2O_8 \rightleftarrows Ca (Fe, Mg)_2Al_2Si_3O_{12}$$
olivine          anorthite          pyrope

This reaction shows the limit of stability of an undersaturated alkali basalt melt.

*Amphibolitic source*

The fact that amphibole is a hydrated mineral with a fair amount of potash could be important in explaining the origin of alkali magma. The work of Bowen (1928) and the later experimental work of Yoder and Tilley (1962) suggested that the resorption of amphibole crystals sinking into a higher-temperature environment may transform a tholeiitic melt into an alkali basalt melt. Yoder and Tilley (1962) demonstrated that amphiboles could be formed from basalts and gabbros by adding water at high pressure, and that basaltic liquids in the presence of water can be quenched to give amphiboles (Fig. 12-5). Amphibole is not stable at a pressure above about 30 kbar ($\simeq$100 km) (Lambert and Wyllie, 1970). Olivine tholeiites do not exist at water pressure greater than 1.4 kbar (Yoder and Tilley, 1962). Above that pressure, amphibole starts to form.

Lacroix (1917) recognized the close resemblance between basalts and a rock of hornblendite composition. Later experimental work by Yoder and Tilley (1962) also suggested the possible similarity between hornblendite and basalts. These authors, in recasting the composition of a Kilauea (Hawaii) olivine tholeiite into the molecular formula of an amphibole (tremolite$_{9.7}$ edinite$_{30.2}$ tschermakite$_{41.6}$ glaucophane$_{18.5}$), have noticed the similarity to a natural amphibole found in altered eclogite. This discovery suggests that natural amphibole exists and approximates the composition of an olivine tholeiite (Yoder and Tilley, 1962).

Since amphiboles are found associated with basalt and serpentinized peridotite in oceanic fracture zones, they may be considered high-pressure equivalents of hydrated basaltic material. Hornblendite, another amphibole-bearing rock, was found in association with alkali basalts in a deep hole drilled in the eastern Indian Ocean during DSDP Leg 22 (Site 211; Hékinian, 1974a) (see Chapter 1).

Among the possible sources of basaltic melt discussed above, a garnet lherzolite seems to be one of the most important rock types which could give rise to the basaltic rocks by partial melting. Many garnet lherzolites on islands and on the continents were brought to the surface as nodule inclusions in basaltic lava. The nodules found in the kimberlite suite of Lesotho (South Africa) constitute two types of garnet lherzolite: (1) the granular type, and (2) the sheared type. Nixon and Boyd (1973) have suggested that the granular lherzolites were derived from mantle material depleted in basaltic constituents. The sheared lherzolites are from a less-depleted mantle zone located in the asthenosphere. Lherzolite, having a strongly sheared appearance with submylonitic texture, is found in abundance on Mid-Atlantic Ridge provinces. The chemical difference between the two types of lherzolite is evidenced by the $TiO_2$-$Al_2O_3$ variation diagram shown in Fig. 12-6. The granular lherzolite is depleted in both $TiO_2$ content (<0.1%) and $Al_2O_3$ content (<1%) with respect to the other type (Fig. 12-6). It is also interesting to note that most

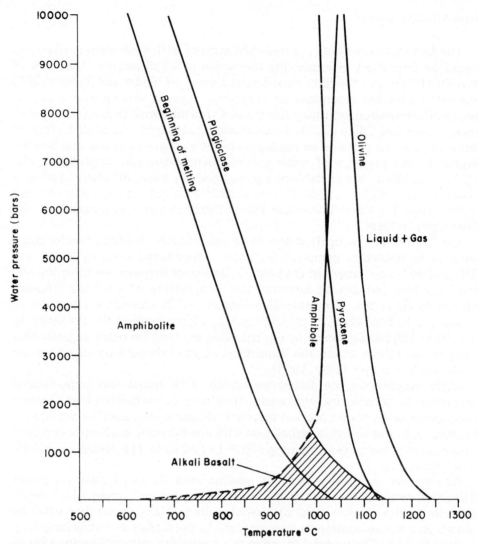

Fig. 12-5. Projection of the pressure-temperature relations of a natural alkali basalt-water system using data on prehistoric flow from Hualalai (Hawaii) (simplified after Yoder and Tilley, 1962, p. 452). The alkali basalt field is restricted to the low-pressure environment (shallow depth); the amphibolite covers a large field of stability.

serpentinized peridotites from the Puerto Rico Trench and from thrust-faulted regions (e.g. Macquarie Ridge) are similar to the granular lherzolite as far as their depletion in $TiO_2$ and $Al_2O_3$ content is concerned (Fig. 12-6). The sheared type of serpentinized lherzolite which is the least depleted in Ti, Al, Ca and Fe corresponds to the serpentinized peridotites commonly found exposed on mid-oceanic ridges and fracture zones.

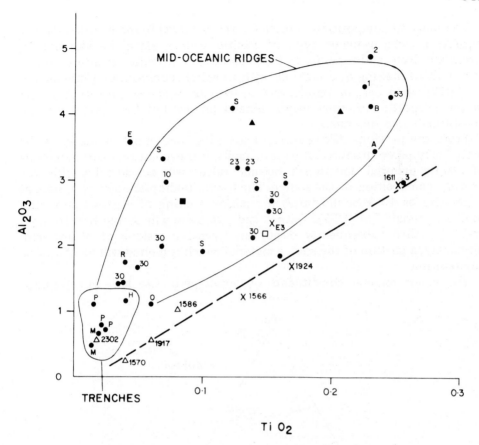

Fig. 12-6. $Al_2O_3$-$TiO_2$ (wt.%) variation diagram of peridotitic rocks from the ocean floor compared to those from subaerial regions. The fields of Mid-Atlantic Ridge samples are from transform faults (Miyashiro et al., 1969a, Hékinian and Aumento, 1973; and this work). The numbers close to the filled circles (0, 10, 23, 30, 53) correspond to the degree of latitude north on the Mid-Atlantic Ridge. ■ = average of 7 harzburgites from the Rift Mountain region near 45°N, Mid-Atlantic Ridge (Aumento and Loubat, 1971). $S$ = peridotite from St. Paul's Rocks (Washington, 1930; Tilley, 1966). The samples from the Mid-Indian Oceanic Ridge are from the Rodriguez transform fault (Hékinian, 1968) and from other various segments of the ridge system (Chernycheva and Bezrukov, 1966; C. G. Engel and Fisher, 1975). $A$ = average of 10 lherzolite, □ = average of 8 harzburgites from the Mid-Indian Oceanic Ridge (Dmitriev, 1974). The field of oceanic trenches (= Puerto Rico Trench) includes samples from thrust-faulted and seamount areas ($M$ = Macquarie Ridge, $H$ = Horseshoe Seamounts). Samples from subaerial regions include the Allende meteorite (*1*), a peridotite from the Salt Lake Crater in Hawaii (*2*, Boyd and Nixon, 1975), and a spinel lherzolite from Hawaii (▲; Mysen and Kushiro, 1977). The dashed line relates sheared garnet lherzolite (×: samples 1924, 1566, 1611, E3) with granular garnet lherzolite nodules (△: samples 1570, 1586, 1917, 2302) (Nixon and Boyd, 1973).

The range in composition of certain trace elements found in basaltic rocks suggests a heterogeneous type of mantle source, since the variation of strontium, lead, and uranium isotopes and the K/Rb ratio are interpreted as the result of melting of a material with a variable composition (Subbarao et al., 1977). Most oceanic basalts vary in their trace element distribution and in the proportion of their major mineral phases, but their major mineral constituents stay the same.

From the previous discussion, and assuming that garnet peridotite is the most likely parental source of basaltic melts, it is evident that partial melting of mantle material underneath oceanic ridges will considerably alter the primary composition of the source. The forsterite-diopside-pyrope system at a pressure of 40 kbar illustrates the partial melting of a garnet-peridotite source material (Fig. 12-7). Garnet and pyroxene will be the first to melt, and they then subsequently disappear, leaving a residue of olivine and, ultimately, a portion of the source material which is melted will be of dunitic composition.

The major mineral constituents of garnet peridotite have large enough

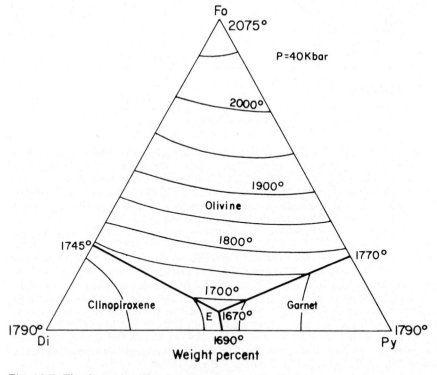

Fig. 12-7. The forsterite (Fo)-diopside (Di)-pyrope (Py) ternary system at 40 kbar (after Davis and Schairer, 1965, p. 124) showing the various isotherms. E = piercing-point (eutectic-like) where olivine, pyroxene and garnet coexist together.

mineral/melt ratio distribution coefficients to have a pronounced effect on trace element distribution. The proportion of the garnet/pyroxene ratio could also vary from place to place in the upper mantle, forming a hetero- geneous parental source of garnet peridotite. Thus, the partial melting of a zone with a high garnet/pyroxene ratio could give rise to a melt rich in plagio- clase. Garnet and pyroxene are also the first phases to disappear during partial melting, and the residual material will have the composition of an olivine peridotite which, upon serpentinization, will be similar to a lherzolite containing relics of pyroxene and olivine. When a melt derived from the partial melting of a garnet peridotite from a depth of about 70—80 km (about 25 kbar) attains a sizeable volume, it will rise within the upper mantle and the crust through fissures and cracks. The magma rising adiabatically could either be stationary within the crust for a certain period of time or continue towards the surface where it erupts. It is often difficult to distinguish between flows that are directly connected with the partial melting source and those which have been differentiated within the crust. A few criteria were enumer- ated in Chapter 2 when referring to the volcanics from the FAMOUS area near $36°50'N$ on the Mid-Atlantic Ridge. Some basaltic glass with very few microlites of plagioclase and olivine and containing high MgO (>9%) and high CaO (>12%) could be considered as derived from primitive melts which have not undergone any shallow-depth differentiation processes.

ORIGIN OF ULTRAMAFICS FROM THE OCEAN FLOOR

In Chapters 2 and 7 the various occurrences of serpentinized peridotites and other ultramafics have been discussed in relationship to the major struc- tural features of the ocean floor. In addition to rift valley regions and fracture zones, ultramafics are found on islands and on some elevated volcanoes (sea- mounts). These are usually in the form of nodules included in alkali basalt lava pouring onto the surface. It is believed that the nodules of peridotites and serpentinized peridotites are solid masses brought up from the upper mantle. Miyashiro et al. (1969a) have suggested that the high-calcium (CaO >2%) serpentinites found in the Mid-Atlantic Ridge have a chemical compo- sition similar to that of the upper mantle nodules found in basalts. It is also likely that ultramafic bodies exist under the cover of relatively recent vol- canoes. Up to now, no ultramafics have been collected from the inner floor of rift valleys or from recent spreading centers of the East Pacific Rise. This fact could be due to the capping of extrusive basaltic flows. Ultramafics are usually associated with relatively old scarpments of the rift valley walls, rift mountain regions, and the older scarps of transform faults. Dredged material from these regions shows sporadic occurrences of ultramafics along fault scarps located at various levels of the exposed crust (Chapter 7). These ultra- mafics are generally minor constituents of dredge hauls which are primarily made up of basaltic rocks. This suggests that the exposure of ultramafics might be limited in most cases (Fig. 1-1). It is not unreasonable to think that such limited and localized occurrences of ultramafic bodies could be due to the exposure by faulting of crystal-fractionated material from a basaltic melt. However, ultramafics on the ocean floor are not always associated with mafic rocks; there are areas such as the Josephine Bank (Horseshoe Seamounts zone off Gibraltar), the King's Trough (between the Peake Deep and the Freen Deep), and the Macquarie Ridge (western Pacific) where abundant serpentinized peridotites have been found (Chapter 4). Hence, these regions are of particular interest since they expose a large volume of ultramafics and related intrusives and metamorphics.

Field observation has shown that serpentinized peridotite occurs at a certain distance from the Rift Valley axis in the Mid-Atlantic Ridge and it is rare to find it within the first 30 km away from the axial zone. However, between 100 and 300 km from the accreting plate boundary region, serpentinized peridotite is commonly found near 45°N and 11°N in the Atlantic Ocean (Aumento and Loubat, 1971; Bonatti and Honnorez, 1976). The sporadic occurrence of serpentinized bodies parallel to the axis of the Rift Valley led Bonatti et al. (1970) to postulate that diapirism is most intense away from the axial zone. However, it is not yet clear why serpentinized peridotite does not occur nearer to the axis of plate boundaries since tectonism is a main factor affecting the walls of the rift valleys. Also, from the known distribution of ultramafics on the ocean floor it is not yet possible to make any correlation between their tectonic setting and distances from accreting plate boundary regions.

The ultramafic complex of the ocean floor also could have been generated by the intrusion of a serpentinite diapir during tectonic activity (Cann, 1971b). The presence of metamorphic minerals often accompanying the serpentinized peridotite suggests that faulting or other tectonic motions are important phenomena in the emplacement of these rocks. The intrusion of serpentinized bodies into the central valley of fracture zones has a deep-seated origin (Miyashiro et al., 1969a). Metamorphosed basalts occurring together with serpentinized peridotite suggested a metamorphic origin for these ultramafics. The serpentinization and metamorphism may have taken place at about the same time during a convective ascent of upper mantle peridotite (Miyashiro et al., 1969a). Other authors (Melson and Thompson, 1971; G. Thompson and Melson, 1972) have also suggested the intrusive nature of peridotitic material in fracture zones. However, in contrast to Miyashiro's interpretation, these authors believe that layer 3 is made up of a complex association of serpentinized peridotite, some basalt, and gabbro. Assuming that the serpentinized peridotite emplaced in oceanic fracture zones has a deep-seated origin, then the conditions of pressure and temperature at which the minerals forming the unaltered original peridotite are in equilibrium must be determined. From experimental work (D. H. Green, 1963), it was found that the substitution of alumina within the plagioclase-peridotite stability field is a function mainly of pressure rather than temperature. Since orthopyroxene is presumably the high-temperature pyroxene which is often found altered into bastite in most serpentinized peridotite, it is likely that the alumina content of the enstatite variety may reflect the depth of origin of a plagioclase-olivine-pyroxene-spinel assemblage. The relatively high $Al_2O_3$ content (1.3%) of pyroxene from the serpentinized peridotite from the Vema Fracture Zone led Melson and Thompson (1971) to conclude that these rocks were formed essentially at high temperature (1200°C) and moderate pressure (<10 kbar). Experimental work by Rotshtein (1960, 1961) on the paragenesis of olivine-pyroxene-spinel also indicates a condition of stability on the order of 10 kbar and 1000°C. Both clinopyroxene and ortho-pyroxene are stable under high pressure and temperature conditions. The fact that only orthopyroxene is found in some serpentinized peridotite may suggest that all the clinopyroxene was melted. It is likely that the most primitive serpentinized peridotite corresponds to the composition of a lherzolite where both abundant orthopyroxene and clinopyroxene co-exist together with olivine and garnet.

Some textural features encountered in serpentinized peridotite suggest the tectonic motions which may have taken place within the crust and the upper mantle. Cataclastic textural features are indicative of oriented stress on the ultramafic flow. Granular mylonites and contorted minerals indicate a deformation under the influence of a hydration process. This is evidenced by the presence of hydrated minerals such as amphiboles, chlorite, serpentine and hematite filling the cleavage planes and interstices of minerals. Differential motions of the oceanic lithosphere and asthenosphere, such as faulting, thrust-faulting, and folding of two bodies with different viscosity, might enhance flow-structural features, relics of bent cleavage planes, and a development of secondary mineral phases. Serpentinization of peridotitic masses will lower the density from about 3.3 g/cm$^3$ to about 2.5 g/cm$^3$. In addition, the serpentinized peridotites are often associated with breccia material containing broken-up fragments of rocks and minerals which are the product of tectonic dislocation. The type of structures and associations in which some serpentinized peridotites are found indicates that they are derived from a relatively deep-seated material and emplaced by tectonism. However, not all the ultramafics need to be considered a residue of a partial melt or deep-seated mantle material; unsheared peridotites or peridotites showing a limited degree of alteration could represent shallow cumulates of a differentiated magma chamber. The association of basalts, gabbros and ultramafics also suggested that a magmatic differentiation, under appropriate conditions in a tectonically quiet zone, produced layering with an upward progression from dunite to peridotite to gabbro and to basalt (Aumento and Loubat, 1971).

354

MAGMATIC PROCESSES

The sporadic sampling of ocean floor volcanics and intrusive sequences has not given a clear picture of magmatic processes involved at accreting plate boundary regions. Ancient ophiolitic complexes which are considered to be old spreading ridge systems are, up to now, the best examples of oceanic crust and upper mantle lithologic sequences, and many geologists refer to the ophiolites when dealing with oceanic crust and upper mantle compositional variations. The sequential appearance of the ophiolitic provinces, which go from an ultramafic to a gabbroic and an interlayered complex with various types of cumulates, indicates a differentiated magma chamber. The different generation of dike complexes which cuts through the sequences and feeds the upper extrusive flows covering the entire unit suggests the complex and the heterogeneous nature of a magma chamber. Obviously, when comparing magmatic models of recent spreading ridges to those of ancient ophiolitic sequences, the late tectonic processes involved during the emplacement of oceanic ridge systems onto a subaerial environment must be considered. However, prior to comparing ancient and recent geological formations, it is important to have a good idea about the paleoenvironments and the type of structure involved. Ophiolitic sequences around the world are not exclusively the ancient expression of modern mid-oceanic ridge systems. Some could represent intra-plate uplifted structures or ridge systems of marginal basins.

Most geochemists have used trace element abundance in order to retrace parental magma and its partially melted derivatives. The advantage of studying the distribution of many trace elements in rocks is that they do not enter into the lattice of mineral phases forming basaltic rocks. The low concentration of these elements means that they are not effectively fractionated during differentiation processes. However, because these elements are not incorporated in the mineral phases, it is also possible for them to be heterogeneously distributed within basaltic rocks. For instance, in highly cumulated rocks, where intercrystalline spaces are limited, trace elements which have not entered into the mineral lattices are likely to migrate elsewhere during the cooling of the melt. Until there is more systematic knowledge about element distribution within a magmatic sequence from its source to its last stage of evolution, any discussion of the various processes exploring the origin of basaltic melts from an upper-mantle composition will remain speculative.

Recently Langmuir et al. (1977) have summarized several hypotheses of magmatic processes explaining the origin of various types of basaltic melts erupted on mid-ocean ridges. For a more in-depth discussion on the physico-chemical constituents of various melts and the thermal energy requirements for melting, the study by Yoder (1978) should be consulted. The physical processes involved in producing magma in the upper mantle are given a general review in the following sections.

*Fractional melting* has been proposed by Presnall (1969) and Roeder (1974) as the process by which a portion of the mantle is melted and the melt is removed from the parent source as it is formed. The mechanism could either be due to a melting of a subsolidus pyrolite composition as suggested by T. H. Green (1969), or due to a direct melting of the initial source region. As the fractional melting of the same source region proceeds, the composition of the melt will vary. If the temperature increases, more melting will occur and give rise to a range of different liquid compositions. At the beginning of fractional melting, the low-temperature mineral assemblages and the large ion lithophile elements will be removed first. This fact makes it clear that basalts which are enriched in the non-incorporated rare earth elements will be formed from the first melting of the source material, and further melting of the same source region cannot give rise to the same type of basalt. Cyclic eruptions of lava with compositions which are repeated each time could not occur except if the source region changes after each eruption.

*Batch melting* is not essentially different from fractional melting. As the name indicates, it involves a process by which the liquid phase stays in contact and in equilibrium condition with the residue until melting is completed, and then the melt is removed all at once, as a batch, from its source. As in the case of fractional melting, the residue will continually change its composition as melting proceeds.

*Zone refining, or zone melting*, has been defined by Harris (1957) as the process by which a magma formed at great depth in the mantle could become enriched in elements which are unable to be incorporated in the mineral phases found in the mantle. The elements expected to enrich the melt during its ascent to the surface are: K, Rb, Cs, Ba, Pb, Zr, Th, U, Nb, Ta, P, C, H and Cl. The removal of material from a melt will involve the migration of a portion of fluid to the adjacent country rock. A liquid of a partially melted material could also mix with the surrounding rock during an upward ascent of the magma. As melting and mixing occur, the composition of the liquid changes accordingly. Precipitation of heavy minerals will occur at the base of the displaced column. As the molten column rises, the accumulated material will increase. Zone melting requires a large volume of melt to start with, in order for the magma to react with certain zones.

*Magma mixing* reflects heterogeneous magma sources of the mantle which, upon melting, would mix together and give rise to different types of melts. This model implies that at least two different magma sources have melted independently and have produced various amounts of mixing. Such a model was used (Schilling et al., 1976) to explain the gradual compositional changes of volcanics erupted along mid-oceanic ridges which are associated with elevated volcanic platforms (e.g. Azores, Iceland, Galapagos; see Chapter 2).

*Dynamic melting* has been proposed by Langmuir et al. (1977) in order to explain the compositional differences in the basaltic flows encountered on the inner floor of the Rift Valley in the FAMOUS area. This model involves upwelling of a portion of the mantle underneath the ridge axis, and when the solidus is reached, melting starts. As melting of the mantle source proceeds, there is a continuous extraction of liquid. The melt removed will rise, and during its ascent might lead to zone refining and fractional crystallization. A positive aspect of this model is that dynamic melting could occur in several places at once. Various degrees of partial melting and different depths of melting could, with this model, explain the heterogeneity of basalts erupted on the ridge axis.

*Fractional crystallization.* The process of fractional crystallization takes place when a melt and a solid are formed and when both phases can either stay in contact with each other or be separated. Evidence of fractional crystallization is found in basaltic and gabbroic formations. Cumulate textures showing crystal growth after settling of olivine and pyroxene are found in troctolitic and normal gabbros (Chapter 1). Also, resorbed textural features are observed in some picritic and highly-phyric plagioclase basalts (Chapters 1 and 2), where the early-formed phases are partially remelted during their ascent towards the surface. The olivine and the plagioclase are often not in equilibrium with their surrounding glassy matrix and could be considered as if they were removed from their original environment of solidification. The composition of most ocean floor basalts falls within the field of olivine tholeiites and quartz tholeiites in the generalized nepheline-diopside-olivine-hypersthene-quartz diagram (Fig. 1-6) and close to the cotectic represented by the olivine-plagioclase of the ternary plagioclase-pyroxene-olivine diagram (Figs. 1-10, 1-11, 1-12). In general, the order of crystallization in most oceanic basalts is as follows: spinel $\rightarrow$ olivine $\rightarrow$ olivine + plagioclase $\rightarrow$ plagioclase + pyroxene. During early crystallization, the liquid will be depleted in Cr and Mg due to the precipitation of spinel and olivine. With the appearance of plagioclase and pyroxene, the liquid will be depleted in $Al_2O_3$ and CaO. The residual melt will become oversaturated with respect to Si, and the concentration of Ti, K and P will increase.

MAGMA RESERVOIR MODELS

As magma rises upward underneath oceanic ridges, it may form stationary pools in the crust prior to its eruption on the ocean floor. The size and shape of these magma pools or chambers are unknown, although some hypothetical models have been suggested. Inferences as to the type of magma reservoir present underneath accreting plate boundary regions are made based only on scattered data localized in a few small segments of oceanic ridges where both petrological and geophysical studies have been carried out (Chapter 2). It is likely that the volume and the extent of magmatic upwelling may vary considerably along various ridge segments as the structure and the morphology of the sea floor changes.

Models for magma chambers in accreting plate boundary regions were considered by Osmaston (1971), Cann (1974), and Sleep (1975). These authors are in favor of the upwelling of hot material which takes place essentially within a narrow zone intersecting the rift valley of slow-spreading mid-oceanic ridges. Sleep (1975) followed Parker and Oldenburg's (1973) assumption that the total heat flux conducted and converted from mid-oceanic ridges equals the amount of heat yielded by narrow dikes at the ridge's axes. According to this model, the temperature gradient for a slow-spreading ridge (<2 cm/yr) is steeper and the heat loss more prominent than for fast-spreading ridges (>5 cm/yr). If temperature is a function of age (McKenzie and Sclater, 1969), heat conduction of slow-spreading ridges is more rapid near the ridge axis (Sleep, 1975). For faster-spreading ridges (>5 cm/yr), the highest temperature gradient computed (Sleep, 1975) extends laterally away from the ridge axis and appears more consistent with the model for a wider source of upwelling (Fig. 12-1). From seismic, petrological and experimental thermal studies of accreting plate boundary regions, several alternative models for magma upwelling and storage in the lithosphere have been proposed.

*Large magma chamber*

Cann (1968, 1974) suggested that magma chambers exist at high levels underneath mid-oceanic ridges. As a ridge spreads, increments of lava are erupted to form a roof on the chamber and a layered crust composed of a top of basaltic pillow flow and dikes, a middle formation of gabbroic rocks, and a lower formation of cumulate gabbro. This model, also called the "infinite onion", was adopted by several authors (Rosendahl et al., 1976; Orcutt et al., 1975; Bryan and Moore, 1977; Hékinian et al., 1976) to explain geophysical and petrological observations on the East Pacific Rise near 8—9°N. The term "infinite onion" was applied to this model because layers of crust are peeled off as the magma chamber is fed with liquid from deeper parts of the lithosphere. In the FAMOUS area near 37°N, Bryan et al. (1976) proposed that differentiation processes existing in the magma chamber could account for the various types of basaltic rocks encountered on the inner floor of the Rift Valley (see Fig. 2-14). The significant chemical and petrographic zonation on the inner floor of the Rift Valley could be explained by a thermal gradient which decreases from the center of the inner floor to the side. A hotter magma enriched in olivine could occur at the center and a cooler lava could be erupted at the side. The fact that there is an increase in age of the eruption towards the margins was explained by the less frequent eruptive phases on the flanks (see Chapter 2).

This model does not appear satisfactory for slow-spreading ridges because it fails to explain the along-strike compositional variation. The argument that flank eruptions are less common than axial eruptions does, however, explain the presence of age differences along the ridge itself. Several other objections have been pointed out concerning the "infinite onion" model. The strongest one has to do with the fact that no seismic shear waves exist in the FAMOUS area underneath the inner floor. Another objection is that the thermal constants for slow-spreading ridges advocated by Sleep (1975) make the existence of a large magma chamber (2—3 km) unlikely (see Figs. 2-16 and 12-1).

*Narrow magma chamber*

This model, also called the "infinite leek" because of its narrow cylinder-shaped reservoir, is fed from beneath via pervasive cracks through which the melt is channeled. This model conforms to Weertman's (1971) suggestion that if the crust behaves as an elastic plate, a crack with melt will form at the bottom of the plate. This crack will increase its length until it pinches at its lower end and shuts, and the pocket of melt trapped inside will rise towards the surface. Several cracks together near the upper crust could form a potential magma pool. A detailed description of the infinite leek model and its advantages is proposed by Nisbet and Fowler (1978). The size of any magma chambers generated by this type of model fits the geophysical data and the theoretical temperature gradients calculated for slow-spreading ridges. From Sleep's (1975) thermal model, a temperature of about 1180°C (within the range of basalt melting) could be attained at a depth of less than 5 km. Using a latent heat calculation, Nisbet and Fowler (1978) have made a thermal model for a magma existing underneath the inner floor in the FAMOUS area at 37°N in the Atlantic Ocean. The temperature used for the liquidus is 1230—1250°C and for the solidus, 1160—1185°C. Independently, it was found that the temperature of equilibrium at the solidus between olivine and the melt varies between 1203 and 1146°C in the inner floor of the Rift Valley (Hékinian et al., 1976). From the above considerations, Nisbet and Fowler (1978) have postulated that the width of the magma chamber is about 2 km for a zone with a half spreading rate of 1 cm/yr. However, on the basis of topographic and compositional variations, it is likely that smaller pools of magma exist underneath the inner floor in the FAMOUS area.

*Columns of magmatic upwelling*

An alternative approach to the model of narrow magmatic upwelling presented above is to consider the presence of several pathways cutting through the lithosphere where melt is channeled. These pathways could have various shapes, and their size will vary depending upon the rate of spreading and the tectonic constraints at the time of magmatic upwelling. The mechanism of spreading generates zones of weakness in the lithosphere where cracks and fissures can extend down to the melted region. This melted area could be represented either by a mantle rising diapir in the sense described by D. H. Green and Ringwood (1967), or by fractional melting of the mantle source as proposed by Presnall (1969). During fast spreading, intense fracturing of the lithosphere takes place with a subsequent upwelling of melt. In crustal zones rendered extremely weak by an abundance of fractures, magmatic pools are expected to form. The area covered by the rise of the magmatic columns is believed to consist of narrow corridors extending underneath and along the strike of the ridge axis where maximum stresses are expected. The lateral extent of these columns is variable depending on the pattern of the plate's motion and of the spreading rates. From surface morphology of flows and from the degree of rock alteration, it is inferred that for both the Mid-Atlantic Ridge (FAMOUS area) and the East Pacific Rise (near 21°N) the uprising columns underneath these ridges should not be more than 1 km in width. The occurrence of dispersed columns of upwelling melt towards the surface implies the presence of dike swarms and sheeted complexes of basaltic composition near the surface. These will later grade downwards into anisotropic and differentiated gabbroic complexes, while mafic cumulates are found at the deepest part of the upwelling zone. A residue of partial melting consisting of depleted peridotite can, upon hydration and further fracturing of the lithosphere, subsequently rise towards the surface.

The advantage of postulating columns of magmatic upwelling is that it explains the episodic and the heterogeneous types of basalt eruption on the accreting plate boundary regions (Chapter 2). Also, the volcanics erupted on a given ridge segment could change their composition as a function of the spreading rate (Chapter 2). Assuming that the degree of fissuring of the lithosphere depends on the mechanism of spreading, any change in the spreading rate will alter the pattern of magmatic production. Therefore, the amount of magma available will be a function of the volume of mantle affected by the melting. However, our present knowledge of the compositional variation of the mantle is not yet well assessed. The relationship between an upwelling of magma in the lithosphere and a mantle convection system underneath accreting plates is unknown. Deep seismic reflection studies are needed in order to have a better understanding of the pattern of magmatic upwelling.

# APPENDIX

## ANALYTICAL PROCEDURES

The chemical analyses presented in the various tables included in the text were performed at the Centre Océanologique de Bretagne unless otherwise indicated.

Both electron scanning microscopy and point analyses (on a surface of less than 3 $\mu$m) were performed with the help of M. Bohn on a computerized electron microprobe unit (Camebax). The electron beam incidence of the Camebax has a take-off angle of about 42° and the limit of detection is about 1000 parts per millions (ppm). The precision of the method used is about 1% of the absolute value of concentration.

X-ray fluorescence analyses of bulk-rock samples were performed at the Centre Océanologique de Bretagne by H. Bougault, P. Cambon and J. Etoubleau using a Siemens instrument. The powdered samples are fused with lithium tetraborate containing lanthanum oxide which form a glass disc. Both major and minor elements are analysed. Comparative studies carried out by different laboratories gave the following average mean values for the major oxides: $SiO_2$, ±0.5; $Al_2O_3$, ±0.2; $Fe_2O_3$, ±0.2; MgO, ±0.2; CaO, ±0.2; $K_2O$, ±0.04; and $TiO_2$, ±0.06 (Bougault, 1980). For major elements the relative accuracy obtained is on the order of 1%, with the exception of MnO and $TiO_2$. The accuracy of trace element concentrations from 0 to 150 ppm is between ±1 and 3 ppm (Bougault, 1980).

## ALTERATION OF BASALTIC ROCKS

The thickness of hydrated glass ($S$) is related to time by the following formula (Friedman and Smith, 1960):

$$S\sqrt{CT}$$

where $S$ = thickness, $T$ = time, and $C$ = a constant depending on the composition of the glass which is equal to 480—2000 $m^2/10^3$ years for Hawaiian basaltic glass.

## CRYSTALLIZATION INDEX

The crystallization index (C.I.) established by Poldervaart and Parker (1964) is used to show the compositional variation of mafic and intermediate types of rocks. The C.I. is expressed in weight percent and is defined as:

$$C.I. = \Sigma \, (An + Di' + Fo' + Sp')$$

where An = normative anorthite; Di' = magnesian diopside, $CaMgSi_2O_6$, recalculated from normative diopside; Fo' = normative forsterite + normative enstatite converted to forsterite; Sp' = magnesian spinel, $MgAl_2O_4$, recalculated from normative corundum in ultramafic rocks.

The conversion factors used in calculating Di' and Fo' are:

$$Di' = 2.157003 \; Endi \; (En \; of \; normative \; diopside)$$

Fo' = Fo + 0.700837 Enhy (En of normative hypersthene)

For more details on the use and calculation of the crystallization index the reader should refer to Poldervaart and Parker (1964).


## DIFFERENTIATION INDEX

The differentiation index (D.I.) has been defined by Thornton and Tuttle (1960) to represent the sum of the normative percentages of quartz, orthoclase, albite, nepheline, leucite and kalsilite:

D.I. = Q + Or + Ab + Ne + Lc + Ks

The differentiation index is a concise way of measuring the relative degree of the "basicity" of the rock and it is best used for felsic types of rocks.


## PHYSICAL PROPERTIES

Specific gravity of oceanic rocks

| Rock type | Specific gravity |
|---|---|
| Basaltic glass | 2.80 — 2.83 |
| Crystalline basalt | 2.70 — 2.90 |
| Metabasalt | 2.6  — 2.75 |
| Serpentinized peridotite | 2.3  — 2.6 |
| Gabbro | 2.9  — 3.0 |
| Metamorphosed gabbro | 2.75 — 2.85 |
| (Fe, Cu, Zn) sulfide | 2.97 — 3.00 |

Refractive index of basaltic glass = 1.595—1.603.


Laboratory measurements of compressional wave velocities of oceanic rocks (after Fox et al., 1976)

| Rock type | Density (g/cm$^3$) | Compressional wave velocity (km/s) | |
|---|---|---|---|
| | | 0.25 kbar  →  | 2 kbar |
| Fresh basalt | 2.68 | 5.19 | 5.58 |
| Metamorphosed basalts | 2.85 | 5.64 | 6.03 |
| Metamorphosed gabbros | 2.89 | 5.55 | 6.04 |
| Serpentinized peridotites | 2.36 | 3.78 | 4.18 |

# REFERENCES

Ailin-Pyzik, I.B. and Sommer, S.E., 1979. Trace element analysis of altered DSDP basalts by selected area X-ray fluorescence and electron microprobe. *30th Pittsburgh Conf. Anal. Chem. Appl. Spectrosc., Cleveland, Ohio, March, 1979* (abstract).

Althaus, E., 1967. The triple point andalusite-sillimanite-kyanite. *Contrib. Mineral. Petrol.*, 16: 29—44.

Amstutz, G.C., 1974. *Spilites and Spilitic Rocks. International Union of Geological Sciences Series, No. 4.* Springer Verlag, New York, N.Y., 482 pp.

Anderson, A.T. and Gottfried, D., 1971. Contrasting behaviour of P, Ti, and Nb in a differentiated high-alumina olivine tholeiite and a calc-alkaline-andesite suite. *Geol. Soc. Am. Bull.*, 82: 1929—1942.

Anderson, D.L. and Archambeau, C.B., 1964. The anelasticity of the earth. *J. Geophys. Res.*, 69: 2071—2084.

Anderson, R.N., McKenzie, D.P. and Sclater, J.G., 1973. Gravity, bathymetry and convection in the earth. *Earth Planet. Sci. Lett.*, 18: 391—407.

Anderson, R.N., Clague, D.A., Klitgord, K.D., Marshall, M. and Mishimori, R.K., 1975. Magnetic and petrological variations along the Galapagos Spreading Center and their relation to the Galapagos melting anomaly. *Geol. Soc. Am. Bull.*, 86: 683—694.

Anderson, R.N., Uyeda, S. and Miyashiro, A., 1976. Geophysical and geochemical constraints of converging plate boundaries, I. Dehydration in the downgoing slab. *Geophys. J. R. Aston. Soc.*, 44: 333—357.

Anderson, T.F., Donnelly, T.W., Drever, J.L., Eslinger, E., Gierkes, J.M., Kastner, M., Lawrence, J.R. and Perry, E.A., 1976. Geochemistry and diagenesis of deep sea sediments from Leg 35 of the Deep Sea Drilling Project. *Nature*, 261: 473—476.

Andrews, A.J., 1978. *Petrology and geochemistry of alteration in layer 2 basalts, DSDP Leg 37.* Ph.D. Thesis, University of Western Ontario, London, Ont.

Andrews, J.E., 1973. Correlations of seismic reflectors. In: *Initial Reports of the Deep Sea Drilling Project, 21.* U.S. Government Printing Office, Washington, D.C., pp. 459—479.

Aramaka, S., 1963. Geology of Asama volcano. *Tokyo Univ. Fac. Sci. J., Sect. 2*, 14: 229—443.

Arculus, R.J. and Johnson, R.W., 1978. Criticism of generalized models for the magmatic evolution of arc-trench systems. *Earth Planet. Sci. Lett.*, 39: 118—126.

Arcyana, 1975. Transform fault and rift valley from bathyscaphe and diving saucer. *Science*, 190: 108—116.

Arcyana, 1977. Rocks collected by bathyscaphe and diving saucer in the FAMOUS area of the Mid-Atlantic Rift Valley: petrological diversity and structural setting. *Deep-Sea Res.*, 24: 565—589.

**Arcyana, 1978.** *FAMOUS, Atlas Photographique.* CNEXO/Gauthiers-Villars, Paris, 128 pp.

Arndt, N.T., 1976. Melting relationship of ultramafic lavas (komotiites) at 1 atm and high pressure. *Carnegie Inst. Washington Yearb.*, 75: 555—562.

Arrhenius, G., 1963. Pelagic sediments. In: M.N. Hill (Editor), *The Sea, 3.* Interscience, New York, N.Y., pp. 655—727.

Aumento, 1968. The Mid-Atlantic Ridge near 45°N, II. Basalts from the area of Confederation Peak. *Can. J. Earth Sci.*, 5: 1—21.

Aumento, F. and Loncarevic, B.D., 1969. The Mid-Atlantic Ridge near 45°N, III. Bald Mountain. *Can. J. Earth Sci.*, 6: 11—23.

Aumento, F. and Loubat, H., 1971. The Mid-Atlantic Ridge near 45°N: serpentinized ultramafic intrusions. *Can. J. Earth Sci.*, 8: 631—663.

Aumento, F. and Milligan, B.A., 1971. Uranium content of Mid-Atlantic Ridge rocks. *EOS, Trans. Am. Geophys. Union*, 52: 375 (abstract).

Auzende, J.M., Olivet, J.L., Charvet, J., Le Lann, A., Le Pichon, X., Monteiro, H.J., Nicolas, A. and Ribeiro, A., 1978. Sampling and observation of oceanic mantle and crust on Gorringe Bank. *Nature*, 273(5657): 45—49.

Baker, I., 1969. Petrology of the volcanic rocks of Saint Helena Islands, South Atlantic. *Geol. Soc. Am. Bull*, 80: 1283—1310.

Baker, P.E., Gass, I.G., Harris, P.G. and Le Maitre, R.W., 1964. Vulcanological report on the Royal Society expedition to Tristan da Cunha, 1962. *Philos. Trans. R. Soc. London, Ser. A*, 256: 439—578.

Balashov, Y.A., Dmitriev, L.V. and Sharaskin, A.Y., 1970. Distribution of rare earths and yttrium in the bedrock of the ocean floor. *Geokhimiyia*, 6: 647—660 (in Russian; translated in *Geochem. Int.*, 7: 456—468).

Ballard, R.D. and Van Andel, T.H., 1977. Morphology and tectonics of the inner rift valley at lat. 36°50′N on the Mid-Atlantic Ridge. *Geol. Soc. Am. Bull.*, 88: 507—530.

Ballard, R.D., Bryan, W.B., Hentzler, J.R., Keller, G., Moore, J.G. and Van Andel, T.H., 1975. Manned submersible observations in the FAMOUS area, Mid-Atlantic Ridge. *Science*, 190: 103—108.

Banghar, A.R. and Sykes, L.R., 1969. Focal mechanisms of earthquakes in the Indian Ocean and adjacent regions. *J. Geophys. Res.*, 74: 632—649.

Banks, H.H., 1972. Iron-rich saponite: additional data on samples dredged from the Mid-Atlantic Ridge, 22°N latitude. *Smithsonian Contrib. Earth Sci.*, 9: 39—42.

Baragar, W., Plant, A.G., Pringle, G.J. and Schau, M., 1977. Petrology and alteration of selected units of Mid-Atlantic Ridge basalts samples from Sites 332 and 335, DSDP. *Can. J. Earth Sci.*, 14: 837—874.

Barnes, I. and O'Neil, J.R., 1969. The relationship between fluids in some fresh Alpine-type ultramafics and possible modern serpentinization Western United States. *Geol. Soc. Am. Bull.*, 80: 1947—1960.

Bass, M.N., 1971. Voluble abyssal basalt populations and their relation to sea-floor spreading rates. *Earth Planet. Sci. Lett.*, 11: 18—22.

Batiza, R., Rosendahl, B.R. and Fisher, R.L., 1977. Evolution of oceanic crust, 3. Petrology and chemistry of basalts from the East Pacific Rise and the Siqueiros transform fault. *J. Geophys. Res.*, 82: 265—276.

Bauer, G.R., 1970. The geology of Tofua Island, Tonga. *Pacif. Sci.*, 24: 333—350.

Beetz, P.R.W., 1934. Geology of southwest Angola between Cunene and Lunda axis. *Geol. Soc. S. Alp. Trans.*, 36: 137—176.

Bender, J.F., Hodges, F.N. and Bence, A.E., 1978. Petrogenesis of basalts from the Project FAMOUS area: experimental study from 0 to 15 kbars. *Earth Planet. Sci. Lett.*, 41: 277—302.

Berthois, L. and Froidefond, J.M., 1978. Relief du versant nord de la Fosse de la Romanche dans la zone comprise entre 18°26′ et 18°31′ de longitude ouest. *Bull. Inst. Geol. Bassin d'Aquitaine*, 24: 107—117.

Bezrukov, P.L. and Kanaev, V.F., 1963. Principal structural features of the bottom of the northeastern part of the Indian Ocean. *Dokl. Akad. Nauk SSSR*, 153: 926—929.

Bezrukov, P.L., Krylov, A.Ya. and Chernysheva, V.I., 1966. Petrography and absolute age of basalts from the floor of the Indian Ocean. *Oceanology*, 6: 210—214.

Birch, F., 1969. Density and composition of the upper mantle: first approximation as an olivine layer. In: P. J. Hart (Editor), *The Earth's Crust and Upper Mantle. Am. Geophys. Union. Geophys. Monogr.*, 13: 18—36.

Birch, F. and LeComte, P., 1960. Temperature-pressure plane for albite composition. *Am. Sci.*, 258: 209—217.

Biscaye, P.E., 1965. Mineralogy and sedimentation of recent deep sea clay in the Atlantic Ocean and adjacent seas and oceans. *Geol. Soc. Am. Bull.*, 76: 803—832.

Bischoff, J.L., 1969. Red Sea geothermal brine deposits: their mineralogy, chemistry and genesis. In: E. T. Degens and D. A. Ross (Editors), *Hot Brines and Recent Heavy Metal Deposits in the Red Sea*. Springer-Verlag, Berlin, pp. 368—401.

Bischoff, J.L., 1972. A ferroan nontronite from the Red Sea geothermal system. *Clays Clay Miner.*, 20: 217—223.

Bischoff, J.L. and Dickson, W., 1975. Seawater-basalt interaction at 200°C and 500 bars: applications for origin of sea-floor heavy-metal deposits and regulation of seawater chemistry. *Earth Planet. Sci. Lett.*, 25: 385—397.

Bishop, A.C. and Wollery, A.R., 1973. A basalt-trachyite-phonolite series from Ua Pu, Marguesas Island, Pacific Ocean. *Contrib. Mineral. Petrol.*, 39: 309—326.

Bogdanov, Y.A. and Ploshko, V.V., 1968. Igneous and metamorphic rocks from the abyssal Romanche depression. *Dokl. Akad. Nauk SSSR*, 177: 173—176.

Bonatti, E., 1968. Fissure basalts and ocean floor spreading on the East Pacific Rise. *Science*, 161: 886—888.

Bonatti, E., 1978. Vertical tectonism in oceanic fracture zones. *Earth Planet. Sci. Lett.*, 37: 369—379.

Bonatti, E. and Honnorez, J., 1976. Sections of the Earth's crust in equatorial Atlantic. *J. Geophys. Res.*, 81: 4104—4116.

Bonatti, E. and Joensuu, O., 1966. Deep-sea iron deposits from the South Pacific. *Science*, 154: 643—645.

Bonatti, E., Honnorez, J. and Ferrara, G., 1970. Equatorial Mid-Atlantic Ridge: petrologic evidence of an alpine-type rock assemblage. *Earth Planet. Sci. Lett.*, 247—256.

Bonatti, E., Honnorez, J. and Ferrara, F., 1971. Peridotite-Gabbro-basalt complex from the equatorial Mid-Atlantic Ridge. *Philos. Trans. R. Soc. London, Ser. A*, 268: 385—402.

Bonatti, E., Honnorez, J. and Ferrara, G., 1974. Ultramafic rocks: peridotite-gabbro-basalt complex from the equatorial Mid-Atlantic Ridge. *Philos. Trans. R. Soc. London, Ser. A*, 268: 247—256.

Bonatti, E., Guerstein-Honnorez, B.M. and Honnorez, J., 1976. Copper-iron-sulfide mineralization from the equatorial Mid-Atlantic Ridge. *Econ. Geol.*, 71: 1515—1525.

Bonatti, E., Chermak, A. and Honnorez, J., 1979. Tectonic and igneous emplacement of crust in oceanic transform zones. In: M. Talwani, C. G. A. Harrison and D. E. Hayes (Editors), *Deep Sea Drilling Results in the Atlantic Ocean: Ocean Crust. Am. Geophys. Union, Maurice Ewing Ser.*, 2: 239—248.

Boström, K. and Peterson, M.N.A., 1966. Precipitates from hydrothermal exhalations on the East Pacific Rise. *Econ. Geol.*, 61: 1258—1265.

Boström, K. and Peterson, M.N.A., 1969. The origin of aluminum-poor ferromanganoan in area of high heat flow on the East Pacific Rise. *Mar. Geol.*, 7: 427—447.

Boström, K., Peterson, M.N.A., Joensuu, O. and Fisher, D.E., 1969. Aluminum-poor ferromanganoan sediments on active oceanic ridges. *J. Geophys. Res.*, 74: 3261—3270.

Bott, M.H.P., Browitt, C.W.A. and Stacey, A.P., 1971. The deep structure of the Iceland-Faroe Ridge. *Mar. Geophys. Res.*, 1: 328—351.

Bougault, H., 1980. *Contribution des éléments de transition à la compréhension de la genèse des basaltes océaniques*. Thèse de Doctorat d'Etat, Université de Paris VII, Paris, 221 pp.

Bougault, H. and Hékinian, R., 1974. Rift Valley in the Atlantic Ocean near 36°50′N: petrology and geochemistry of basaltic rocks. *Earth Planet. Sci. Lett.*, 24: 249—261.

Bougault, H., Joron, L. and Treuil, M., 1979a. Alteration, fractional crystallization, partial melting, mantle properties from trace elements in basalts recovered in the North Atlantic, In: M. Talwani, C. G. A. Harrison and D. E. Hayes (Editors), *Deep Sea Drilling Results in the Atlantic Ocean: Ocean Crust. Am. Geophys. Union, Maurice Ewing Ser.*, 2: 352—368.

Bougault, H., Cambon, P., Corre, O., Joron, J.L. and Treuil, M., 1979b. Evidence for variability of magmatic processes and upper mantle heterogeneity in axial region of the Mid-Atlantic Ridge near 22° and 36°N. In: J. Francheteau (Editor), *Processes at Mid-Ocean Ridges. Tectonophysics*, 55: 11—34.

Bowen, N.L., 1915. The crystallization of haplobasaltic, haplodioritic, and related magmas. *Am. J. Sci.*, 40: 161—185.

Bowen, N.L., 1928. *The Evolution of the Igneous Rocks.* Princeton University Press, Princeton, N.J., 334 pp.

Bowen, N.L. and Tuttle, O.F., 1949. The system $MgO-SiO_2-H_2O$. *Geol. Soc. Am. Bull.*, 60: 439—460.

Bowin, C., 1973. Origin of the Ninety East Ridge from studies near the equator. *J. Geophys. Res.*, 78: 6029—6043.

Bowin, C.O., Nalwalk, A.J. and Hersey, J.B., 1966. Serpentinized peridotite from the north wall of the Puerto Rico trench. *Geol. Soc. Am. Bull.*, 77: 257—270.

Boyd, F.R. and Nixon, P.H., 1975. Origins of the ultramafic nodules from some kimberlites of northern Lesotho and the Monastery Mine, South Africa. In: L. H. Ahrens, J. B. Dawson, A. R. Duncan and A. J. Erlank (Editors), *Physics and Chemistry of the Earth*, 9. Pergamon, Oxford, pp. 431—454.

Brongniart, A., 1913. Tableau de la classification des roches mélangées. *J. Mines*, 34: 31—48.

Brothers, R.N. and Martin, K.R., 1970. The geology of Macauley Island, Kermadec group, Southwest Pacific. *Bull. Volcanol.*, 34: 330—346.

Bryan, W.B., 1967. Geology and petrology of Clarion Island, Mexico. *Geol. Soc. Am. Bull.*, 78: 1461—1476.

Bryan, W.B., 1979. Regional variation and petrogenesis of basalt glasses from the FAMOUS area, Mid-Atlantic Ridge. *J. Petrol.*, 20: 293—325.

Bryan, W.B. and Moore, J.G., 1977. Compositional variations of young basalts in the Mid-Atlantic Ridge Rift Valley near lat. 36°44'N. *Geol. Soc. Am. Bull.*, 88: 556—570.

Bryan, W.B., Stice, G.D. and Ewart, A., 1972. Geology, petrography and geochemistry of the volcanic islands of Tonga. *J. Geophys. Res.*, 77: 1566—1585.

**Bryan, W.B., Thompson, G., Frey, F.A. and Dickey, J.S., 1976. Inferred geologic setting and differentiation in basalts from the Deep Sea Drilling Project.** *J. Geophys. Res.*, 81: 4285—4304.

**Bryan, W.B., Thompson, G. and Frey, F.A., 1979. Petrological character of the Atlantic crust from DSDP and IPOD drill sites.** In: M. Talwani, C.G.A. Harrison and D.E. Hayes (Editors), *Deep Sea Drilling Results in the Atlantic Ocean: Ocean Crust. Am. Geophys. Union, Maurice Ewing Ser.*, 2: 278—284.

Buddinger, T.F. and Enbysk, B.J., 1967. Late Tertiary date from the East Pacific Rise. *J. Geophys. Res.*, 72: 2271—2274.

Bunch, T.E. and LaBorde, R., 1976. Mineralogy and compositions of selected basalts from DSDP Leg 34. In: *Initial Reports of the Deep Sea Drilling Project, 34.* U.S. Government Printing Office, Washington, D.C., pp. 263—275.

Bunce, T.E., Bowin, C.O. and Chase, R.L., 1966. Preliminary results of the 1964 cruise of R.V. "Chain" to the Indian Ocean. *Philos. Trans. R. Soc. London, Ser. A.*, 259: 218—226.

Campbell, A.S. and Fyfe, W.S., 1965. Analcime-albite equilibra. *Am. J. Sci.*, 263: 807—816.

Campsie, J., Bailey, J.C. and Rasmussen, M., 1973a. Chemistry of tholeïites from the Galapagos Islands and adjacent ridges. *Nature*, 245: 122—124.

Campsie, J., Bailey, J.C., Rasmussen, M. and Dittmer, F., 1973b. Chemistry of tholeiites from the Reykjanes and Charlie Gibbs fracture zone. *Nature Phys. Sci.*, 244: 71—73.

Cann, J.R., 1968. Geological processes at the Mid-Ocean Ridge crest. *Geophys. J. R. Astron. Soc.*, 15: 331—341.

Cann, J.R., 1969. Spilites from the Carlberg Ridge, Indian Ocean. *J. Petrol.*, 10: 1—19.

Cann, J.R., 1970a. Rb, Sr, Y, Zr and Nb in some ocean floor basaltic rocks. *Earth Planet. Sci. Lett.*, 10: 7—11.

Cann, J.R., 1970b. New model for the structure of the ocean crust. *Nature*, 226: 928—930.

Cann, J.R., 1971a. Major element variations in ocean floor basalts. *Philos. Trans. R. Soc. London, Ser. A*, 268: 495—505.

Cann, J.R., 1971b. Petrology of basement rocks from Palmer Ridge, northeast Atlantic. *Philos. Trans. R. Soc. London, Ser. A*, 268: 605—617.

Cann, J.R., 1974. A model for oceanic crustal structure developed. *Geophys. J. R. Astron. Soc.*, 39: 169—187.

Cann, J.R., 1979. Metamorphism in the ocean floor. In: M. Talwani, C.G.A. Harrison and D.E. Hayes (Editors), *Deep Sea Drilling Results in the Atlantic Ocean: Ocean Crust. Am. Geophys. Union, Maurice Ewing Ser.*, 2: 230—238.

Cann, J.R. and Funnell, B.M., 1967. Palmer Ridge: a section through the upper part of the ocean crust? *Nature*, 213: 661—664.

Cann, J.R. and Vine, F.J., 1966. An area on the crest of the Carlsberg Ridge: petrology and magnetic survey. *Philos. Trans. R. Soc. London, Ser. A*, 259: 198—217.

Cann, J.R., Winter, C.K. and Pritchard, R.G., 1977. A hydrothermal deposit from the floor of the Gulf of Aden. *Mineral. Mag.*, 41: 193—199.

Carmichaël, J.S.E., 1964. The petrology of Thingmuli, a Tertiary volcano in eastern Iceland. *J. Petrol.*, 5: 435—460.

**Chaigneau, M., Hékinian, R. and Chéminée, J.L., 1980. Magmatic gases extracted and** analysed from ocean floor volcanics. *Bull. Volcanol.*, 43: 241—253.

Chase, C.G., Menard, H.W., Larson, R.L., Sharmon, G.F., III and Smith, S.M., 1970. History of sea-floor spreading west of Baja California. *Geol. Soc. Am. Bull.*, 81: 491—498.

Chase, R.L. and Hersey, J.B., 1968. Geology of north slope of the Puerto Rico Trench. *Deep-Sea Res.*, 15: 297—317.

Chase, T.E., Menard, H.W. and Mammerickx, J., 1970. *Bathymetry of the North Pacific*. Scripps Institution of Oceanography, and Institute of Marine Resources, University of California (chart).

Chernysheva, V.I. and Bezrukov, P.L., 1966. Serpentinite from the crest of the Indo-Arabian ridge. *Dokl. Akad. Nauk S.S.S.R., Earth Sci. Sect.*, 166: 207—210.

Chernysheva, V.I., Dmitriev, L.V. and Udintsev, G.B., 1975. Geological-petrographic description of bedrock in rift zones of the world oceans. In: A.P. Vinogradov and G.B. Udintsev (Editors), *Rift Zones of the World Oceans*. Wiley, New York, N.Y., pp. 87—119 (translated from Russian by N. Kanev).

Choukroune, P., Francheteau, J. and Le Pichon, X., 1978. In situ structural observations along Transform Fault "A" in the FAMOUS area, Mid-Atlantic Ridge. *Geol. Soc. Am. Bull.*, 89: 1013—1029.

Christensen, N.I. and Salisbury, M.H., 1975. Structure and constitution of the lower oceanic crust. *Rev. Geophys. Space Phys.*, 13: 57—86.

Cifelli, R., 1966. Late Tertiary planktonic foraminifera associated with a basaltic boulder from the Mid-Atlantic Ridge. *J. Mar. Res.*, 23: 73—87.

**Clague, D.A., 1970. Petrology of basaltic and gabbroic rocks dredged from the Danger Island through Manihiki plateau. In:** *Initial Reports of the Deep Sea Drilling Project, 33*. U.S. Government Printing Office, Washington, D.C., pp. 891—911.

Clague, D.A. and Bunch, T.E., 1976. Formation of ferrobasalt at East Pacific Mid-Ocean spreading centers. *J. Geophys. Res.*, 81: 4247—4256.

Clark, S.P., 1966. *Handbook of Physical Constants. Geol. Soc. Am., Mem.*, 97: 75—96.

Clark, W.B., Beg, M.A. and Craig, H., 1969. Excess $^3$He in the sea: evidence for terrestrial primordial helium. *Earth Planet. Sci. Lett.*, 6: 213—220.

**Cohen, L.H., Ito, K. and Kennedy, G.C., 1967. Melting and phase relations in an anhydrous** basalt to 40 kilobars. *Am. J. Sci.*, 265: 475—518.

Coleman, R.G., 1977. *Ophiolites.* Springer-Verlag, Berlin, 229 pp.

Connary, S.A., 1972. *Investigation of the Walvis Ridge and environs.* Ph.D. Thesis, Columbia University, Palisades, N.Y., 228 pp.

Coombs, D.S., 1961. Some recent work on the lower grades of metamorphism. *Aust. J. Sci.,* 24: 203—215.

Coombs, D.S., Ellis, A.E., Fyfe, W.S. and Taylor, A.M., 1960. Lower grade mineral facies in New Zealand. *21st Int. Geol. Congr., Copenhagen, Rep.,* 13: 339—351.

Corliss, J.B. and Ballard, R.D., 1977. Oasis of life in cold abyss. *Natl. Geogr.,* 152: 441—453.

Corliss, J.B., Lyle, M., Dymond, J. and Crane, K., 1978. The chemistry of hydrothermal mounds near the Galapagos Rift. *Earth Planet. Sci. Lett.,* 40: 12—24.

Cornen, G. and Maury, R.C. 1980. Petrology of the volcanic island of Annobon, Gulf of Guinea. *Mar. Geol.,* 36: 253—267.

Craig, H. and Lupton, J.E., 1981. Helium-3 and mantle volatiles in the ocean and the oceanic crust. In: C. Emiliani (Editor), *The Sea,* 7. Wiley, New York, N.Y.

Craig, H., Clark, W.B. and Beg, M.A., 1975. Excess $^3$He in deep water on the East Pacific Rise. *Earth Planet. Sci. Lett.,* 26: 125—132.

Crane, K. and Ballard, R.D., 1980. The Galapagos Rift at 86°W, 4. Structure and morphology of hydrothermal fields and their relationship to the volcanic and tectonic processes of the rift valley. *J. Geophys. Res.,* 85: 1443—1454.

Cronan, D.S., 1975. Manganese nodules and other ferromanganese oxide deposits from the Atlantic Ocean. *Geophys. Res.,* 80: 3831—3837.

Cyamex, 1978. First submersible study of the East Pacific Rise: RITA (Rivera-Tamayo) Project, 21°N. *EOS, Trans. Am. Geophys. Union,* 52: 1198.

Cyamex, 1981. First manned submersible dives on the East Pacific Rise, 21°N: general results. *Mar. Geophys. Res.,* 4: 345—379.


Dalrymple, B.G. and Cox, A., 1968. Paleomagnetism, potassium-argon ages and petrology of some volcanic rocks. *Nature,* 217: 323—326.

Davis, B.T.C. and Schairer, J.F., 1965. Melting relations in the join diopside-forsterite-pyrope at 40 kilobars and at one atmosphere. *Carnegie Inst. Washington Yearb.,* 64: 123—126.

Deer, W.A., Howie, R.A. and Zussman, J., 1963. *Rock-Forming Minerals, 2. Chain Silicates.* Longmans, Green and Co., London, 379 pp.

Detrick, R.S. and Klitgord, K.D., 1973. Sound velocity correction table and revised Matthews' tables, Southtow G. *Scripps Inst. Oceanogr., Mar. Phys. Lab. Tech. Memo,* 240: 7 pp.

Detrick, R.S., Mudie, J.D., Lyendyke, B.P. and Macdonald, R.C., 1973. Near bottom observations of an active transform fault (Mid-Atlantic Ridge at 37°N). *Nature,* 246: 59—61.

Dickinson, W.R. and Hatherton, T., 1967. Andesite volcanism and seismicity around the Pacific. *Science,* 157: 801—803.

Dietrich V., Emmerman, R., Oberhansli, R. and Puchelt, M., 1978. Geochemistry of basaltic and gabbroic rocks from West Mariana Basin and the Mariana Trench. *Earth Planet. Sci. Lett.,* 39: 127—144.

Dietz, R.S. and Holden, J.C., 1970. Reconstruction of Pangea; break up and dispersion of continents, Permian to present. *J. Geophys. Res.,* 75: 4939—4956.

Dingle, R.V. and Simpson, E.S.W., 1976. The Walvis Ridge: a review. In: C.L. Drake (Editor), *Geodynamics: Progress and Prospects.* American Geophysical Union, Washington, D.C., pp. 160—176.

Dmitriev, L.V., 1974. Petrochemical study of the basaltic basement of the Mid-Indian Ridge. In: *Initial Reports of the Deep-Sea Drilling Project,* 24. U.S. Government Printing Office, Washington, D.C., pp. 767—779.

Dmitriev, L.V. and Sharaskin, A.Y., 1975. Petrology and petrochemistry of bedrock from the Arabian-Indian Ridge. In: A.P. Vinogradov and G.B. Udintsev (Editors), *Rift Zones of the World Oceans.* Wiley, New York, N.Y., pp. 393—430 (translated from Russian by N. Kanev).

Donaldson, C.H., Brown, R.W. and Reid, A.M., 1976. Petrology and chemistry of basalts from the Nazca plate, 1. Petrology and mineral chemistry. In: *Initial Reports of the Deep Sea Drilling Project, 34.* U.S. Government Printing Office, Washington, D.C., pp. 227—238.

Dongo, G., 1950. Eclogites and glaucophane amphibolites in Venezuela. *Trans. Am. Geophys. Union*, 31: 873—878.

Donnelly, T.W., 1966. Geology of St. Thomas and St. John, U.S. Virgin Islands. *Geol. Soc. Am., Mem.*, 98: 85—176.

Donnelly, T.W., Thompson, G. and Robinson, P.T., 1979. Very-low-temperature hydrothermal alteration of the oceanic crust and the problem of fluxes of potassium and magnesium. In: M. Talwani, C.G.A. Harrison and D.E. Hayes (Editors), *Deep Sea Drilling Results in the Atlantic Ocean: Ocean Crust. Am. Geophys. Union, Maurice Ewing Ser.*, 2: 369—382.

Douaran, S., 1979. *Caracteristique structurale et géophysique de la dorsale médio-Atlantique de 10° à 50° Nord.* Thèse de 3° cycle, Université de Nancy, Nancy.

Duke, N.A. and Hutchinson, R.W., 1974. Geological relationship between massive sulfide bodies and ophiolitic volcanic rocks near York Harbour, Newfoundland. *Can. J. Earth Sci.*, 11: 53—69.

Dymond, J. and Deffeyes, K., 1968. K-Ar ages of deep-sea rocks and their relations to sea-floor spreading. *Trans. Am. Geophys. Union*, 49: 364 (abstract).

Dymond, J., Corliss, J.B., Cobler, R., Muratli, C.M., Chou, C. and Conard, R., 1980. Composition and origin of sediments recovered by deep drilling of sediment mounds, Galapagos spreading center. In: *Initial Reports of the Deep Sea Drilling Project, 54.* U.S. Government Printing Office, Washington, D.C., pp. 377—385.

Einarsson, T., 1967. The extent of the Tertiary basalt formation and the structure of Iceland. In: S. Bjornsson (Editor), *Iceland and Mid-Ocean Ridges. Soc. Sci. Islandica, Rit.*, 38: 170—179.

Emslie, R.F., 1971. Liquidus relations and subsolidus reactions in some plagioclase-bearing systems. *Carnegie Inst. Washington Yearb.*, 69: 148—155.

Engel, A.E.J. and Engel, C.G., 1964. Igneous rocks of the East Pacific Rise. *Science*, 146: 477—485.

Engel, C.G. and Chase, T.E., 1965. Composition of basalts from seamounts off the west coast of Central America. *U.S. Geol. Surv. Prof. Paper*, 525-C: 161—163.

Engel, C.G. and Engel, A.E.J., 1963. Basalts dredged from the northeastern Pacific Ocean. *Science*, 140: 1321—1324.

**Engel, C.G. and Engel, A.E.J., 1966. Volcanic rocks dredged southwest of the Hawaiian Islands.** *U.S. Geol. Surv. Prof. Paper*, 550-D: 104—108.

Engel, C.G. and Fisher, R.L., 1969. Lherzolite, anorthosite, gabbro and basalt dredged from the Mid-Indian Ocean Ridge. *Science*, 166: 1136—1141.

Engel, C.G. and Fisher, R.L., 1975. Granitic to ultramafic rock complexes of the Indian Ocean Ridge system, western Indian Ocean. *Geol. Soc. Am. Bull.*, 86: 1553—1578.

Engel, C.G., Fisher, R.L. and Engel, A.E.J., 1965. Igneous rocks of the Indian Ocean. *Science*, 150: 605—610.

Engel, C.G., Bingham, E. and Fisher, R.L., 1974. Trace element composition of Leg 24 basalts and one diabase. In: *Initial Reports of the Deep Sea Drilling Project, 24.* U.S. Government Printing Office, Washington, D.C., pp. 781—786.

Erlank, A.J. and Reid, D.L., 1974. Geochemistry, mineralogy and petrology of basalts. In: *Initial Reports of the Deep Sea Drilling Project, 25.* U.S. Government Printing Office, Washington, D.C., pp. 543—551.

Eskola, P., 1920. The mineral facies of rocks. *Nor. Geol. Tidsskr.*, 6: 143—194.

Eskola, P., 1921. On the eclogites of Norway. *Kristiana Vidensk. Skr. I, Math.-Natur-vitensk. Kl.*, 8: 118 pp.

Eskola, P., 1939. Die metamorphen Gesteine. In: T.F.W. Barth, C.W. Correns and P. Eskola, *Die Entstehung der Gesteine*. Springer-Verlag, Berlin, pp. 263—407.

Evans, B.W., 1965. Application of reaction rate method to the breakdown equilibria of muscovite and muscovite plus quartz. *Am. J. Sci.*, 263: 647—667.

Fermor, L.L., 1913. Preliminary note on garnet as a geological barometer and on an infra-plutonic zone in earth's crust. *Rec. Geol. Surv. India*, 43: 41—47.

Fisher, D.E., 1978. Terrestrial uranium, heat flow and cosmochemistry. In: R.E. Zartman (Editor), *Short Papers of the Fourth International Conference, Geochronology, Cosmo-chronology, and Isotope Geology. U.S. Geol. Surv. Prof. Paper*, 708-711: 109—111.

Fisher, R.L., Engel, C.G. and Hilde, T.W.C., 1968. Basalts dredged from the Amirante shore flank of the Tonga Trench. *Geol. Soc. Am. Bull.*, 80: 1373—1378.

Fisher, R.L., Engel, C.G. and Hilde, T.W.C., 1968. Basalts dredged from the Mirante Ridge, western Indian Ocean. *Deep-Sea Res.*, 15: 521—534.

Fleet, A.J., Henderson, P. and Kempe, D.R.C., 1976. Rare earth element and related chemistry of some drilled southern Indian Ocean basalts and volcanogenic sediments. *J. Geophys. Res.*, 81: 4257—4268.

Fleisher, U., 1971. Gravity surveys over the Reykjanes Ridge and between Iceland and the Faeroe Islands. *Mar. Geophys. Res.*, 1: 314—327.

Fleming, H.S., Cherkis, N.Z. and Heirtzler, J.R., 1970. The Gibbs Fracture Zone: double fracture at 52.30′N in the Atlantic Ocean. *Mar. Geophys. Res.*, 1: 37—45.

Fodor, R.V., 1977. Petrology of basalt recovered during DSDP Leg 39B. In: *Initial Reports of the Deep Sea Drilling Project, 39*. U.S. Government Printing Office, Washington, D.C., pp. 513—523.

Fodor, R.V. and Hékinian, R., 1981. Petrology of basaltic rocks from the Ceara and the Sierra Leone aseismic rises in the equatorial Atlantic Ocean. *Oceanol. Acta*, 4: 223—228.

Fodor, R.V. and Thiede, J., 1977. Volcanic breccia from DSDP Site 357: implications for the composition and origin of the Rio Grande Rise. In: *Initial Reports of the Deep Sea Drilling Project, 39*. U.S. Government Printing Office, Washington, D.C., pp. 537—543.

Fodor, R.V., Husler, J.W. and Kumar, N., 1977. Petrology of volcanic rocks from an aseismic rise: implications for the origin of the Rio Grande Rise, South Atlantic Ocean. *Earth Planet. Sci. Lett.*, 35: 225—233.

Forbes, R.B. and Hoskin, C.M., 1969. Dredged trachyte and basalt from Kodiak seamount and the adjacent Aleutian Trench, Alaska. *Science*, 166: 502—504.

Forbes, R.B., Dugdale, R.C., Katsura, R.C., Matsumoto, T. and Haramura, H., 1969. Dredged basalts from Giacomini Seamount. *Nature*, 221: 849—850.

Ford, A.B., 1975. Antarctic deep-sea basalt, Southeast Indian Ocean and Balleny Basin, DSDP Leg 28. In: *Initial Reports of the Deep Sea Drilling Project, 28*. U.S. Government Printing Office, Washington, D.C., pp. 835—850.

Forsyth, D.W., 1977. The evolution of the upper mantle beneath mid-ocean ridges. *Tectonophysics*, 38: 89—118.

Fowler, C.M.R. and Matthews, D.H., 1974. Seismic refraction experiments on the Mid-Atlantic Ridge in the FAMOUS area. *Nature*, 249: 752—754.

Fox, P.J., Lowrie, A. and Heezen, B., 1969. Oceanographer fracture zone. *Deep-Sea Res.*, 16: 53—66.

Fox, P.J., Schreider, E. and Peterson, J.J., 1973. The geology of the oceanic crust: com-pressional wave velocities of oceanic rocks. *J. Geophys. Res.*, 78: 5155—5172.

Fox, P.J., Schreider, E., Rowlett, H. and McCarry, K., 1976. The geology of the Oceanog-rapher fracture zone: a model for fracture zones. *J. Geophys. Res.*, 81: 4117—4128.

Francheteau, J. and Le Pichon, X., 1972. Marginal fracture zones as structural framework of continental margins in South Atlantic Ocean. *Bull. Am. Assoc. Pet. Geol.*, 56: 991—1007.

Francheteau, J., Choukroune, P., Hékinian, R., Le Pichon, X. and Needham, H.D., 1976. Oceanic fracture zones do not provide deep sections in the crust. *Can. J. Earth Sci.*, 13: 1223—1235.

Francheteau, J., Needham, H.D., Choukroune, P., Juteau, T., Seguret, M., Ballard, R.D., Fox, J.P., Normark, W., Carranza, A., Cordoba, D., Guerrero, J., Bougault, H., Cambon, P. and Hékinian, R., 1979. Massive deep-sea sulphide ore deposits discovered on the East Pacific Rise. *Nature*, 277: 523—528.

Francis, T.J.G. and Raitt, R.W., 1967. Seismic refraction measurements in the southern Indian Ocean. *J. Geophys. Res.*, 72: 3015—3041.

Francis, T.J.G. and Shor, G.G., 1966. Seismic refraction measurements in the northwest Indian Ocean. *J. Geophys. Res.*, 71: 427—449.

Francis, T.J.G., Porter, I.T. and McGrath, J.R., 1977. Ocean bottom seismograph observations on the Mid-Atlantic Ridge near lat. 37°N. *Geol. Soc. Am. Bull.*, 88: 664—677.

Franckel, J.J., 1968. Forms and structure of intrusive basaltic rocks. In: H.H. Hess and A. Poldervaart (Editors), *Poldervaart Treatise on Rocks of Basaltic Composition, 2*. Interscience, New York, N.Y., pp. 63—100.

Frey, F.A., Bryan, W.B. and Thompson, G., 1974. Atlantic Ocean floor: geochemistry and petrology of basalts from Legs 2 and 3 of the Deep Sea Drilling Project. *J. Geophys. Res.*, 79: 5507.

Frey, F.A., Dickey, J.R., Thompson, G. and Bryan, W.B., 1977. Eastern Indian Ocean DSDP sites: correlations between petrography, geochemistry and tectonic setting. In: J.R. Heirtzler, H.M. Bolli, T.A. Davies, J.B. Saunders and J.G. Sclater (Editors), *Indian Ocean Geology and Biostratigraphy. Am. Geophys. Union, Geophys. Monogr.*, pp. 189—257.

Friedman, I.I. and Smith, R.L., 1960. A new dating method using obsidian, 1. The development of the method. *Am. Antiquity*, 25: 476—522.

Fryer, B.J. and Hutchinson, R.W., 1976. Generation of metal deposits on the sea floor. *Can. J. Earth. Sci.*, 13: 126—135.

Furnes, H., 1975. Experimental palagonitization of basaltic glasses of varied compositions. *Contrib. Mineral. Petrol.*, 50: 105.

Furumoto, A.S., Woollard, G.P., Campbell, J.F. and Hussong, D.M., 1968. Variation in the thickness of the crust in the Hawaiian archipelago. In: L. Knopoff, C.L. Drake and P.J. Hart (Editors), *The Crust and Upper Mantle of the Pacific Area. Am. Geophys. Union, Geophys. Monogr.*, 12: 94—111.

Gass, J.G., Mallick, D.I.J. and Cox, K.C., 1973. Volcanic islands of the Red Sea. *J. Geol. Soc. London*, 129: 275—310.

Gast, P., 1965. Terrestrial ratio of potassium to rubidium and the composition of earth's mantle. *Science*, 147: 858—860.

Gavasci, A.T., Fox, P.J. and Ryan, W.B.F., 1970. Petrology of rocks from the crestal area of the Gorringe Bank. In: *Initial Reports of the Deep Sea Drilling Project, 13*. U.S. Government Printing Office, Washington, D.C., pp. 749—752.

Girod, M., 1972. A propos des andesites des Açores. *Contrib. Mineral. Petrol.*, 35: 159—167.

Girod, M., Camus, G. and Valette, Y., 1971. Diversity of tholeiitic rocks at St. Paul Island (Indian Ocean). *Contrib. Mineral. Petrol.*, 33: 108—117.

Gladkikh, V.S. and Chernysheva, V.I., 1966. Rare elements in suboceanic mafic extrusives. *Geochem. Int.*, 3: 786—789.

Glennie, K.W., Boeuf, M.G.A., Hughes-Clark, M.W., Moody-Stuart, M., Pilaar, W.F.H. and Reinhardt, B.M., 1974. Geology of the Oman Mountains, 1—3. *Verh. K. Ned. Geol. Mijnbouwkd. Genoot., Geol. Ser.*, 31: 423 pp.

Gorini, M., 1976. *The tectonic fabric of the equatorial Atlantic adjoining continental margins*. Ph.D. Thesis, Columbia University, New York, N.Y.

Goslin, J. and Sibuet, J.C., 1975. Geophysical study of the easternmost Walvis Ridge, South Atlantic: deep structure. *Geol. Soc. Am. Bull.*, 86: 1713—1724.

most Walvis Ridge, South Atlantic: morpho... *Bull.*, 85: 619—632.

Green, D.H., 1963. Alumina content of enstatite in a V... dotite. *Geol. Soc. Am. Bull.*, 74: 1397—1402.

Green, D.H. and Ringwood, A.E., 1963. Mineral assemblages in a model mantle composition. *J. Geophys. Res.*, 68: 937—945.

Green, D.H. and Ringwood, A.E., 1967. The genesis of basaltic magmas. *Contrib. Mineral. Petrol.*, 15: 103—190.

Green, D.H. and Ringwood, A.E., 1970. Mineralogy of peridotite composition under upper-mantle conditions. *Phys. Earth Planet. Inter.*, 3: 359—371.

Green, T.H., 1969. High pressure experimental studies on the origin of anorthosite. *Can. J. Earth Sci.*, 427—440.

Greenbaum, D., 1972. Magmatic processes of ocean ridges evidence from the Troodos Massif, Cyprus. *Nature, Phys. Sci.*, 238: 18—21.

Greenewalt, D. and Taylor, P.T., 1974. Deep-Tow magnetic measurements across the axial valley of the Mid-Atlantic Ridge. *J. Geophys. Res.*, 79: 4401—4405.

Gutenberg, B. and Richter, C.F., 1954. *Seismicity of the Earth*. Princeton University Press, Princeton, N.J., 2nd ed., 273 pp.

Harker, R.I., 1932. *Metamorphism — A Study of the Transformations of Rock-Masses*. Methuen, London, 362 pp.

Harris, P.G., 1957. Zone refining and the origin of potassic basalts. *Geochim. Cosmochim. Acta*, 12: 195—208.

Hart, R., 1973. A model for chemical exchange in the basalt-seawater system of oceanic layer II. *Can. J. Earth Sci.*, 10: 799—816.

Hart, R., 1976. Chemical variance in deep ocean basalts. In: *Initial Reports of the Deep Sea Drilling Project, 34*. U.S. Government Printing Office, Washington, D.C., pp. 301—335.

Hart, S.R., 1971. K, Rb, Cs and Ba contents and Sr isotope ratios of ocean floor basalts. *Philos. Trans. R. Soc. London, Ser. A*, 268: 573—587.

Hart, S.R. and Staudigel, H., 1978. Ocean crust—seawater interaction. *EOS, Trans. Am. Geophys. Union*, 54: 409 (abstract).

Heath, G.R. and Van Andel, T.H., 1971. Tectonic and sedimentation in the Panama Basin: geological results of Leg 16, Deep Sea Drilling Project. In: *Initial Reports of the Deep Sea Drilling Project, 26*. U.S. Government Printing Office, Washington, D.C., pp. 899—913.

Hedge, C.E. and Peterman, Z.E., 1970. The strontium isotopic composition basalts from the Gorda and Juan de Fuca Rises, northeastern Pacific Ocean. *Contrib. Mineral. Petrol.*, 27: 114—120.

Hedge, C.E., Watkins, N.D., Hildreth, R.A. and Doering, W.P., 1973. $^{87}Sr/^{86}Sr$ ratios in basalts from islands in the Indian Ocean. *Earth Planet. Sci. Lett.*, 21: 29—34.

Heezen, B.C. and Rawson, M., 1977. Visual observations of the sea floor subduction line in the Middle America Trench. *Science*, 196: 423—426.

Heezen, B.C. and Tharp, M., 1966. Physiography of the Indian Ocean. *Philos. Trans. R. Soc. London, Ser. A*, 259: 137—149.

Heezen, B.C., Bunce, E.T., Hersey, J.B. and Tharp, M., 1964. Chain and Romanche fracture zone. *Deep-Sea Res.*, 11: 11—33.

Hékinian, R., 1968. Rocks from the Mid-Oceanic Ridge in the Indian Ocean. *Deep-Sea Res.*, 15: 195—213.

Hékinian, R., 1970. Gabbro and pyroxenite from a deep-sea core in the Indian Ocean. *Mar. Geol.*, 9: 287—294.

Hékinian, R., 1971a. Chemical and mineralogical differences between abyssal hill basalts and ridge tholeiites in the eastern Pacific Ocean. *Mar. Geol.*, 11: 77—91.

Hékinian, R., 1971b. Petrological and geochemical study of spilites and associated rocks from St. John, Virgin Islands. *Geol. Soc. Am. Bull.*, 82: 659—682.

Hékinian, R., 1972. Volcanics from the Walvis Ridge in the southeast Atlantic Ocean. *Nature*, 239: 91—93.

Hékinian, R., 1974a. Petrology of igneous rocks from Leg 22 in the northeastern Indian Ocean. In: *Initial Reports of the Deep Sea Drilling Project, 22.* U.S. Government Printing Office, Washington, D.C., pp. 413—447.

Hékinian, R., 1974b. Petrology of the Ninety-East Ridge (Indian Ocean) compared to other aseismic ridges. *Contrib. Mineral. Petrol.*, 43: 125—147.

Hékinian, R. and Aumento, F., 1973. Rocks from the Gibbs fracture zone and the Minia Seamount near 53°N in the Atlantic Ocean. *Mar. Geol.*, 14: 47—72.

Hékinian, R. and Fevrier, M., 1979. Comparison between deep-sea hydrothermal deposits recovered from recent spreading ridges. In: *La Génèse des Nodules de Manganese, Coll. Int. C.N.R.S., Gif-sur-Yvette, 25—30 September, 1978*, pp. 167—178.

Hékinian, R. and Hoffert, M., 1975. Rate of palagonitization and manganese coating on basaltic rocks from the rift valley in the Atlantic Ocean near 36°50'N. *Mar. Geol.*, 19: 91—109.

Hékinian, R. and Thompson, G., 1976. Comparative geochemistry of volcanics from rift valleys, transform faults and aseismic ridges. *Contrib. Mineral. Petrol.*, 57: 145—162.

Hékinian, R., Bougault, H. and Pautot, G., 1973. Atlantique Nord: étude préliminaire des roches de la fracture Gibbs (53° Nord) et de la zone de fracture Açores-Gibraltar. *C.R. Acad. Sci. Paris, Ser. D*, 276: 3281—3284.

Hékinian, R., Moore, J.G. and Bryan, W.B., 1976. Volcanic processes of the Mid-Atlantic Ridge Rift Valley near 36°49'N. *Contrib. Mineral. Petrol.*, 5: 83—110.

Hékinian, R., Rosendahl, B.R., Cronan, D.S., Dmitriev, Y., Fodor, R.V., Goll, R.M., Hoffert, M., Humphris, S.E., Mattey, D.P., Natland, J., Petersen, N., Roggenthen, W., Schrader, E.L., Srivastava, R.K. and Warren, M., 1978. Hydrothermal deposits and associated basement rocks from the Galapagos spreading center. *Oceanol. Acta*, 1: 473—481.

Hékinian, R., Fevrier, M., Bischoff, J.L., Picot, P. and Shanks, W.C., 1980. Sulfide deposits from the East Pacific Rise near 21°N. *Science*, 207: 1433—1444.

Herron, E.M., 1972. Sea-floor spreading and the Cenozoic history of the east central Pacific. *Geol. Soc. Am. Bull.*, 83: 1671—1692.

Hey, R.N., 1977. Tectonic evolution of the Cocos-Nazca spreading center. *Geol. Soc. Am. Bull.*, 88: 1404—1420.

Hill, M.N., 1957. Recent geophysical exploration of the ocean floor. *Phys. Chem. Earth*, 2: 129—163.

Hodges, F.N. and Papike, J.J., 1977. Petrology of basalts, gabbros and peridotites from DSDP Leg 37. In: *Initial Reports of the Deep Sea Drilling Project, 37.* U.S. Government Printing Office, Washington, D.C., pp. 711—719.

Hoffert, M., Perseil, A., Hékinian, R., Choukroune, P., Needham, H.D., Francheteau, J. and Le Pichon, X., 1978. Hydrothermal deposits sampled by diving saucer in Transform Fault "A" near 37°N on the Mid-Atlantic Ridge, FAMOUS area. *Oceanol. Acta*, 1: 73—86.

Holcombe, T.L., Vogt, P.R., Matthews, J.E. and Murchison, R.R., 1973. Evidence for sea floor spreading in the Ceyman Trough. *Earth Planet. Sci. Lett.*, 20: 357—371.

Holden, J.C. and Dietz, R.S., 1972. Galapagos gore, NazCoPac triple junction and Carnegie/Cocos Ridge. *Nature*, 235: 266—269.

Honnorez, J., 1969. Sur l'origine artificielle de la coloration du verre basaltique altérée des hyaloclastites de Palagonia (Sicile). *Schweiz. Mineral. Petrogr. Mitt.*, 49: 65—76.

Honnorez, J., 1972. *La Palagonitization: l'Altération sous-marine du Verre volcanique basique de Popogonia (Sicile). Vulkaninstitut I. Friedlander, Publ., 9.* Birkhausen, Basel, 132 pp.

374

Honnorez, J. and Kirst, P., 1975. Petrology of rodingites from the equatorial Mid-Atlantic fracture zones and their geotectonic significance. *Contrib. Mineral. Petrol.*, 49: 233—257.
Hutchinson, R.W. and Searle, D.L., 1971. Stratabound pyrite deposits in Cyprus and relations to other sulphide ores. *Soc. Min. Geol. Jpn. Spec. Issue*, 3: 198—205.

Iddings, J.P., 1892. The origin of igneous rocks. *Bull. Philos. Soc. Washington*, 12: 89—213.
Irwin, W.P. and Coleman, R.G., 1974. North polar projection showing the principal ophiolite belts of the world. In: R.G. Coleman (Editor), *Ophiolites*. Springer-Verlag, Berlin, 229 pp.
Isacks, B. and Molnar, P., 1971. Distribution of stresses in descending lithosphere from global survey of focal-mechanism solutions of mantle earthquakes. *Rev. Geophys. Space Phys.*, 9: 103—174.
Ito, K. and Kennedy, G.C., 1967. Melting and phase relations in natural peridotite to 40 kilobars. *Am. J. Sci.*, 265: 519.

Jehl, V., 1975. *Metamorphisme et les fluides associés des roches océaniques de l'Atlantique Nord*. Thèse de Docteur Ingenieur, Université de Nancy, Nancy, 242 pp.
Johannes, W., 1968. Experimental investigation of the reaction forsterite + $H_2O$ ⇌ serpentine + brucite. *Contrib. Mineral. Petrol.*, 19: 309—315.
Johansen, A., 1939. *A Descriptive Petrography of the Igneous Rocks, 1*. University of Chicago Press, Chicago, Ill., 318 pp.
Johnson, G.L. and Vogt, P.R., 1973. The Mid-Atlantic Ridge from 47° to 51°N. *Geol. Soc. Am. Bull.*, 84: 3443—3462.
Johnson, J.R., 1979. Transitional basalts and tholeiites from the East Pacific Rise, 9°N. *J. Geophys. Res.*, 84: 1635—1652.
Jolly, W., 1970. Zeolite and prehnite-pumpellyite facies in south central Puerto Rico. *Contrib. Mineral. Petrol.*, 27: 204—224.
Joplin, G.A., 1968. The shoshonitic association: a review. *J. Geol. Soc. Aust.*, 15: 275—294.
Joron, J.L., Brigueu, L., Bougault, H. and Treuil, M., 1980. East Pacific Rise, Galapagos spreading center and Siqueiros fracture zone hydromagmatophile elements: a comparison with the North Atlantic. In: *Initial Reports of the Deep Sea Drilling Project, 54*. U.S. Government Printing Office, Washington, D.C., pp. 725—735.
Juteau, T., Eissen, J.P., Francheteau, J., Needham, H.D., Choukroune, P., Rangin, C., Seguret, M., Ballard, R.D., Fox, P.J., Normark, W.R., Carranza, A., Cordoba, D. and Guerrero, J., 1980. Homogeneous basalts from the East Pacific Rise at 21°N: steady state magma reservoirs at moderately fast spreading centers. *Oceanol. Acta*, 9: 487—503.

Kanamori, H., 1963. Study on the crustal-mantle structure in Japan, 2. *Tokyo Univ. Earthq. Res. Inst. Bull.*, 41: 761—779.
Kashintsev, G.L. and Rudnik, B., 1974. New data on basalts of the East Indian Ocean Ridge. *Int. Geol. Rev.*, 18: 1165—1172.
Kauzel, E.G., 1972. *Regionalisation of the lithosphere and asthenosphere of the Pacific Ocean*. Ph.D. Thesis, Columbia University, New York, N.Y., 147 pp.
Kay, R., Hubbard, N.J. and Gast, P.W., 1970. Chemical characteristic and origin of oceanic ridge volcanic rocks. *J. Geophys. Res.*, 75: 1585—1613.
Kempe, D.R.C., 1974. The petrology of the basalts from Leg 26. In: *Initial Reports of the Deep Sea Drilling Project, 26*. U.S. Government Printing Office, Washington, D.C., pp. 465—501.

Kennedy, W.Q., 1933. Trends of differentiation in basaltic magmas. *Am. J. Sci.*, 25: 239—256.

Kharin, G.N., 1974. The petrology of magmatic rocks, DSDP Leg 38. In: *Initial Reports of the Deep Sea Drilling Project, 38*. U.S. Government Printing Office, Washington, D.C., pp. 685—715.

Klitgord, K.D. and Mudie, J.D., 1974. The Galapagos spreading centre: a near bottom geophysical survey. *Geophys. J. R. Astron. Soc.*, 38: 563—586.

Komar, P.D., 1972. Mechanical interactions of phenocrysts and flow differentiation of igneous dikes and sills. *Geol. Soc. Am. Bull.*, 83: 973—988.

Kumar, N. and Embley, R.W., 1977. Evolution and origin of Ceara Rise: an aseismic rise in the western equatorial Atlantic. *Geol. Soc. Am. Bull.*, 88: 683—694.

Kuno, H., 1959. Origin of cenozoic provinces of Japan and surrounding areas. *Bull. Volcanol., Ser. 2*, 20: 37—76.

Kuno, H., 1960. High-alumina basalt. *J. Petrol.*, 1: 121—145.

Kuno, H., 1966. Lateral variation of basalt magma across continental margins and island arcs. *Bull. Volcanol.*, 29: 195—222.

Kuno, H., Yamasaki, K., Iida, C. and Nagashima, K., 1957. Differentiation of Hawaiian magma. *Jpn. J. Geol. Geogr.*, 28: 179—218.

Kurasawa, H., 1959. Petrology and chemistry of the Amogi volcanic rocks, Izu Peninsula, Japan. *Chikyu-Kagaku*, 44: 1—18 (in Japanese with English abstract).

Kushiro, I., 1960. Si-Al relation in clinopyroxenes from igneous rocks. *Am. J. Sci.*, 258: 548—554.

Kushiro, I., 1970. Stability of amphibole and phlogopite in the upper mantle. *Carnegie Inst. Washington Yearb.*, 68: 245—247.

Kushiro, I., 1972. Partial melting of synthetic and natural peridotites at high pressure. *Carnegie Inst. Washington Yearb.*, 71: 357—362.

Kushiro, I. and Kuno, H., 1963. Origin of primary basalt magmas and classification of basaltic rocks. *J. Petrol.*, 4: 75—89.

Kushiro, I. and Thompson, R.N., 1972. Origin of some abyssal tholeiites from the Mid-Atlantic Ridge. *Carnegie Inst. Washington Yearb.*, 71: 403—406.

Kushiro, I. and Yoder, H.S., 1966. Anorthite-forsterite and anorthite-enstatite reactions and their bearing on the basalt-eclogite transformation. *J. Petrol.*, 7: 337—362.

Lacroix, A., 1917. Sur la transformation de quelques roches éruptives basiques en amphibolites. *C. R. Acad. Sci. Paris*, 164: 969—974.

Lambert, K. and Wyllie, P.J., 1970. Melting in the deep crust and upper mantle and the nature of the low-velocity layer. *Phys. Earth Planet. Inter.*, 3: 316—322.

Langmuir, L.H., Bender, J.F., Bence, A.E., Hanson, G.H. and Taylor, S.R., 1977. Petrogenesis of basalts from the FAMOUS area, Mid-Atlantic Ridge. *Earth Planet. Sci. Lett.*, 36: 133—156.

Lapido-Laureiro, F.E. de V., 1968. Sub-volcanic carbonatite structures of Angola. *23rd Int. Geol. Congr., Prague, Rep.*, 2: 147—161.

Larson, R.L., 1971. Near-bottom geologic studies of the East Pacific Rise crest. *Geol. Soc. Am. Bull.*, 82: 823—842.

Larson, R.L., Menard, H.W. and Smith, S.M., 1968. Gulf of California: a result of ocean-floor spreading and transform faulting. *Science*, 161: 781.

Laughton, A.S., 1966. The Gulf of Aden. *Philos. Trans. R. Soc. London, Ser. A*, 259:150 pp.

Laughton, A.S., Whitmarsh, R.B. and Jones, M.T., 1970. The evolution of the Gulf of Aden. *Philos. Trans. R. Soc. London, Ser. A*, 267: 227—266.

Laughton, A.S., Matthews, D.H. and Fisher, R.L., 1971. The structure of the Indian Ocean. In: A.E. Maxwell (Editor), *The Sea 4*. Wiley-Interscience, New York, N.Y., pp. 543—586.

LeBas, M.J., 1962. The role of aluminum in igneous clinopyroxenes with relation to their parentage. *Am. J. Sci.*, 260: 267—288.

~~~~~ spreading. *J. Geophys. Res.*, 73: 210.

Le Pichon, X. and Langseth, M.G., Jr., 1969. Heat flow from the mid-ocean ridges and sea-floor spreading. *Tectonophysics*, 8: 319—344.

Le Pichon, X., Houtz, R.E., Drake, C.L. and Nute, J.E., 1965. Crustal structure of the mid-ocean ridges. *J. Geophys. Res.*, 70: 319—339.

Le Pichon, X., Ewing, J.I. and Houtz, R.E., 1970. The Gibraltar end of the Azores-Gibraltar plate boundary: an example of compressive tectonics. In: *Tectonics, Upper Mantle Committee Symp., Flagstaff, Ariz., July* (abstract).

Liou, J.G., 1970. Synthesis and stability relations of wairakite, $CaAl_2Si_4O_{12} \cdot 2H_2O$. *Contrib. Mineral. Petrol.*, 27: 259—282.

Loncarevic, B.D., Mason, C.S. and Matthews, D.H., 1966. Mid-Atlantic Ridge near 45°N. 1. The Median Valley. *Can. J. Earth Sci.*, 3: 327—349.

Lonsdale, P., 1977. Deep-tow observations at the mounds abyssal hydrothermal field, Galapagos Rift. *Earth Planet. Sci. Lett.*, 36: 92—110.

Ludden, J.N., Thompson, G., Bryan, W.B. and Frey, F.A., 1980. The origin of lavas from the Ninety-East Ridge, eastern Indian Ocean: an evolution of fractional crystallization models. *J. Geophys. Res.*, 85: 4405—4420.

Luyendyk, B.P. and Engel, C.G., 1969. Petrological, magnetic and chemical properties of basalts dredged from an abyssal hill in the northeast Pacific. *Nature*, 223: 1049—1050.

Macdonald, G.A. and Katsura, T., 1964. Chemical composition of Hawaiian lavas. *J. Petrol.*, 5: 82—133.

Macdonald, K.C., 1977. Near-bottom magnetic anomalies, asymmetric spreading, oblique spreading and tectonics of the Mid-Atlantic Ridge near lat. 37°N. *Geol. Soc. Am. Bull.*, 88: 541—555.

McKenzie, D. and Sclater, J.G., 1969. Heat flow in the eastern Pacific and sea floor spreading. *Bull. volcanol.*, 33: 101—118.

Martin, H., Mathias, M. and Simpson, E.S., 1960. The Damaraland subvolcanic ring complexes in South West Africa. *21st. Int. Geol. Congr., Copenhagen, Rep.*, 13: 156—174.

Mas, M.J., 1962. The role of aluminum in igneous clinopyroxenes with relation to their parentage. *Am. J. Sci.*, 260: 267—288.

Matthews, D.H., 1971. Altered basalts from Swallow Bank; an abyssal hill in the N.E. Atlantic and from a nearby seamount. *Philos. Trans. R. Soc. London, Ser. A*, 268: 551—571.

Maxwell, A.E., Von Herzen, R.P., Hsü, K.J., Andrews, J.E., Saito, T., Percival, S.F., Milow, E.D., Jr. and Boyce, R.E., 1970. Deep sea drilling in the South Atlantic. *Science*, 168: 1047—1049.

Maxwell, J.C., 1970. The Mediterranean ophiolites and continental drift. In: H. Johnson and B.L. Smith (Editors), *Megatectonics of Continents and Oceans*. Rutgers University, New Brunswick, N.J., pp. 167—193.

Mazzullo, L.J., Bence, A.E. and Papike, J.J., 1976. Petrology and phase chemistry of basalts from DSDP Leg 34; Nazca plate. In: *Initial Reports of the Deep Sea Drilling Project, 34*. U.S. Government Printing Office, Washington, D.C., pp. 245—261.

Melson, W.G. and Thompson, G., 1971. Petrology of a transform fault and adjacent segments. *Philos. Trans. R. Soc. London, Ser. A*; 268: 423—441.

Melson, W.G. and Thompson, G., 1973. Glassy abyssal basalts, Atlantic sea floor near St. Paul's Rocks: petrography and composition of secondary clay minerals. *Geol. Soc. Am. Bull.*, 84: 703—716.

**Melson, W.G. and O'Hearn, T., 1979. Basaltic glass erupted along the Mid-Atlantic Ridge** between 0—37°N: relationship between composition and latitude. In: M. Talwani, C.G.A. Harrison and D.E. Hayes (Editors), *Deep Sea Drilling Results in the Atlantic Ocean: Ocean Crust. Am. Geophys. Union, Maurice Ewing Ser.*, 2: 249—261.

Melson, W.G. and Van Andel, T.H., 1966. Metamorphism in the Mid-Atlantic Ridge, 22° latitude. *Mar. Geol.*, 4: 165—186.

Melson, W.G., Bowen, V.T., Van Andel, T.H. and Siever, R., 1966. Greenstone from the Central Valley of the Mid-Atlantic Ridge. *Nature*, 209: 604—605.

Melson, W.G., Jorosewitch, E., Bowen, V.T. and Thompson, G., 1967. St. Peter and St. Paul Rocks: a high temperature, mantle-derived intrusion. *Science*, 155: 1532—1535.

Melson, W.G., Thompson, G. and Van Andel, T.H., 1968. Volcanism and metamorphism in the Mid-Atlantic Ridge, 22°N latitude. *J. Geophys. Res.*, 73: 5925—5941.

Melson, W.G., Vallier, T., Wright, T.L., Byerly, G. and Nelen, J.A., 1976. Chemical diversity of abyssal volcanic glass erupted along Pacific, Atlantic and Indian Ocean sea-floor spreading centers. *Am. Geophys. Union, Geophys. Monogr.*, 19: 354—368.

Melson, W.G., Byerly, G.R., Nelen, J.A., O'Hearn, T., Wright, T.L. and Vallier, T., 1977. A catalog of the major element chemistry of abyssal volcanic glasses. *Smithsonian Contrib. Earth Sci.*, 19: 31—60.

Menard H.W., 1964. *Marine Geology of the Pacific*. McGraw-Hill, New York, N.Y., 271 pp.

Menard, H.W. and Dietz, R.S., 1951. Submarine geology of the Gulf of Alaska. *Geol. Soc. Am. Bull.*, 62: 1263—1285.

Minster, J.B., Jordon, T.H., Molnar, P. and Haines, E., 1974. Numerical modelling of instantaneous plate tectonics. *Geophys. J. R. Astron. Soc.*, 36: 541—576.

Miyashiro, A., 1972. Metamorphism and related magmatism in plate tectonics. *Am. J. Sci.*, 272: 629—656.

Miyashiro, A., 1973. *Metamorphism and Metamorphic Belts*. George Allen and Unwin Ltd., London, pp. 293—309.

Miyashiro, A., 1974. Volcanic rock series in island arcs and active continental margins. *Am. J. Sci.*, 274: 321—355.

Miyashiro, A., 1978. Nature of alkalic rock series. *Contrib. Mineral. Petrol.*, 66: 91—104.

Miyashiro, A., Shido, F. and Ewing, M., 1969a. Composition and origin of serpentinites from the Mid-Atlantic Ridge near 24° and 30° north latitude. *Contrib. Mineral. Petrol.*, 23: 117—127.

Miyashiro, A., Shido, F. and Ewing, M., 1969b. Diversity and origin of abyssal tholeiite from the Mid-Atlantic Ridge near 24° or 30°N latitude. *Contrib. Mineral Petrol.*, 23: 38—52.

Miyashiro, A., Shido, F. and Ewing, M., 1970a. Crystallization and differentiation in abyssal tholeiites and gabbros from mid-oceanic ridges. *Earth Planet. Sci. Lett.*, 7: 261—265.

Miyashiro, A., Shido, F. and Ewing, M., 1970b. Petrologic models for the Mid-Atlantic Ridge. *Deep-Sea Res.*, 17: 109—123.

Miyashiro, A., Shido, F. and Ewing, M., 1971. Metamorphism in the Mid-Atlantic Ridge near 24° and 30°N. *Philos. Trans. R. Soc. London, Ser. A*, 268: 589—603.

Molnar, P. and Aggarwal, Y.P., 1971. A microearthquake survey in Kenya. *Seismol. Soc. Am. Bull.*, 61: 195—201.

Molnar, P. and Sykes, L.R., 1969. Tectonics of the Caribbean and Middle America regions from focal mechanism and seismicity. *Geol. Soc. Am. Bull.*, 80: 1639—1684.

Moore, J.G. and Schilling, J.G., 1973. Vesicles, water and sulphur in Reykjanes Ridge basalts. *Contrib. Mineral. Petrol.*, 41: 105—118.

Moore, J.G., Fleming, H.S. and Phillips, J.D., 1974. Preliminary model for extrusion and rifting at the axis of the Mid-Atlantic Ridge, 36°48′ north. *Geology*, 2: 437—440.

378

of recent spreading ridges. *Contrib. Mineral. Petrol.*, 72: 425—436.

Morgan, W.J., 1971. Convection plumes in the lower mantle. *Nature*, 230: 42—43.

Morgan, W.J., 1972a. Deep convection plumes and plate motions. *Bull. Am. Assoc. Pet. Geol.*, 56: 203—213.

Morgan, W.J., 1972b. Plate motion and deep mantle convection. *Geol. Soc. Am., Mem.*, 132: 1—7.

Morgan, W.J., 1981. Hotspot tracks and the opening of the Atlantic and Indian Oceans. In: C. Emiliani (Editor), *The Sea*, 7. Wiley, New York, N.Y., pp. 443—487.

Morgenstein, M., 1967. Authigenic cementation of scoriaceous deep-sea sediments west of the Society Ridge; South Pacific. *Sedimentology*, 9: 105—118.

Muehlenbachs, K., 1980. The alteration and aging of the basaltic layer of the sea floor: oxygen isotope evidence from DSDP-IPOD Legs 51, 52 and 53. In: *Initial Reports of the Deep Sea Drilling Project, 51, 52, 53.* U.S. Government Printing Office, Washington, D.C., Part 2, pp. 1159—1167.

Muehlenbachs, K. and Clayton, R.N., 1972. Oxygen isotope studies of fresh and weathered submarine basalts. *Can. J. Earth Sci.*, 9: 172—184.

Muir, J.D., 1951. The clinopyroxenes of the Skaergaard intrusion, eastern Greenland. *Mineral. Mag.*, 29: 690.

Muir, J.D. and Tilley, C.E., 1961. Mugearites and their place in alkali igneous rock series. *J. Geol.*, 69: 186—203.

Muir, J.D., Tilley, C.E. and Scoon, J.H., 1964. Basalts from the northern part of the rift zone of the Mid-Atlantic Ridge. *J. Petrol.*, 5: 409—434.

Muir, J.D., Tilley, C.E. and Scoon, J.H., 1966. Basalts from the northern part of the Mid-Atlantic Ridge, 2. The Atlantis collection near 30°N. *J. Petrol.*, 7: 193—201.

Murase, T. and McBirney, A.R., 1973. Properties of some common igneous rocks and their melts at high temperature. *Geol. Soc. Am. Bull*, 84: 3563—3592.

Murray, J. and Renard, A.F., 1891. *Report on Deep-sea Deposits Based on the Specimens Collected During the Voyage of H.M.S. "Challenger" in the Years 1872 to 1876.* H.M. Stationery Office, London, pp. 34—147, 451—488.

Mysen, B.O., 1976. Experimental determination of some geochemical parameters relating to conditions of equilibrium of peridotite in the upper mantle. *Am. Mineral.*, 61: 677—683.

Mysen, B.O. and Kushiro, I., 1977. Compositional variation of coexisting phases with degree of melting of peridotie in the upper mantle. *Am. Mineral.*, 62: 843—857.

Needham, H.D. and Francheteau, J., 1974. Some characteristics of the Rift Valley in the Atlantic Ocean near 36°48′ north. *Earth Planet. Sci. Lett.*, 22: 29—43.

Nicholls, G.D., 1965. Basalts from the deep ocean floor. *Mineral Mag. (Tilley Volume)*, 34: 373—388.

Nicholls, G.D. and Islam, M.R., 1971. Geochemical investigations of basalts and associated rocks from the ocean floor and their implications. *Philos. Trans. R. Soc. London, Ser. A*, 268: 469—486.

Nicholls, G.D., Nalwalk, A.J. and Hays, E.E., 1964. The nature and composition of rock samples dredged from the Mid-Atlantic Ridge between 22°N and 52°N. *Mar. Geol.*, 1: 333—343.

Nicholls, I.A. and Ringwood, A.E., 1972. Production of silica-saturated tholeiitic magma in island arcs. *Earth Planet. Sci. Lett.*, 17: 243—246.

Nisbet, E.G. and Fowler, C.M.R., 1978. The Mid-Atlantic Ridge at 37° and 45°N: some geophysical and petrological constraints. *Geophys. J. R. Astron. Soc.*, 54: 631—660.

**Nixon, P.H. and Boyd, F.R., 1973. The discrete nodule (megacryst) association in kimberlites from Northern Lesotho.** In: P.H. Nixon (Editor), *Lesotho Kimberlites.* Lesotho National Development Corporation, Maseru, pp. 67—75.

Noe-Nygaard, A., 1949. Samples of volcanic rocks from the sea bottom between the Faeroes and Iceland. *Geografics. Ann.*, 31: 348—356.

Noe-Nygaard, A., 1962. The geology of the Faeroes. *Q. J. Geol. Soc. London*, 118: 375—383.

Noe-Nygaard, A. and Rasmussen, J., 1968. Petrology of a 3000 meter sequence of basaltic lavas in the Faeroes Islands. *Lithos*, 1: 286—304.

O'Hara, M.J. and Mercy, E.L.P., 1963. Petrology and petrogenesis of some garnetiferous peridotites. *Trans. R. Soc. Edinburgh*, 45: 251—313.

Oliver, J., 1969. Structure and evolution of the mobile seismic belts. *Phys. Earth Planet. Inter.*, 2: 350—362.

Oliver, J. and Isacks, B., 1967. Deep earthquake zones: anomalous structures in the upper mantle and lithosphere. *J. Geophys. Res.*, 72: 4259—4275.

Olivet, J.L., Sichler, B., Thonon, P., Le Pichon, X., Martinais, J. and Pautot, G., 1970. La faille transformante Gibbs entre le rift et la marge du Labrador. *C.R. Acad. Sci. Paris, Sér. D*, 271: 949—952.

Olivet, J.L., Le Pichon, X., Monti, S., Sichler, B. and Pautot, G., 1974. Charlie Gibbs Fracture Zone. *J. Geophys. Res.* 79: 2059—2072.

Orcutt, J.B., Kennett, B., Dorman, L. and Protherow, W.A., 1975. Evidence for a low velocity zone underlying a fast-spreading rise crest. *Nature*, 256: 475.

Osborn, E.F. and Tait, D.B., 1952. The system diopside-forsterite-anorthite. *Am. J. Sci. (Bowen Volume)*, 413: 413—433.

Osmaston, M.F., 1971. Genesis of ocean ridge median valleys and continental rift valleys. *Tectonophysics*, 11: 387—407.

Palmason, G., 1970. *Crustal Structure of Iceland from Explosion Seismology.* Science Institute, University of Iceland, Reykjanes, 239 pp.

Parker, R.L. and Fleisher, M., 1968. Geochemistry of niobium and tentalum. *U.S. Geol. Surv. Prof. Paper*, 612: 1—43.

Parker, R.L. and Oldenburg, D.W., 1973. Thermal model of ocean ridges. *Nature Phys. Sci.*, 242: 137—139.

Parmentier, E.M. and Spooner, E.T.C., 1978. A theoretical study of hydrothermal convection and the origin of the ophiolitic sulphides ore deposits of Cyprus. *Earth Planet. Sci. Lett.*, 40: 33—44.

Paster, T.P., 1968. *Petrological variations within submarine basalt pillows of the South Pacific and Antarctic Oceans.* Ph.D. Thesis, Florida State University, Tallahassee, Fla., 108 pp.

Peacock, M., 1926. The petrology of Iceland, I. The basic tuffs. *Trans. R. Soc. Edinburgh*, 55: 51—76.

Pearce, J.A. and Cann, J.R., 1971. Ophiolite origin investigated by discriminant analysis using Ti, Zr and Y. *Earth Planet. Sci. Lett.*, 12: 339—349.

Pearce, J.A. and Cann, J.R., 1973. Tectonic setting of basic volcanic rocks determined using trace element analyses. *Earth Planet. Sci. Lett.*, 19: 290—300.

Pearce, J.A. and Norry, M.J., 1979. Petrogenetic implications of Ti, Zr, Y and Nb variations in volcanic rocks. *Contrib. Mineral. Petrol.*, 69: 33—47.

380

Petelin, V.P., 1964. Hard rocks from the deep sea trenches of the south-west part of the Pacific Ocean. In: *Bottom Geology of the Oceans and Seas, Doklady Soviet Geologists. 22nd Sess. Int. Geol. Congr.*, pp. 78—86.

Peterman, Z.E. and Hedge, C.E., 1971. Related Sr isotopic and chemical variations in oceanic basalt. *Geol. Soc. Am. Bull.*, 82: 493—500.

Peterson, M.N.A. and Goldberg, E.D., 1962. Feldspar distribution in South Pacific sediments. *J. Geophys. Res.*, 67: 3477.

Phillips, J.D. and Fleming, H.S., 1977. Multi-beam sonar study of Mid-Atlantic Ridge rift valley, 36°—37°N. *Geol. Soc. Am. Bull.*, 88: 1—5.

Poehls, K.A., 1974. Seismic refraction on the Mid-Atlantic Ridge at 37°N. *J. Geophys. Res.*, 79: 337—373.

Poldervaart, A. and Green, J., 1965. Chemical analysis of submarine basalts. *Am. Mineral.*, 50: 1723—1727.

Poldervaart, A. and Parker, A.B., 1964. The Crystallization Index as a parameter of igneous differentiation in binary variation diagrams. *Am. J. Sci.*, 262: 281—289.

Powell, J.L., Faure, G. and Hurley, P.H., 1965. $^{87}$Sr abundances in a suite of Hawaiian volcanic rocks of varying silica content. *J. Geophys. Res.*, 70: 1509—1513.

Presnall, D.C., 1969. The geometrical analysis of partial fusion. *Am. J. Sci.*, 267: 1178—1194.

Prévot, M. and Lecaille, A., 1976. Sur le caractère épisodique du fonctionnement des zones d'accrétions: critique des arguments géomagnétiques. *Bull. Soc. Géol. Fr. (7)*, 18: 903—911.

Purdy, G.M., 1975. The eastern end of the Azores-Gibraltar plate boundary. *Geophys. J. R. Astron. Soc.*, 43: 973—1000.

Raitt, R.W., 1963. The crustal rocks. In: M.N. Hill (Editor), *The Sea*, 3. Wiley, New York, N.Y., pp. 85—102.

Rankama, K., 1948. On the geochemistry of niobium. *Ann. Acad. Sci. Fenn.*, Ser A3, 13: 1—57.

Rea, D.K., 1975. Model for the formation of topographic features of the East Pacific Rise crest. *Geology*, 3: 77—80.

Reid, I., Orcutt, J.A. and Protherow, W.A., 1977. Seismic evidence for a narrow zone of partial melting underlying the East Pacific Rise at 21°N. *Geol. Soc. Am. Bull.*, 88: 678—682.

Reid, J. and Macdonald, K.G., 1973. Microearthquake study of the Mid-Atlantic Ridge near 37°N using sonobuoys. *Nature*, 246: 88—90.

Renard, V., Schrumpf, B., Sibuet, J.C. and Carré, D., 1975. *Bathymétrie détaillée d'une partie de la vallée du Rift et de la faille transformante près de 36°50'N dans l'Océan Atlantique.* CNEXO/Bureau de Recherches Géologiques et Minières, Orléans.

Rhodes, J.M., Blanchard, D.R., Rodgers, K.V., Jacobs, J.W. and Brannon, J.C., 1976. Petrology and chemistry of basalts from the Nazca plate, 2. Major and trace element chemistry. In: *Initial Reports of the Deep Sea Drilling Project*, 34. U.S. Government Printing Office, Washington, D.C., pp. 239—244.

Rhodes, J.M., Dungan, M.A., Blanchard, D.P. and Long, P.E., 1979. Magma mixing at mid-ocean ridges: evidences from basalts drilled near 22°N on the Mid-Atlantic Ridge. In: J. Francheteau (Editor), *Processes at Mid-Ocean Ridges. Tectonophysics*, 55: 35—61.

Richardson, S.W., Gilbert, M.C. and Bell, P.M., 1969. Experimental determination of kyanite-andalusite and andalusite-sillimanite equilibria; the alumina-silicate triple point. *Am. J. Sci.*, 267: 259—272.

Richey, J.E. and Thomas, H.H., 1930. The geology of Ardnamurchan, northwest Mull and Coll. *Geol. Surv. Scotl., Mem., Edinburgh.* (Reference in: Poldervaart, A., 1962. Aspect of basalt petrology. *J. Geol. Soc., India*, 3: 1—4.)

Ridley, W.J., Perfit, M.R. and Adams, M.L., 1974. Petrology of basalts from the Deep Sea Drilling Project, Leg 38. In: *Initial Reports of the Deep Sea Drilling Project, 38*. U.S. Government Printing Office, Washington, D.C., pp. 731—739.

Ringwood, A.E., 1967. The pyroxene-garnet transformation in the earth's mantle. *Earth Planet. Sci. Lett.*, 2: 255—263.

Ringwood, A.E., MacGregor, I.D. and Boyd, F.R., 1964. Petrological constitution of the upper mantle. *Carnegie Inst. Washington Yearb.*, 63: 147—152.

Robinson, P.T. and Whitford, D.J., 1974. Basalts from the eastern Indian Ocean. In: *Initial Reports of the Deep Sea Drilling Project, 27*. U.S. Government Printing Office, Washington, D.C., pp. 551—559.

Roeder, P.L., 1974. Paths of crystallization and fusion in systems showing ternary solution. *Am. J. Sci.*, 274: 48—60.

Roeder, P.L. and Emslie, R.F., 1970. Olivine-liquid equilibrium. *Contrib. Mineral. Petrol.*, 29: 275—289.

Rogers, A.F. and Kerr, P.F., 1942. *Optical Mineralogy*. McGraw-Hill, New York, N.Y., 390 pp.

Rona, P.A., 1976. Pattern of hydrothermal mineral deposition: Mid-Atlantic Ridge crest at latitude 26°N. *Mar. Geol.*, 21: 59—66.

Rona, P.A., 1977. Plate tectonics, energy and mineral resources: basic research leading to pay-off. *EOS Trans. Am. Geophys. Union*, 58: 629—639.

Rona, P.A., 1978. Criteria for the recognition of hydrothermal mineral deposits in oceanic crust. *Econ. Geol.*, 73: 135—160.

Rosendahl, B.R., 1976. Evolution of oceanic crust, 2. Constraints, implications and inferences. *J. Geophys. Res.*, 81: 5305—5314.

Rosendahl, B.R., Raitt, R.W., Dorman, L.M., Bibee, L.D., Hussong, D.M. and Sutton, G.H., 1976. Evolution of oceanic crust, 1. A physical model of the East Pacific Rise crest derived from seismic refraction data. *J. Geophys. Res.*, 81: 5294—5304.

Rotshtein, A.A., 1961. Phase relationships in the peridotite of Davros and Belkhelv. *Izv. Akad. Nauk SSSR, Ser. Geol.*, 3 (in Russian).

Rotshtein, A.A., 1962. Magmatic facies of ultrabasic igneous rocks of the tholeiite series. *Izv. Akad. Nauk SSSR* (in Russian). (Cited in: Vinogradov, A.P. and Udintsev, G.B. (Editors), 1975. *Rift Zones of the World Ocean.* Wiley, New York, N.Y., 503 pp.)

Ruegg, W., 1962. Rasgos morfologicos — geologicos intramarinos y sus contrapartes en el suel continental peruano. *Bol. Soc. Geol. Peru*, 38: 97—142.

Sayles, F.L. and Bischoff, J.L., 1973. Ferromanganoan sediments in the equatorial East Pacific Rise. *Earth Planet. Sci. Lett.*, 19: 330—336.

Scarfe, C.M. and Smith, D.G.W., 1977. Mineralogy and chemistry of secondary phases in some basaltic rocks from DSDP Leg 37. In: *Initial Reports of the Deep Sea Drilling Project, 37*. U.S. Government Printing Office, Washington, D.C., pp. 825—828.

Scheidegger, K.F., 1973. Temperature and composition of magmas ascending along mid-ocean ridges. *J. Geophys. Res.*, 78: 3340—3355.

Scheidegger, K.F. and Stakes, D.S., 1977. Mineralogy, chemistry and crystallization sequence of clay minerals in altered tholeiitic basalts from the Peru Trench. *Earth Planet. Sci. Lett.*, 36: 413—422.

Schilling, J.G., 1973. Iceland mantle plume: geochemical study of Reykjanes Ridge. *Nature*, 242: 565—571.

382

Schilling, J.G., 1975a. Rare earth variations across "normal segments" of the Reykjanes Ridge, 60°—53°N, Mid-Atlantic Ridge, 29°S, and East Pacific Rise, 2°—19°S, and evidence on the composition of the underlying low-velocity layer. *J. Geophys. Res.*, 80: 1459—1473.

Schilling, J.G., 1975b. Azores mental blob: rare earth evidence. *Earth Planet. Sci. Lett.*, 25: 103—115.

Schilling, J.G. and Noe-Nygaard, A., 1974. Faeroe-Iceland mantle plume: rare earth evidence. *Earth Planet. Sci. Lett.*, 24: 1—14.

Schilling, J.G., Anderson, R.N. and Vogt, P., 1976. Rare earth, Fe and Ti variations along the Galapagos Spreading Centre, and their relationship to the Galapagos mantle plume. *Nature*, 26: 108—113.

Schrader, E.L., Rosendahl, B.R., Furbish, W.J. and Meadows, G., 1980. Picritic basalts from the siqueiros transform fault. In: *Initial Reports of the Deep Sea Drilling Project, 54*. U.S. Government Printing Office, Washington, D.C., pp. 71—78.

Schweitzer, E.L., Papike, J.J. and Bence, E.A., 1979. Statistical analysis of clinopyroxenes from deep-sea basalts. *Am. Mineral.*, 64: 501—513.

Scientific Party, DSDP Leg 26, 1974. Site 251. In: *Initial Reports of the Deep Sea Drilling Project, 26*. U.S. Government Printing Office, Washington, D.C., pp. 21—73.

Scientific Party, DSDP Leg 34, 1976. Sites 319, 320 and 321. In: *Initial Reports of the Deep Sea Drilling Project, 34*. U.S. Government Printing Office, Washington, D.C., pp. 17—153.

Scientific Party, DSDP Leg 38, 1976a. Sites 336 and 352. In: *Initial Reports of the Deep Sea Drilling Project, 38*. U.S. Government Printing Office, Washington, D.C., pp. 23—116.

**Scientific Party, DSDP Leg 45, 1976b. "Challenger" drills on Leg 45. *Geotimes*, 21(4): 20—23.**

Scientific Party, DSDP Leg 46, 1976c. Drilling into ocean crust. *Geotimes*, 21(9): 21—23.

Scientific Party, DSDP Leg 37, 1977a. Introduction and site reports. In: *Initial Reports of the Deep Sea Drilling Project, 37*. U.S. Government Printing Office, Washington, D.C., pp. 15—326.

Scientific Party, DSDP Leg 39, 1977b. Site 354: Ceara Rise. In: *Initial Reports of the Deep Sea Drilling Project, 39*. U.S. Government Printing Office, Washington, D.C., pp. 45—49.

**Scientific Party, DSDP Leg 54, 1977c. On the East Pacific Rise "Glomar Challenger" completes 54th cruise. *Geotimes*, 22(11): 19—22.**

Scientific Party, DSDP Leg 46, 1978. Site 386: 23°N, Mid-Atlantic Ridge. In: *Initial Reports of the Deep Sea Drilling Project, 46*. U.S. Government Printing Office, Washington, D.C., pp. 265—303.

Scientific Party, DSDP Leg 51, 1980a. Introduction and explanatory notes. In: *Initial Reports of the Deep Sea Drilling Project, 51*. U.S. Government Printing Office, Washington, D.C., pp. 5—22.

Scientific Party, DSDP Leg 54, 1980b. Sites 419—423 and 426—429: ocean crust drilling on the East Pacific Rise and in the Siqueiros fracture zone near 9°N. In: *Initial Reports of the Deep Sea Drilling Project, 54*. U.S. Government Printing Office, Washington, D.C., pp. 81—232.

Scientific Party, DSDP Leg 54, 1980c. Sites 424 and 425: geothermal drilling on the Galapagos Rift. In: *Initial Reports of the Deep Sea Drilling Project, 54*. U.S. Government Printing Office, Washington, D.C., pp. 233—304.

**Scientific Party, DSDP Leg 70, 1980d. Off Galapagos, metals traced to exchange in sea water. *Geotimes*, 25(4): 16—17.**

Sclater, J.G., Anderson, R.N. and Lee Bell, M., 1971. The elevation of ridges and the evolution of the central eastern Pacific. *J. Geophys. Res.*, 76: 78—88.

Sclater, J.G., Von der Borch, C., Veevers, J.J., Hékinian, R., Thompson, R.W., Pimm, A., McGowran, B., Gartner, S. and Johnson, D.A., 1974. *Initial Reports of the Deep Sea Drilling Project, 22*. U.S. Government Printing Office, Washington, D.C., pp. 815—837.

Scott, M.R., Scott, R.B., Rona, P.A., Butler, L.W. and Nalwalk, A.J., 1974. Rapidly accumulating manganese deposit from the Median Valley of the Mid-Atlantic Ridge. *Geophys. Res. Lett.*, 1: 355.

Scott, R.B. and Hajash, A., Jr., 1976. Initial submarine alteration of basaltic pillow lavas: a microprobe study. *Am. J. Sci.*, 276: 480—501.

Scott, R.B., Burkett, D.H. and Leaird, J.D., 1970. Experimental alteration of basaltic glass in low temperature aqueous environments. *Geol. Soc. Am. Abstr.*, 2: 509.

Scott, R.B., Rona, P.A., Butler, L.W., Nalwalk, A.J. and Scott, M.R., 1972. Manganese crusts of the Atlantis fracture zone. *Nature*, 239: 77—79.

Scott, R.B., Rona, P.A., McGregor, B.A. and Scott, M.R., 1974. The TAG hydrothermal field. *Nature*, 251: 301—302.

Seyfried, W.E., Shanks, W.C. and Bischoff, J.C., 1976. Alteration and vein formation in Site 321 basalt. In: *Initial Reports of the Deep Sea Drilling Project*, 34. U.S. Government Printing Office, Washington, D.C., pp. 385—392.

Seymour-Sewell, R.B., 1925. Geographic and oceanographic research in Indian waters, 1. The geography of the Andaman Sea Basin. *Mem. Asiat. Soc. Bengal*, 9: 1—26.

Shand, S.J., 1949. Rocks of the Mid-Atlantic Ridge. *J. Geol.*, 57: 89—91.

Shibata, T., 1976. Phenocryst-bulk rock composition relations of abyssal tholeiites and their petrogenetic significance. *Geochim. Cosmochim. Acta*, 40: 1407—1417.

Shibata, T. and Fox, P.J., 1975. Fractionation of abyssal tholeiites: samples from the Oceanographer fracture zone (35°N, 35°W). *Earth Planet. Sci. Lett.*, 27: 62—72.

Shibata, T., Delong, S.E. and Walker, D., 1979a. Abyssal tholeiites from the Oceanographer fracture zone, 1. Petrology and fractionation. *Contrib. Mineral. Petrol.*, 70: 89—108.

Shibata, T., Thompson, G. and Frey, F.A., 1979b. Tholeiitic and alkali basalts from the Mid-Atlantic Ridge at 43°N. *Contrib. Mineral. Petrol.*, 70: 127—141.

Shido, F., Miyashiro, A. and Ewing, M., 1971. Crystallization of abyssal tholeiites. *Contrib. Mineral. Petrol.*, 31: 251—266.

Shor, G.G. and Pollard, D.D., 1964. Mohole site selection studies north of Maui. *J. Geophys. Res.*, 69: 1626—1637.

Sibuet, J.C. and Mascle, J., 1978. Plate kinematic implications of Atlantic equatorial fracture zone trends. *J. Geophys. Res.*, 83: 3401—3421.

Sigurdson, H., 1977. Generation of Icelandic rhyolites by melting of plagiogranites in the oceanic layer. *Nature*, 269: 25—28.

Simpson, E.S.W., 1971. The geology of the southwest African continental margin: a review. In: F. Delanny (Editor), *The Geology of the East Continental Margin. Inst. Geol. Sci., Rep.*, 70/16: 153—170.

Simpson, E.S.W., Schlich, R., Gieskes, J.M., Girdley, W.A., Leclaire, L., Marshall, B.V., Moore, C., Müller, C., Sigal, J., Vallier, J.C., White, S.M. and Zobel, B., 1974. *Initial Reports of the Deep Sea Drilling Project*, 25. U.S. Government Printing Office, Washington, D.C., pp. 187—208.

Sleep, N.H., 1975. Formation of oceanic crust: some thermal constraints. *J. Geophys. Res.*, 80: 4037—4042.

Smith, R.E., 1968. Redistribution of major elements in the alteration of some basic lavas during burial metamorphism. *J. Petrol.*, 9: 191—219.

Solomon, S.C., 1973. Shear wave attenuation and melting beneath the Mid-Atlantic Ridge. *J. Geophys. Res.*, 78: 6044—6058.

Spiess, F.N., Macdonald, K.C., Atwater, T., Ballard, R., Carrenza, D., Cordoba, D., Cox, V., Diaz Garcia, V.M., Francheteau, J., Guerrero, J., Hawkins, J., Haymon, R., Hessler, R., Juteau, T., Kastner, M., Larson, R., Luyendyk, B., Macdougall, J.D., Miller, S., Normark, W., Orcutt, J. and Ranger, C., 1980. East Pacific Rise: hot springs and geophysical experiments. *Science*, 207: 1421—1433.

Spindel, R.C., Davis, S.B., Macdonald, K.C., Porter, R.P. and Phillips, J.D., 1974. Microearthquake survey of median valley of the Mid-Atlantic Ridge at 36°30′N. *Nature*, 248: 577—579.

Spooner, E.T.C. and Fyfe, W.S., 1973. Sub-sea floor metamorphism, heat and mass transfer. *Contrib. Mineral. Petrol.*, 42: 287—304.

Stebbins, J. and Thompson, G., 1978. The nature and petrogenesis of intra-oceanic plate alkaline eruptive and plutonic rocks: King's Trough, northeast Atlantic. *J. Volcanol. Geotherm. Res.*, 4: 333—361.

Stocks, Th., 1960. Zur Bodenstalt des Indischen Ozeans Erdkünde. *Arch. Wiss. Geogr.*, 14: 161—170.

Stoffa, R.L., Buhl, P., Herron, T.D., Kan, T.K. and Ludwig, W.I., 1980. Mantle reflection beneath the crestal zone of the East Pacific Rise from multi-channel seismic data. *Mar. Geol.*, 35: 83—97.

Störzer, D. and Selo, M., 1974. Ages par la méthode des traces de fission de basaltes prélevés dans la vallée axiale de la dorsale medio-Atlantique aux environs de 37°N. *C.R. Acad. Sci., Paris, Sér. D*, 279: 1649—1651.

Störzer, D. and Selo, M., 1976. Uranium and fission track ages of some basalts from the FAMOUS area. *Bull. Soc. Géol. Fr.*, 18(4): 807—810.

Subbarao, K.V., 1972. The strontium isotopic composition of basalts from the East Pacific and Chile rises and abyssal hills in the eastern Pacific Ocean. *Contrib. Mineral. Petrol.*, 37: 111—120.

Subbarao, K.V. and Hedge, C.E., 1973. K, Rb, Sr, and $^{87}Sr/^{86}Sr$ in rocks from the Mid-Indian Oceanic Ridge. *Earth Planet Sci. Lett.*, 18: 223—228.

Subbarao, K.V. and Hékinian, R., 1978. Alkali-enriched rocks from the central eastern Pacific Ocean. *Mar. Geol.*, 26: 249—268.

Subbarao, K.V., Clark, G.S. and Forbes, R.B., 1973. Strontium isotopes and some seamount basalts from the northeastern Pacific Ocean. *Can. J. Earth Sci.*, 10: 1479—1484.

Subbarao, K.V., Reddy, V.V., Hékinian, R. and Chandrasekharam, D., 1977. Large ion lithophile elements and Sr and Pb isotopic variation in volcanic rocks from the Indian Ocean. In: J.R. Heirtzler, H.M. Bolli, T.A. Davies, J.B. Saunders and J.G. Sclater (Editors), *Indian Ocean Geology and Biostratigraphy*. American Geophysical Union, Washington, D.C., pp. 259—278.

Subbarao, K.V., Kempe, D.R.C., Reddy, V.V., Reddy, G.R. and Hékinian, R., 1979a. Review of the geochemistry of Indian and other oceanic rocks. In: L.H. Ahrens (Editor), *Origin and Distribution of the Elements*. Pergamon Press, Oxford, pp. 367—399.

Subbarao, K.V., Reddy, V.V., Reddy, G.R. and Hékinian, R., 1979b. Rare earth geochemistry of basalts from the FAMOUS area, Mid-Atlantic Rift Valley — a preliminary study. *Geol. Soc. India Bull.*, 20: 565—569.

Summerhayes, C.P., 1967. Note on Macquerie Ridge and the Tonga-Kermadec complex. *N.Z. J. Sci.*, 10: 808—812.

Sun, S.S., Tatsumoto, M. and Schilling, J.G., 1975. Mantle plume mixing along the Reykjanes Ridge axis: lead isotopic evidence. *Science*, 190: 143—147.

Swindale, L.D. and Pow-Foong Fan, 1967. Transformation of gibbsite to chlorite in ocean bottom sediments. *Science*, 157: 799—800.

Sykes, L.R., 1967. Mechanism of earthquakes and nature of faulting on the mid-oceanic ridges. *J. Geophys. Res.*, 72: 2131—2153.

Sykes, L.R., 1969. Seismicity of the Mid-Oceanic Ridge system. In: J.P. Hart (Editor), *Earth Crust and Upper Mantle. Am. Geophys. Union, Geophys. Monogr.*, 13: 148—160.

Sykes, L.R. and Sbar, M.L., 1974. Focal mechanism solutions of intraplate earthquakes and stresses in the lithosphere. In: C. Jansson (Editor), *Geodynamics of Iceland and the North Atlantic Area*. Reidel, Dordrecht, pp. 207—224.

Symes, R.F., Bevan, J.C. and Hutchison, R., 1977. Phase chemistry studies on gabbro and peridotite rocks from Site 334, DSDP Leg 37. In: *Initial Reports of the Deep Sea Drilling Project, 37*. U.S. Government Printing Office, Washington, D.C., pp. 841—846.

Talwani, M. and Udintsev, G., 1976. In: *Initial Reports of the Deep Sea Drilling Project, 38*. U.S. Government Printing Office, Washington, D.C., pp. 1213—1242.

Talwani, M., Windish, C.C. and Langseth, M.G., Jr., 1971. Reykjanes Ridge crest: a detailed geophysical study. *J. Geophys. Res.*, 76: 473—517.

Talwani, M., Udintsev, G. and White, S.M., 1976. Introduction and explanatory notes, Leg 38. In: *Initial Reports of the Deep Sea Drilling Project, 38*. U.S. Government Printing Office, Washington, D.C., pp. 3—19.

Tarling, D.H. and Gale, N.H., 1968. Isotopic dating and paleomagnetic polarity in the Faeroe Islands. *Nature*, 218: 1043—1044.

Tatsumoto, M., Hedge, C.E. and Engel, A.E.J., 1965. Potassium, strontium, thorium, uranium and $^{87}Sr/^{86}Sr$ in oceanic tholeiitic basalt. *Science*, 150: 886—888.

Taylor, F.B., 1910. *Geol. Soc. Am. Bull.*, 21: 217. (Reference from Washington, H.S., 1930. The origin of the Mid-Atlantic Ridge. *J. Md. Acad. Sci.*, 1: 20—29.)

Taylor, P.T. and Hékinian, R., 1971. Geology of a newly discovered seamount in the New England seamount chain. *Earth Planet. Sci. Lett.*, 11: 73—82.

Thiede, J., 1964. Sedimentary structures in pelagic and hemipelagic sediments from the central and southern Atlantic Ocean. In: *Initial Reports of the Deep Sea Drilling Project, 39*. U.S. Government Printing Office, Washington, D.C., pp. 407—415.

Thiede, J., 1977. The subsidence of aseismic ridges: evidence from sediments on Rio Grande Rise (S.W. Atlantic Ocean). *Am. Assoc. Pet. Geol. Bull.*, 61: 920—940.

Thompson, A.B., 1970. Laumontite equilibria and the zeolite facies. *Am. J. Sci.*, 269: 267—275.

Thompson, G. and Melson, W.G., 1972. The petrology of oceanic crust across fracture zones in the Atlantic Ocean: evidence of a new kind of sea floor spreading. *J. Geol.*, 80: 526—538.

Thompson, G., Bryan, W.B., Frey, F.A. and Suny, C.M., 1974. Petrology and geochemistry of basalts and related rocks from Sites 214, 215, 216. DSDP Leg 22, Indian Ocean. In: *Initial Reports of the Deep Sea Drilling Project, 22*. U.S. Government Printing Office, Washington, D.C., pp. 459—468.

Thompson, G., Bryan, W.B., Frey, F.A., Dickey, J.S. and Suen, C.J., 1976. Petrology and geochemistry of basalts from DSDP Leg 34, Nazca plate. In: *Initial Reports of the Deep Sea Drilling Project, 34*. U.S. Government Printing Office, Washington, D.C., pp. 215—226.

Thompson, G., Bryan, W.B., Frey, F.A. and Dickey, J.S., Jr., 1978. Basalts and related rocks from deep-sea drilling sites in the central and eastern Indian Ocean. *Mar. Geol.*, 26: 119—138.

Thompson, G., Bryan, W.B. and Melson, W.G., 1980. Geological and geophysical investigation of the Mid-Cayman Rise spreading center: geochemical variation and petrogenesis of basalt glasses. *J. Geol.*, 88: 41—55.

Thompson, R.N., 1972. Melting behavior of two Snake River lavas at pressure up to 35 kb. *Carnegie Inst. Washington Yearb.*, 71: 400—410.

Thornton, C.P. and Tuttle, O.F., 1960. Chemistry of igneous rocks, 1. Differentiation Index. *Am. J. Sci.*, 258: 664—684.

Tilley, C.E., 1950. Some aspects of magmatic evolution. *Q. J. Geol. Soc. London*, 106: 37—61.

Tilley, C.E., 1966. Note on a dunite (peridotite) mylonite of St. Paul's Rocks (Atlantic). *Geol. Mag.*, 103: 120—123.

Tilley, C.E. and Long, J.V.P., 1967. The porphyroclast minerals of the peridotite-mylonites of St. Paul's Rocks (Atlantic). *Geol. Mag.*, 104: 46—48.

Tobin, D.G., Ward, P.L. and Drake, C.L., 1969. Microearthquake in the Rift Valley of Kenya. *Geol. Soc. Am. Bull.*, 80: 2034—2046.

Turner, F.J. and Verhoogen, J., 1960. *Igneous and Metamorphic Petrology*. McGraw-Hill, New York, N.Y., 694 pp.

386

Uchupi, I., 1971. Bathymetric map of the Atlantic, Carribean and Gulf of Mexico. *Woods Hole Oceanogr. Inst.*, Ref. No. 71-72.
Udintsev, G.B., 1975. "Vityaz" expedition to the rift zone of the Indian Ocean (36th cruise). In: A.P. Vinogradov and G.B Udintsev (Editors), *Rift Zones of the World Ocean.* Wiley, New York, N.Y., pp. 5—31 (translated from Russian by N. Kanev).
Udintsev, G.B. and Chernysheva, V.I., 1965. Rock samples from the upper mantle of the Indian Ocean rift zone. *Dokl. Akad. Nauk SSSR, Ser. Geol.*, 165: 85—88.
Uyeda, S. and Horai, K., 1964. Terrestrial heat flow in Japan. *J. Geophys. Res.*, 69: 2121—2141.

Vacquier, V., Raft, A.D. and Warren, R.E., 1961. Horizontal displacements in the floor of the N.E. Pacific Ocean. *Geol. Soc. Am. Bull.*, 72: 1251—1258.
Van Andel, T.H. and Bowin, C., 1968. Mid-Atlantic Ridge between 22° and 23° north latitude and the tectonics of mid-ocean rises. *J. Geophys. Res.*, 73: 1279—1298.
Van Andel, T.H., Heath, G.R., Malfait, B.T., Heinrichs, D.E. and Ewing, J.I., 1971. Tectonics of the Panama Basin, eastern equatorial Pacific. *Geol. Soc. Am. Bull.*, 82: 1489—1508.
Varne, R. and Rubenach, M.J., 1972. Geology of Macquarie Island and its relationship to oceanic crust. In: D.E. Hayes (Editor), *Antarctic Oceanology, II. The Australian-New Zealand Sector. Am. Geophys. Union, Antarct. Res. Ser.*, 19: 251—266.
Varne, R., Gee, R.D. and Quilty, P.G., 1969. Macquarie Island and the cause of oceanic linear magnetic anomalies. *Science*, 166: 230—233.
Veevers, J.J., Jones, J.G. and Talent, J.A., 1971. Indo-Australian stratigraphy and the configuration and dispersal of Gondwanaland. *Nature*, 229: 383—388.
Vogt, P.R. and Avery, D.E., 1974. Detailed magmatic surveys in the northeast Atlantic and Labrador Sea. *J. Geophys. Res.*, 79: 363—384.
Von Herzen, R.P., 1964. Indian Ocean heat flow. *Univ. Calif., Scripps Inst. Oceanogr.*, SIO Ref. No. 64-19.

Wager, L.R., 1960. The major element variation of the layered series of the Skaergaard intrusion and a re-estimation of the average composition of the Nidden layered series and the successive residual magma. *J. Petrol.*, 1: 364—398.
Wager, L.R. and Deer, W.A., 1939. The petrology of the Skaergaard intrusion, Kangedlugssuaq, East Greenland. *Medd. Grönland*, 105(4): 352 pp.
Walker, D., Powell, M.A., Lofgren, G.E. and Hays, J.F., 1978. Dynamic crystallization of a eucrite basalt. *Proc. 9th Lunar Planet. Sci. Conf.*, pp. 1369—1391.
Walker, G.P.L., 1960. Zeolite zones and dikes distribution on relation to the structure of the basalts of eastern Iceland. *J. Geol.*, 68: 515—528.
Washington, H.S., 1930. The origin of the Mid-Atlantic Ridge. *J. Md. Acad. Sci.*, 1: 20—29.
Watkins, M.D. and Gunn, B.M., 1971. Petrology geochemistry and magnetic properties of some rocks dredged from the Macquerie Ridge. *N.Z. J. Geol. Geophys.*, 14: 153—168.
Watts, A.B., Bodine, J.H. and Ribe, N.M., 1980. Observations of flexure and the geological evolution of the Pacific Ocean basin. *Nature*, 283: 532—537.
Weertman, J., 1971. Theory of water-filled crevasse in glaciers applied to vertical magma transport beneath oceanic ridges. *J. Geophys. Res.*, 76: 1171—1183.
White, W.M., Hart, S.R. and Schilling, J.G., 1975. Geochemistry of the Azores and the Mid-Atlantic Ridge: 29°N to 60°N. *Carnegie Inst. Washington Yearb.*, 74: 224—234.
White, W.M. and Bryan, W.B., 1977. Sr-isotope, K, Rb, Cs, Sr, Ba, and rare earth geochemistry of basalts from the FAMOUS area. *Geol. Soc. Am. Bull.*, 88: 571—576.
Whitmarsh, R.B., 1973. Median Valley refraction line, Mid-Atlantic Ridge at 37°N. *Nature*, 246: 297—299.
Whitmarsh, R.B., Wesser, O.E., Syed Ali, Boudreaux, J.E., Fleischer, R.L., Jipa, D., Kidd, R.B., Mallik, T.K., Nigrini, C., Matter, A., Siddiquie, H.N., Stoffers, P., Hamilton, N. and Hunziker, J., 1974. Arabian Sea. In: *Initial Reports of the Deep Sea Drilling Project, 23.* U.S. Government Printing Office, Washington, D.C., pp. 33—383.

Williams, D.L., Von Herzen, R.P., Sclater, J.G. and Anderson, R.N., 1974. The Galapagos spreading center: lithospheric cooling and hydrothermal circulation. *Geophys. J. R. Astron. Soc.*, 38: 587—608.

Williams, D.L., Green, K., Van Andel, T.H., Von Herzen, R.P., Dymond, J.R. and Crane, K., 1979. The hydrothermal mounds of the Galapagos Rift: observations with D.S.R.V. "Alvin" and detailed heat flow studies. *J. Geophys. Res.*, 84: 7467—7484.

Williams, H. and Stevens, R.K., 1974. The ancient continental margin of eastern North America. In: C.A. Burk and C.L. Drake (Editors), *The Geology of Continental Margins*. Springer-Verlag, Berlin, pp. 781—796.

Wilshire, H.G., 1959. Deuteric alteration of volcanic rocks. *J. Proc. R. Soc. N.S.W.*, 98: 105—120.

Wilson, J.T., 1965. Submarine fracture zones, aseismic ridges and the International Council of Scientific Unions line: proposed western margin of the East Pacific Ridge. *Nature*, 207: 907—911.

Winchell, A.N. and Winchell, H., 1961. *Elements of Optical Mineralogy, II. Descriptions of Minerals*. Wiley, New York, N.Y., 551 pp.

Wiseman, J.D.H., 1937. Geological and mineralogical investigations, 1. Basalts from the Carlsberg Ridge, Indian Ocean. *Rep. John Marray Exped. Br. Mus. Nat. Hist.*, 3: 1—28.

Wiseman, J.D.H., 1966. St. Paul's Rocks and the problem of the upper mantle. *Geophys. J. R. Astron. Soc.*, 11: 519—525.

Wolery, T.J. and Sleep, N.H., 1976. Hydrothermal circulation and geochemical flux at mid-ocean ridges. *J. Geol.*, 84: 249—275.

Wyllie, P.J., 1970. Ultramafic rocks and the upper mantle. *Mineral. Soc. Am. Spec. Paper*, 3: 3—32.

Yagi, K., 1960. A delenite block dredged from the bottom of the "Vitiaz Deep," Mariana Trench. *Proc. Jpn. Acad.*, 36: 213—216.

Yeats, R.S., Forbes, W.C., Heath, G.R. and Scheidegger, F., 1973. Petrology and geochemistry of DSDP Leg 16 basalts, eastern equatorial Pacific. In: *Initial Reports of the Deep Sea Drilling Project, 16*. U.S. Government Printing Office, Washington, D.C., pp. 617—640.

Yoder, H.S., Jr., 1975. Heat of melting of simple systems related to basalts and eclogites. *Carnegie Inst. Washington Yearb.*, 74: 515—519.

Yoder, H.S., Jr., 1976. *Generation of Basaltic Magma*. National Academy of Science, Washington, D.C., 265 pp.

Yoder, H.S., Jr., 1978. Basic magma generation and aggregation. *Bull. Volcanol.*, 41: 301—316.

Yoder, H.S., Jr. and Tilley, C.E., 1962. Origin of basalt magmas: an experimental study on natural and synthetic rock systems. *J. Petrol.*, 3: 342—532.

# SUBJECT INDEX